This book comes with access to more content online.

Create custom quizzes from hundreds
of practice problems!

Register your book or ebook at
www.dummies.com/go/getaccess.

Select your product, and then follow the prompts
to validate your purchase.

You'll receive an email with your PIN and instructions.

1,001 Algebra I Practice Problems FOR DUMMIES®

by Mary Jane Sterling

John Wiley & Sons, Inc.

1,001 Algebra I Practice Problems For Dummies®

Published by
John Wiley & Sons, Inc.
111 River St.
Hoboken, NJ 07030-5774
www.wiley.com

Published by John Wiley & Sons, Inc., Hoboken, New Jersey

Published simultaneously in Canada

For general information on our other products and services, please contact our Customer Care Department within the U.S. at 877-762-2974, outside the U.S. at 317-572-3993, or fax 317-572-4002.

For technical support, please visit www.wiley.com/techsupport.

Wiley publishes in a variety of print and electronic formats and by print-on-demand. Some material included with standard print versions of this book may not be included in e-books or in print-on-demand. If this book refers to media such as a CD or DVD that is not included in the version you purchased, you may download this material at http://booksupport.wiley.com. For more information about Wiley products, visit www.wiley.com.

Library of Congress Control Number: 2012956402

ISBN 978-1-118-44671-3 (pbk); ISBN 978-1-118-44665-2 (ebk); ISBN 978-1-118-44666-9 (ebk); ISBN 978-1-118-44669-0 (ebk)

Manufactured in the United States of America

SKY10022498_111220

About the Author

Mary Jane Sterling is the author of six *For Dummies* titles: *Algebra I For Dummies, Algebra II For Dummies, Trigonometry For Dummies, Math Word Problems For Dummies, Business Math For Dummies,* and *Linear Algebra For Dummies.* She has also written many supplementary workbooks and study aids.

Mary Jane continues doing what she loves best: teaching mathematics. When not teaching or writing *For Dummies* books, she loves spending her time reading murder mysteries and fishing for her dinner.

Dedication

The author dedicates this book to her son, Sgt. James T. Sterling, USMC, and the other members of the 4th Air/Naval Gunfire Liaison Company, Det Juliet (part of Operation Enduring Freedom 2012). Jim and the others in his unit, as well as all military serving the U.S., have our utmost respect and appreciation.

Author's Acknowledgments

The author issues a big thank you to project editor Tim Gallan, who has taken on the huge challenge of creating this new product. He has taken a lot of raw material and made it into this wonderful, finished workbook. Thank you so much for your hard work and patience.

Also, a heartfelt thank you to the math editors, Suzanne Langebartels and Stephanie Long. As careful as I may be to do all the problems correctly, there is always that chance of a silly error. The editors keep me honest!

And, of course, a grateful thank you to acquisitions editor Lindsay Lefevere, who again found me another interesting project.

Publisher's Acknowledgments

We're proud of this book; please send us your comments at `http://dummies.custhelp.com`. For other comments, please contact our Customer Care Department within the U.S. at 877-762-2974, outside the U.S. at 317-572-3993, or fax 317-572-4002.

Some of the people who helped bring this book to market include the following:

Acquisitions, Editorial, and Vertical Websites

Senior Project Editor: Tim Gallan

Executive Editor: Lindsay Sandman Lefevere

Copy Editor: Suzanne Langebartels

Assistant Editor: David Lutton

Editorial Program Coordinator: Joe Niesen

Technical Editor: Stephanie Long

Editorial Manager: Michelle Hacker

Editorial Assistants: Rachelle S. Amick, Alexa Koschier

Composition Services

Senior Project Coordinator: Kristie Rees

Layout and Graphics: Carrie A. Cesavice, Melissa Smith, Erin Zeltner

Proofreaders: Lindsay Amones, Debbye Butler

Indexer: Potomac Indexing, LLC

Publishing and Editorial for Consumer Dummies

Kathleen Nebenhaus, Vice President and Executive Publisher

David Palmer, Associate Publisher

Kristin Ferguson-Wagstaffe, Product Development Director

Publishing for Technology Dummies

Andy Cummings, Vice President and Publisher

Composition Services

Debbie Stailey, Director of Composition Services

Contents at a Glance

Table of Contents

Introduction

One-thousand-one algebra problems: That's a *lot* of algebra problems.

It will take you seven days to do all of them, if you do 143 each day. Whew! It will take you 91 days to do all of them, if you manage to do 11 each day. And, of course, it will take you 1,001 days to do all the problems if you do just one each day. Whatever your game plan, this is still a lot of problems. You may want to start at the beginning and do each problem in turn, or you may want to jump around and do the problems in an order that suits you best. Either plan is doable. Either plan is fine. Just watch out for topics that build on one another — you may need the information from one skill to succeed in another.

Practice makes perfect. Unlike other subjects where you can just read or listen and absorb the information sufficiently, mathematics takes practice. The only way to figure out how the different algebraic rules work and interact with one another is to get into the problems — get your hands dirty, so to speak. Many problems appear to be the same, on the surface, but different aspects and challenges have been inserted to make the different problems unique. The concepts become more set in your mind when you work with the problems and have the properties confirmed with your solutions.

Yes, one-thousand-one algebra problems are a lot of problems. But you may find that this just whets your appetite for more. Enjoy!

What You'll Find

This book has 1,001 algebra problems divided up among 23 chapters. Each chapter has many different sets of questions. The sets of questions are sometimes in a logical, sequential order, going from one part of a topic to the next and then to the next. Other times the sets of questions represent the different ways a topic can be presented. In any case, you're given instructions on doing the problems. And sometimes you're given a particular formula or format to use.

Instead of just having answers to each of the problems, you find a worked-out solution for each and every one. Flip to the back of the book for the step-by-step process needed to solve the problems. The solutions include verbal explanations inserted in the work where necessary. Sometimes an alternate procedure may be offered. Not everyone does algebra exactly the same way, but this book tries to provide the most understandable and success-promoting process to use when solving the algebra problems presented.

How This Workbook Is Organized

This workbook is divided into two main parts: questions and answers. But you probably figured that out already.

Part I: Questions

The questions chapters cover many different topics:

- **Basic operations:** The first six chapters cover the types of numbers and the types of operations on those numbers that are essential to working in algebra. The natural numbers and whole numbers are fine for elementary arithmetic, but you need to broaden your horizons with signed numbers and decimals and fractions and exponential expressions. All these types of numbers are added, subtracted, multiplied, and divided. The rules for the different types of numbers have similarities and differences. The problems can help you come to grips with these situations and recognize what's the same and what's different.

  Also important in algebra are the operations involving radicals, absolute value, and factorial. And, tying together all the numbers and operations are the rules on how to deal with them: the order in which you perform the operations, and then the effect of grouping symbols on the whole process.

- **Algebraic expressions:** An algebraic expression can consist of one or more terms — separated by addition and subtraction — or it can be in factored form. The factored form has everything connected by multiplication and division. Each of these forms is useful in some process or another, so it's important to be able to change from one form to another and back again. Multiply out the factors if you want a listing of terms from highest exponent to lowest. Or, factor many terms to make them all just one if you want to solve for a root or reduce some fraction.

  You'll find techniques for multiplying by one term or two — or more. There are some helpful tricks for raising binomials to higher powers. And then you find the factoring techniques — from rules of divisibility to factoring by grouping. One of the challenges of factoring expressions is deciding which technique to use. You find lots of practice to help you make those decisions.

- **Solving equations:** What is the point of learning all those algebra basics and then going through the factoring process? One of the favorite and most common goals for all that practice is to use the techniques to solve an equation. Solving an equation means identifying the number or numbers you can replace the variable with to make a true statement.

  You'll find factoring and the multiplication property of zero to be your first approach, and then you'll also have the quadratic formula to use on some of the more challenging second-degree equations. Polynomials can be solved using synthetic division to help with the factoring. And then you have radical and absolute value equations — with their particular challenges. Finish the section off with inequalities, and you'll have run the gamut of solving for what variables can represent.

- **Applications:** Mention the words *story problem*, and you'll see either a shudder or a brightening smile. People either love them or they don't. But story problems (practical applications) are a main goal of learning to use algebra effectively.

  The practical applications found in this section of the workbook are broken into many different types. You find some that are based on an established formula: area, perimeter, simple interest, and so on. Other applications have to do with relationships between numbers or sizes of objects. The trick to doing those applications is understanding the wording, which is why you come armed with all the basics under your belt. Get to work on the work problems before you age too much with the age problems. Just write yourself a simple algebraic equation, and you're almost finished.

- **Graphing:** Most of us are very visual — we understand things better when a picture is drawn. I usually draw pictures when working on word problems; it helps me focus on what type of equation to write. But the pictures in this section are a bit more structured. The pictures here involve the *Cartesian coordinate* system, which involves placing points, segments, and lines in their proper positions. Graphing lines is often used when solving systems of equations. And graphing is found in pretty much all the mathematics that follows algebra. This is where you can get a good start on the topic.

Part II: Answers

This part provides not only the answers to all the questions but explanations of the answers as well. So you get the solution, and you see how to arrive at that solution.

Beyond the Book

This book is chock-full of algebra goodness — I've given you enough problems to significantly improve your confidence with all things algebra. But maybe you want to track your progress as you tackle the problems, or maybe you're stuck on a few types of algebra problems and wish they were all presented in one place where you could methodically make your way through them. No problem! Your book purchase comes with a free one-year subscription to all 1,001 practice problems online. You get on-the-go access any way you want it — from your computer, smartphone, or tablet. Track your progress and view personalized reports that show where you need to study the most. And then do it. Study what, where, when, and how you want.

What you'll find online

The online practice that comes free with this book offers you the same 1,001 questions and answers that are available here, presented in a multiple-choice format. The beauty of the online problems is that you can customize your online practice to focus on the topic areas that give you the most trouble. So if you aren't yet a whiz at factoring polynomials and solving quadratic equations, then select these problem types and BAM! — just those types of problems appear for your solving pleasure. Or, if you're short on time but want to get a mixed bag of a limited number of problems, you can plug in the quantity of problems you want to practice and that many — or few — of a variety of algebra problems appears. Whether you practice a couple hundred problems in one sitting or a couple dozen, and whether you focus on a few types of problems or practice every type, the online program keeps track of the questions you get right and wrong so that you can monitor your progress and spend time studying exactly what you need.

You can access this online tool using a PIN code, as described in the next section. Keep in mind that you can create only one login with your PIN. Once the PIN is used, it's no longer valid and is nontransferable. So you can't share your PIN with other users after you've established your login credentials.

How to register

To gain access to additional tests and practice online, all you have to do is register. Just follow these simple steps:

1. **Register your book or ebook at Dummies.com to get your PIN. Go to `www.dummies.com/go/getaccess`.**
2. **Select your product from the dropdown list on that page.**
3. **Follow the prompts to validate your product, and then check your email for a confirmation message that includes your PIN and instructions for logging in.**

If you do not receive this email within two hours, please check your spam folder before contacting us through our Technical Support website at `http://support.wiley.com` or by phone at 877-762-2974.

Now you're ready to go! You can come back to the practice material as often as you want — simply log on with the username and password you created during your initial login. No need to enter the access code a second time.

Your registration is good for one year from the day you activate your PIN.

Where to Go for Additional Help

The written directions given with the individual problems are designed to tell you what you need to do to get the correct answer. Sometimes the directions may seem vague if you aren't familiar with the words or the context of the words. Go ahead and look at the solution to see if that helps you with the meaning. But if the vocabulary is still unrecognizable, you may want to refer to the glossary in an algebra book, such as *Algebra I For Dummies,* written by yours truly and published by the fine folks at Wiley.

The solution to each problem is given at the end of its respective chapter. But you may not be able to follow from one step to the next. Is something missing? This book is designed to provide you with enough practice to become very efficient in algebra, but it isn't intended to give the step-by-step explanation on how and why each step is necessary. You may need to refer to *Algebra I For Dummies* or *Algebra I Essentials For Dummies* (also written by me and published by Wiley) to get more background on a problem or to understand why a particular step is taken in the solution of the problem.

Part I
The Questions

In this part . . .

One thousand and one algebra problems. That's a lot of work. But imagine how much work it was for me to write them. Don't get me started. Anyway, here are the general types of questions you'll be dealing with:

- Performing basic operations (Chapters 1 through 6)
- Changing the format of algebraic expressions (Chapters 7 through 12)
- Solving Equations (Chapters 13 through 17)
- Applying algebra by using formulas and solving word problems (Chapters 18 through 20)
- Graphing (Chapters 21 through 23)

Chapter 1

Signing on with Signed Numbers

Signed numbers include all real numbers, positive or negative, except 0. In other words, signed numbers are all numbers that have a positive or negative sign. You usually don't put a plus sign in front of a positive number, though, unless you're doing math problems. When you see the number 7, you just assume that it's +7. The number 0 is the only number that isn't either positive or negative and doesn't have a plus or minus sign in front of it; it's the dividing place between positive and negative numbers.

The Problems You'll Work On

As you work with signed numbers (and positive and negative values), here are the types of problems you'll do in this chapter:

- Placing numbers in their correct position on the number line — starting from smallest to largest as you move from left to right
- Performing the absolute value operation — determining the distance from the number to 0
- Adding signed numbers — finding the sum when the signs are the same, and finding the difference when the signs are different
- Subtracting signed numbers — changing the second number to its opposite and then using the rules for addition
- Multiplying and dividing signed numbers — counting the number of negative signs and assigning a positive sign to the answer when an even number of negatives exist and a negative sign to the answer when an odd number of negatives exist

What to Watch Out For

Pay careful attention to the following items when working on the signed number problems in this chapter:

- Keeping track of the order of numbers when dealing with negative numbers and fractions
- Working from left to right when adding and subtracting more than two terms
- Determining the sign when multiplying and dividing signed numbers, being careful not to include numbers without signs when counting how many negatives are present
- Reducing fractions correctly and dividing only by common factors

Placing Real Numbers on the Number Line

1–6 *Determine the correct order of the numbers on the real number line.*

1. Determine the order of the numbers:

$-3, 4, -1, 0, -4$

2. Determine the order of the numbers:

$-3, 3, -2, 0, 1$

3. Determine the order of the numbers:

$-1, 2, -5, \frac{3}{7}, -\frac{7}{3}$

4. Determine the order of the numbers:

$\frac{5}{6}, -\frac{6}{5}, -2, -4, 0$

5. Determine the order of the numbers:

$\sqrt{3}, -\sqrt{2}, 0, 3, -4$

6. Determine the order of the numbers:

$-3, \sqrt{3}, 0, 2, 4, -\frac{7}{2}$

Using the Absolute Value Operation

7–10 *Evaluate each expression involving absolute value.*

7. $|-4|$

8. $|-7.6|$

9. $-|-2|$

10. $-\left|-\frac{2}{3}\right|$

Adding Signed Numbers

11–20 *Find the sum of the signed numbers.*

11. $-4 + (-2) =$

12. $2 + (-4) =$

13. $-2 + 4 =$

14. $-5 + 3 =$

15. $-6 + 6 =$

16. $7 + (-2) =$

17. $5 + (-4) + (-2) =$

18. $-1 + 2 + (-3) + 4 =$

19. $-67 + 68 + (-69) + 70 =$

20. $-4 + (-5) + (-6) + (-7) + 7 + 4 =$

Subtracting Signed Numbers

21–30 *Find the difference between the signed numbers.*

21. $-4 - 6 =$

22. $7 - (-8) =$

23. $6 - 3 =$

24. $-9 - (-4) =$

25. $-7 - 7 =$

26. $-7 - (-7) =$

27. $3-(-2)=$

28. $-[-2]-3=$

29. $-[-4]-(-4)=$

30. $0-(-5)=$

Multiplying and Dividing Signed Numbers

31 – 50 *Find the products and quotients involving signed numbers.*

31. $2(-3)=$

32. $-4(-5)=$

33. $-5(6)=$

34. $3(-1)=$

35. $(-7)(-7)=$

36. $(-8)(8)=$

37. $-6\left(-\frac{5}{3}\right)=$

38. $20\left(-\frac{3}{4}\right)=$

39. $-2(0)=$

40. $(-1)(-1)(-1)(-1)=$

41. $\frac{-6}{2}=$

42. $\frac{-8}{-4} =$

43. $\frac{12}{-3} =$

44. $\frac{-60}{-15} =$

45. $\frac{0}{-2} =$

46. $\frac{-5}{1} =$

47. $\frac{-16}{2(-4)} =$

48. $\frac{2(-6)(-1)}{4(-3)} =$

49. $\frac{-4(-3)(-2)(-1)}{6(-1)(-1)(-1)} =$

50. $\frac{2(2)(-3)(-3)}{(-2)(-2)(3)(3)} =$

Chapter 2

Recognizing Algebraic Properties and Notation

The properties used in mathematics were established hundreds of years ago. Mathematicians around the world wanted to be able to communicate with one another; more specifically, they wanted to get the same answers when working on the same questions. To help with that, they developed and adopted rules such as the commutative property of addition and multiplication, the associative property of addition and multiplication, and the distributive property.

The Problems You'll Work On

To strengthen your skills with algebraic properties and notation, you'll practice doing the following in this chapter:

- Using the distributive property of multiplication over addition and subtraction
- Paying attention to the order of operations
- Simplifying radicals and radical expressions
- Reassociating terms for easier computation
- Regrouping and commuting for ease and accuracy

What to Watch Out For

Here are a few things to keep in mind while you work in this chapter:

- Distributing a negative number over several terms and being sure to apply the negative sign to each term
- Recognizing the fraction line as a grouping symbol
- Performing the absolute value operation when it's used as a grouping symbol
- Applying the correct exponent when multiplying or dividing variables

Applying Traditional Grouping Symbols

51–58 *Simplify the expressions.*

51. $6-(5-3)=$

52. $(4-3)-5=$

53. $5[6+(3-5)]=$

54. $8\{3-[4+(5-6)]\}=$

55. $\frac{8-4}{2}=$

56. $\frac{12}{9-11}=$

57. $\frac{3-(6-2)}{7-8}=$

58. $\frac{(5-6)-(9-3)}{3-10}=$

Introducing Some Non-Traditional Grouping Symbols

59–64 *Simplify the expressions involving radicals and absolute value.*

59. $\sqrt{12-8}=$

60. $\frac{11-3}{\sqrt{9-5}}=$

61. $\frac{\sqrt{11-10}}{3-4}=$

62. $|5-6|-7=$

63. $5-|4-7|=$

64. $\frac{\sqrt{20-4}}{|9-11|}=$

Distributing Multiplication over Addition and Subtraction

65–72 *Perform the distributions over addition and subtraction.*

65. $2(7 - y) =$

66. $-6(x + 4) =$

67. $-3\left(x - \frac{1}{3}\right) =$

68. $\frac{3}{4}(8 - 16y) =$

69. $x(y - 6) =$

70. $-4x(x - 2y + 3) =$

71. $12\left(\frac{1}{6} - \frac{2}{3} + \frac{3}{4}\right) =$

72. $-5\left(4x - \frac{2}{5}\right) =$

Associating Terms Differently with the Associative Property

73–78 *Use the associative property to simplify the expressions.*

73. $47 + (-47 + 90) =$

74. $(-6 + 23) - 23 =$

75. $\frac{3}{5}\left(\frac{5}{3} \cdot 29\right) =$

76. $\left(811 \cdot \frac{1}{15}\right)15 =$

77. $(16 + 19) + (-19 + 4) =$

78. $(77 - 53.2) + 53.2 =$

Rearranging with the Commutative Property

79–84 *Use the commutative property to simplify the expressions.*

79. $-16 + 47 + 16 =$

80. $\frac{3}{11}+\frac{4}{3}-\frac{3}{11}=$

81. $432 + 673 - 432 =$

82. $\frac{31}{17}\left(-\frac{12}{13}\right)\left(\frac{17}{31}\right)=$

83. $\frac{1}{9}\cdot\frac{25}{11}\cdot 18=$

84. $-3 + 4 + 23 + 3 - 23 =$

Applying More Than One Property to an Expression

85–90 *Simplify each expression using the commutative, associative, and distributive properties.*

85. $-32 + 4(8 - x) =$

86. $-5(x - 2) - 10 =$

87. $48\left(\frac{x}{6}+1\right)-8x=$

88. $-\frac{1}{4}(4x-20)-5=$

89. $-2(3 + y) + 3(y + 2) =$

90. $\frac{1}{8}\cdot\frac{5}{4}\left(-\frac{8}{5}\right)=$

Chapter 3

Working with Fractions and Decimals

Fractions and decimals are closely related. A *fraction* can be expressed as either a repeating or terminating decimal. A *decimal* is a special type of fraction — it always has a denominator that's some power of ten. Decimal numbers are often written with a *lead zero*. You'll see 0.031 instead of .031. The lead zero helps keep the decimal point from getting overlooked.

The Problems You'll Work On

In this chapter, you'll work with fractions and decimals in the following ways:

- Adding and subtracting fractions by finding a common denominator
- Multiplying and dividing fractions by changing to improper fractions and reducing where possible
- Simplifying complex fractions
- Adding and subtracting decimals by aligning decimal points
- Multiplying decimals by assigning the decimal place last
- Dividing decimals by assigning the decimal place first
- Changing fractions to decimals — repeating or terminating
- Changing decimals to fractions and then reducing
- Rounding decimals to designated places

What to Watch Out For

Don't let common mistakes trip you up; remember the following when working with fractions and decimals:

- Finding the least common denominator of fractions before adding or subtracting
- Recognizing the numerators and denominators in the numerator and denominator of a complex fraction
- Reducing fractions correctly by dividing by factors, not terms
- Recognizing the correct decimal place when rounding

Adding and Subtracting Fractions

91–96 *Find the sums and differences of the fractions.*

91. $\frac{3}{5}+\frac{5}{6}=$

92. $\frac{24}{25}-\frac{8}{15}=$

93. $4\frac{3}{8}+6\frac{5}{6}=$

94. $6\frac{3}{10}-2\frac{5}{6}=$

95. $\frac{4}{5}+\frac{3}{8}-\frac{1}{3}-\frac{4}{7}=$

96. $3\frac{1}{8}-1\frac{5}{12}+4-\frac{1}{2}=$

Multiplying Fractions

97–100 *Multiply the fractions and mixed numbers.*

97. $\frac{16}{25}\times\frac{65}{64}=$

98. $-\frac{70}{93}\times\frac{39}{35}=$

99. $3\frac{3}{7}\times 5\frac{4}{9}=$

100. $-3\frac{1}{3}\times 2\frac{2}{5}\times\left(-4\frac{2}{7}\right)=$

Dividing Fractions

100–104 *Divide the fractions and mixed numbers.*

101. $\frac{48}{35}\div\frac{20}{21}=$

102. $7\frac{1}{5}\div 5\frac{1}{7}=$

103. $12 \div 5\frac{3}{5} =$

104. $-45 \div 4\frac{1}{6} =$

Simplifying Complex Fractions

105–110 *Simplify the complex fractions.*

105. $\dfrac{1}{1+\frac{2}{3}} =$

106. $\dfrac{\frac{1}{2}}{2+\frac{1}{3}} =$

107. $\dfrac{\frac{3}{4}}{\frac{3}{5}-\frac{1}{2}} =$

108. $\dfrac{4\frac{1}{2}}{6-\frac{2}{3}} =$

109. $\dfrac{x}{1+\frac{1}{x}} =$

110. $\dfrac{\frac{1}{x+3}-\frac{1}{x}}{3} =$

Adding and Subtracting Decimals

111–114 *Find the sums and differences of the decimal numbers and variable expressions.*

111. $432.04 + 6.0001 =$

112. $15.4 - 5.123 =$

113. $x + 0.043x =$

114. $5.3y - 4.712y =$

Multiplying Decimals

115–118 *Find the products of the decimal numbers and variable expressions.*

115. $4.3 \times 0.056 =$

116. $6.21(-5.5) =$

117. $8.3x(0.004x) =$

118. $3.7y(-4.5y)(-0.1y) =$

Dividing Decimals

119–124 *Find the quotients of the decimal numbers. Round the answer to three decimal places, if necessary.*

119. $36.5 \div 0.05 =$

120. $0.143 \div 1.1 =$

121. $6 \div 0.0123 =$

122. $-72 \div 3.06 =$

123. $1.45 \div 0.03 =$

124. $67.4 \div 0.037 =$

Changing Fractions to Decimals

125–132 *Rewrite each fraction as an equivalent decimal.*

125. $\frac{1}{8} =$

126. $\frac{7}{25} =$

127. $\frac{99}{1,250} =$

128. $\frac{6}{10,000} =$

129. $\frac{4}{9} =$

130. $\frac{8}{11} =$

131. $\frac{30}{37} =$

132. $\frac{9}{101} =$

Changing Decimals to Fractions

133–140 *Rewrite each decimal as an equivalent fraction.*

133. 0.75

134. 0.875

135. 0.0008

136. 0.1525

137. 0.888...

138. 0.636363...

139. 0.261261...

140. 0.285714285714...

Chapter 4

Making Exponential Expressions and Operations More Compatible

An exponential expression consists of a base and a power. The general format of an exponential expression is b^n, where b is the base and n is the power or exponent. The base, b, has to be a positive number, and the power, n, is a real number. Positive powers, negative powers, and fractional powers all have special meanings and designations.

The Problems You'll Work On

Here are some of the things you do in this chapter:

- Multiplying and dividing exponential factors with the same base
- Raising a power to a power — putting an exponent on an exponential expression
- Combining operations— deciding what comes first when multiplying, dividing, and raising to powers
- Changing numbers to the same base so they can be combined
- Writing numbers using scientific notation

What to Watch Out For

Be sure you also remember the following:

- Writing fractional expressions by using the correct power of a base
- Recognizing a common base in different numbers
- Remembering when to add, subtract, and multiply the exponents
- Using the correct power of ten in scientific notation expressions

Multiplying and Dividing Exponentials with the Same Base

141–150 *Perform the operations and simplify.*

141. $3^2 \cdot 3^3$

142. $2^{-1} \cdot 2^6$

143. $4 \cdot 4^2 \cdot 4^{-1}$

144. $5 \cdot 5^{-3} \cdot 5^5$

145. $\frac{6^4}{6^2}$

146. $\frac{2^4}{2}$

147. $\frac{3^2}{3^{-1}}$

148. $\frac{5}{5^{-2}}$

149. $\frac{4^{-1}}{4^2}$

150. $\frac{3^{-2}}{3^{-3}}$

Raising a Power to a Power

151–160 *Compute the powers and simplify your answers.*

151. $(2^2)^3$

152. $(3^2)^2$

153. $(4^4)^{1/2}$

154. $(3^{1/3})^6$

155. $(5^{-2})^{-1}$

156. $(2^{-3})^{-2}$

157. $(3^{9})^{-1/3}$

158. $(4^{-2/5})^{-10}$

159. $(5^{-2/7})^{7}$

160. $(6^{-4/3})^{-3/4}$

Combining Different Operations on Exponentials

161–170 *Use the order of operations to compute the final answers.*

161. $\frac{2^2 \cdot 2^3}{2^4}$

162. $\frac{3^5 \cdot 3^2}{3^9}$

163. $4^{-3}(4^{2})^{2}$

164. $(5^{-1})^{3}(5^{4})^{1/2}$

165. $(6^{-3})^{-1}(6^{2})^{-2}$

166. $4^{3} \div (4^{8})^{3/4}$

167. $(2^{-3})^{-4} \div (2^{3})^{2}(2^{4})^{1/4}$

168. $(4^{2})^{-1/2}(2^{3})^{2}$

169. $(6^{1/2})^{4} \div (4^{-1/2})^{-2}$

170. $(2^{2})^{3}(2^{3})^{2} \div (2^{2})^{2}(2^{3})^{3}$

Changing the Base to Perform an Operation

171–180 *Perform the operations by changing the numbers to the same base.*

171. $2^4 \cdot 4^{-1}$

172. $3^{-2} \cdot 27^2$

173. $4^{1/3} \cdot 8^2 \cdot 2^{4/3}$

174. $5^3 \cdot 25^{-2}$

175. $\left(4^{-3}\right)^2\left(8^{1/3}\right)$

176. $\left(9^2\right)^{-1}(27)^3$

177. $6^{1/3} \cdot 2^{2/3} \cdot 3^{-4/3}$

178. $12^{-1} \cdot 3^2 \cdot 4$

179. $32^{1/2}\left(8^3\right)^{-1/2}$

180. $49^{-1/3}\left(7^2\right)^{-1/6}$

Working with Scientific Notation

181–190 *Perform the operations on the numbers written in scientific notation. Write your answer in scientific notation.*

181. $\left(2\times10^2\right)\left(4\times10^4\right)$

182. $\left(3\times10^4\right)\left(1.7\times10^{-2}\right)$

183. $\left(6\times10^3\right)\left(8\times10^7\right)$

184. $\left(5\times10^{-3}\right)\left(9\times10^{-4}\right)$

185. $(6.4\times10^{10})(5.2\times10^{-10})$

186. $(9\times10^{-2})\div(3\times10^{-4})$

187. $(1.8\times10^{4})\div(3.6\times10^{-2})$

188. $(5.1\times10^{-2})\div(3\times10^{-2})$

189. $(1.44\times10^{5})\div(1.6\times10^{-7})$

190. $(1\times10^{-17})\div(8\times10^{-15})$

Chapter 5

Raking in Radicals

Radical expressions are characterized by radical symbols and an *index* — a small number written in front of the radical symbol that indicates whether you have a cube root, a fourth root, and so on. When no number is written in front of the radical, you assume it's a square root.

The Problems You'll Work On

In this chapter, you get plenty of practice working with radicals in the following ways:

- Simplifying radical expressions by finding a perfect square factor
- Rationalizing denominators with one term
- Rationalizing denominators with two terms, using a conjugate
- Rewriting radicals with fractional exponents
- Dividing with radicals
- Solving operations involving fractional exponents
- Estimating the values of radical expressions

What to Watch Out For

As you get in your groove, solving one radical problem after another, don't overlook the following:

- Choosing the largest perfect square factor when simplifying a radical expression
- Multiplying correctly when writing equivalent fractions, using conjugates
- Performing operations correctly when fractions are involved
- Checking radical value estimates by comparing to nearest perfect square values

Simplifying Radical Expressions

191–196 *Simplify the radical expressions.*

191. $\sqrt{18}$

192. $\sqrt{360}$

193. $\sqrt{32}$

194. $\sqrt{640}$

195. $\sqrt{50}+\sqrt{72}$

196. $\sqrt{80}-\sqrt{125}$

Rationalizing Denominators

197–210 *Simplify the fractions by rationalizing the denominators.*

197. $\frac{5}{\sqrt{5}}$

198. $\frac{12}{\sqrt{6}}$

199. $\frac{18}{\sqrt{3}}$

200. $\frac{\sqrt{20}}{\sqrt{40}}$

201. $\frac{\sqrt{12}}{\sqrt{60}}$

202. $\frac{4}{8+\sqrt{6}}$

203. $\frac{10}{6+\sqrt{5}}$

204. $\frac{14}{7+\sqrt{7}}$

205. $\frac{4}{3-\sqrt{2}}$

206. $\frac{9}{5+\sqrt{10}}$

207. $\frac{2+\sqrt{3}}{4+\sqrt{6}}$

208. $\frac{4-\sqrt{10}}{8+\sqrt{6}}$

209. $\frac{3-\sqrt{20}}{9-\sqrt{5}}$

210. $\frac{4+\sqrt{24}}{1-\sqrt{6}}$

Using Fractional Exponents for Radicals

211–216 *Rewrite each radical expression using a fractional exponent.*

211. $\sqrt[3]{a^5}$

212. $\sqrt[7]{4^3}$

213. $\sqrt[5]{a^{25}}$

214. $\sqrt[3]{x^{15}}$

215. $\frac{2}{3\sqrt{y}}$

216. $\frac{x}{y\sqrt[3]{z}}$

Evaluating Expressions with Fractional Exponents

217–226 *Compute the value of each expression.*

217. $8^{2/3}$

218. $16^{3/4}$

219. $27^{4/3}$

220. $9^{5/2}$

221. $64^{5/6}$

222. $125^{2/3}$

223. $1000^{-2/3}$

224. $32^{-1/5}$

225. $\left(\frac{4}{9}\right)^{3/2}$

226. $\left(\frac{9}{100}\right)^{-1/2}$

Operating on Radicals

227–234 *Perform the operations on the radicals.*

227. $\left(\sqrt[4]{16}\right)^2$

228. $\left(\sqrt[6]{81}\right)^9$

229. $\left(\sqrt[4]{64}\right)^6$

230. $\left(\sqrt[3]{100}\right)^{-3}$

231. $\left(\sqrt[6]{256}\right)\left(\sqrt[3]{4}\right)$

232. $(\sqrt{27})(\sqrt[3]{81})(\sqrt[6]{3})$

233. $\frac{\sqrt[4]{64}}{\sqrt{32}}$

234. $\frac{\sqrt[6]{81}}{\sqrt[3]{243}}$

Operating on Factors with Fractional Exponents

235–242 *Perform the operations on the expressions.*

235. $6^{10/3} \cdot 6^{1/3}$

236. $5^{3/4} \cdot 5^{15/4}$

237. $\frac{8^{12/7}}{8^{3/7}}$

238. $\frac{4^{11/4}}{4^{3/2}}$

239. $\frac{27^{5/6}}{9^{1/4}}$

240. $\frac{16^{3/4}}{32^{3/5}}$

241. $\frac{8^{1/6} \cdot 2^{2/3}}{4^{1/12}}$

242. $\frac{25^{3/2}}{5^{1/2} \cdot 125^{1/6}}$

Estimating Values of Radicals

243–250 *Estimate the value of the radicals to the nearer tenth after simplifying the radicals. Use:* $\sqrt{2} \approx 1.4$, $\sqrt{3} \approx 1.7$, $\sqrt{5} \approx 2.2$

243. $\sqrt{20}$

244. $\sqrt{300}$

245. $\sqrt{75}$

246. $\sqrt{288}$

247. $\sqrt{18}+\sqrt{8}$

248. $\sqrt{125}-\sqrt{80}$

249. $\sqrt{6}$

250. $\sqrt{10}$

Chapter 6

Creating More User-Friendly Algebraic Expressions

Algebraic expressions involve *terms* (separated by addition and subtraction) and *factors* (connected by multiplication and division). Part of the challenge of working with algebraic expressions is in using the correct rules: the order of operations, rules of exponents, distributing, and so on. Function notation helps simplify some expressions by providing a rule and inviting evaluation.

The Problems You'll Work On

In this chapter, you get to put some of those algebraic rules to practice with the following types of problems:

- Finding the sums and differences of like terms
- Multiplying and dividing terms and performing the operations logically
- Applying the order of operations when simplifying expressions
- Evaluating algebraic expressions when variables are assigned specific values
- Using the factorial operation
- Getting acquainted with function notation

What to Watch Out For

Here are a few more things to keep in mind:

- Recognizing *like terms* and the processes involved when combining them
- Reducing fractions correctly — dividing by factors of *all* the terms
- Evaluating expressions within grouping symbols before applying the order of operations
- Reducing fractions involving factorials correctly

Adding and Subtracting Like Terms

251–258 *Simplify by combining like terms.*

251. $4a + 6a$

252. $9xy + 4xy - 5xy$

253. $5z - 3 - 2z + 7$

254. $6y + 4 - 3 - 8y$

255. $7a + 2b + ab - 3 + 4a - 2b - 5ab$

256. $3x^2 + 2x - 1 + 4x^2 - 5x + 3$

257. $9 - 3z + 4 - 7ab + 6b - ab - 4$

258. $x + 3 - y + 4 - z^2 + 5 - 2$

Multiplying and Dividing Factors

259–266 *Multiply or divide, as indicated.*

259. $4(3x)$

260. $-9(5y)$

261. $\frac{12x^2}{3x}$

262. $\frac{24y^2}{-6y^2}$

263. $3xy(4xy^2)$

264. $-5yz^2(3y^2z)$

265. $\frac{42a^2b}{6ab}$

266. $\frac{8ab^2cd}{4abcd}$

Simplifying Expressions Using the Order of Operations

267–286 *Simplify, applying the order of operations.*

267. $2+6\cdot 3$

268. $9-8\div 4$

269. $3\cdot 2+4(-3)$

270. $-6\div 3-4\div 2$

271. $5\sqrt{4}+3\sqrt{9}$

272. $7-2\sqrt{25}$

273. $6-4\cdot 2+8\div 4-1$

274. $20\div 5+4\cdot 3-1$

275. $10-4-8+6\cdot 2$

276. $36\div 6\div 3+3\cdot 2\cdot 5$

277. $4(6-3)$

278. $5(-3+2)$

279. $\frac{4+8}{6}$

280. $\frac{40}{3+7}$

281. $5\sqrt{4+12}$

282. $7\sqrt{10-1}$

283. $3+2(6-4)$

284. $8 - 7(1 + 3)$

285. $4(6 + 1) - 8(3 + 2)$

286. $\frac{6}{2+4} + \frac{12}{7-6}$

Evaluating Expressions Using the Order of Operations

287–296 *Evaluate the expressions.*

287. What is $3x^2$ if $x = -2$?

288. What is $-5x - 1$ if $x = -3$?

289. What is $x(2 - x)$ if $x = 4$?

290. What is $\frac{x-2}{x+3}$ if $x = -2$?

291. What is $2(l + w)$ if $l = 4$ and $w = 3$?

292. What is $\frac{1}{2}bh$ if $b = 9$ and $h = 4$?

293. What is $a_0 + (n-1)d$ if $a_0 = 4$, $n = 11$, and $d = 3$?

294. What is $\frac{9}{5}C + 32$ if $C = 40$?

295. What is $A\left(1+\frac{r}{n}\right)^{nt}$ if $A = 100$, $r = 2$, $n = 1$, and $t = 3$?

296. What is $\sqrt{x(x-a)(x-b)(x-c)}$ if $x = 6$, $a = 4$, $b = 3$, and $c = 5$?

Operating with Factorials

297–300 *Evaluate the factorial expressions.*

297. $3!$

298. $6! - 3!$

299. $\frac{4!}{2!}$

300. $\frac{6!}{3!}$

Focusing on Function Notation

301–310 *Evaluate the functions for the input value given.*

301. If $f(x) = x^2 + 3x + 1$, then $f(2) =$

302. If $g(x) = 9 - 3x^2$, then $g(-1) =$

303. If $h(x) = \sqrt{5 - x}$, then $h(-4) =$

304. If $k(x) = \frac{x-4}{2}$, then $k(10) =$

305. If $n(x) = x^3 + 2x^2$, then $n(2) =$

306. If $p(x) = \frac{(x+1)x^2}{4}$, then $p(3) =$

307. If $q(x) = x! + (x - 1)!$, then $q(4) =$

308. If $r(x) = \frac{x^2 - 4}{3x + 6}$, then $r(8) =$

309. If $t(x) = \frac{5 - x}{4 + x}$, then $t(-3) =$

310. If $w(x) = \sqrt{2x^2 - 7}$, then $w(4) =$

Chapter 7

Multiplying by One or More Terms

Multiplying algebraic expressions is much like multiplying numbers, but the introduction of variables makes the process just a bit more interesting. Products involving variables call on the rules of exponents. And, because of the commutative property of addition and multiplication, arrangements and rearrangements of terms and factors can make the process simpler.

The Problems You'll Work On

When multiplying by one or more terms, you deal with the following in this chapter:

- Distributing terms with one or more factors over two or more terms — multiplication over sums and differences
- Distributing division over sums and differences and dividing each term in the parentheses
- Distributing binomials over binomials or trinomials and then combining like terms
- Multiplying binomials using FOIL: First, Outer, Inner, Last
- Using Pascal's triangle to find powers of binomials
- Finding products of binomials times trinomials that create sums and differences of cubes

What to Watch Out For

With all the distributing and multiplying, don't overlook the following:

- Applying the rules of exponents to all terms when distributing variables over several terms
- Changing the sign of each term when distributing a negative factor over several terms
- Combining the outer and inner terms correctly when applying FOIL
- Starting with the zero power when assigning powers of the second term to the pattern in Pascal's triangle

Distributing One Term Over Sums and Differences

311–315 *Distribute the number over the terms in the parentheses.*

311. $3(2x + 4)$

312. $-4(5y - 6)$

313. $7(x^2 - 2x + 3)$

314. $-8\left(z - \frac{1}{2}\right)$

315. $12\left(4 + \frac{1}{6} - \frac{5}{4}\right)$

Distributing Using Division

316–320 *Perform the division by dividing each term in the numerator by the term in the denominator.*

316. $\frac{18 + 60 + 9}{3}$

317. $\frac{-50 + 75 + 200}{-5}$

318. $\frac{a^2 + 2a^3 - 3a^4}{a^2}$

319. $\frac{6x - 8y + 12z}{2}$

320. $\frac{20x^2y - 24xy^2 + 32y^3}{4y}$

Multiplying Binomials Using Distributing

321–325 *Distribute the first binomial over the second binomial and simplify.*

321. $(a + 1)(x - 2)$

322. $(y - 4)(z^2 + 7)$

323. $(x + 2)(y - 2)$

324. $(x^2 - 7)(x^3 - 8)$

325. $(x^2 + y^4)(x^2 - y^4)$

Multiplying Binomials Using FOIL

326–335 *Multiply the binomials using "FOIL."*

326. $(x - 3)(x + 2)$

327. $(y + 6)(y + 4)$

328. $(2x - 3)(3x - 2)$

329. $(z - 4)(3z - 8)$

330. $(5x + 3)(4x - 2)$

331. $(3y - 4)(7y + 4)$

332. $(x^2 - 1)(x^2 + 1)$

333. $(2y^3 + 1)(3y^3 - 2)$

334. $(8x - 7)(8x + 7)$

335. $(2z^2 + 3)(2z^2 - 3)$

Distributing Binomials Over Trinomials

336–340 *Distribute the binomial over the trinomial and simplify.*

336. $(x + 3)(x^2 - 2x + 1)$

337. $(y - 2)(y^2 + 3y + 4)$

338. $(2z + 1)(z^2 + z + 7)$

339. $(4x - 3)(2x^2 + 2x + 1)$

340. $(y + 7)(3y^2 - 7y + 5)$

Squaring Binomials

341–345 *Square the binomials.*

341. $(x + 5)^2$

342. $(y - 6)^2$

343. $(4z + 3)^2$

344. $(5x - 2)^2$

345. $(8x + y)^2$

Raising Binomials to the Third Power

346–350 *Raise the binomials to the third power.*

346. $(x + 2)^3$

347. $(y - 4)^3$

348. $(3z + 2)^3$

349. $(2x^2 + 1)^3$

350. $(a^2 - b)^3$

Using Pascal's Triangle

351–360 *Raise the binomial to the indicated power.*

351. $(x + 3)^4$

352. $(y - 2)^5$

353. $(z + 1)^6$

354. $(a + b)^7$

355. $(x-2)^7$

356. $(4z+1)^4$

357. $(3y-2)^5$

358. $(2x+3)^6$

359. $(3x+2y)^4$

360. $(2z-3w)^5$

Finding Special Products of Binomials and Trinomials

361–365 *Distribute the binomial over the trinomial to determine the "special" product.*

361. $(x-1)(x^2+x+1)$

362. $(y+2)(y^2-2y+4)$

363. $(z-4)(z^2+4z+16)$

364. $(3x-2)(9x^2+6x+4)$

365. $(5z+2w)(25z^2-10zw+4w^2)$

Chapter 8

Dividing Algebraic Expressions

Division is the opposite or inverse of multiplication. Instead of adding exponents, you subtract the exponents of like variables. When dividing an expression containing several terms by an expression containing just one term, you have two possible situations: the divisor evenly divides each term, meaning fractions formed from each term and the divisor reduce to denominators of 1, or the divisor doesn't evenly divide one or more of the terms. What you do with the second situation depends on the application you're working on at the time. The problems in this chapter present various options.

The Problems You'll Work On

The problems in this chapter are all about division and include the following:

- Dividing several terms by a number
- Dividing several terms by a term containing numerical and variable factors
- Dividing several terms by a binomial, using long division
- Dividing several terms by a binomial, using synthetic division
- Dividing by trinomials

What to Watch Out For

As you work through dividing one expression after another, watch out for the following:

- Assigning the correct sign to each term in the result
- Remembering to change the sign of each term when dividing by a negative term
- Changing the signs of all products of quotient term times divisor term in each step of a long division problem
- Inserting zeros for missing terms when using synthetic division
- Changing the sign of the number in the binomial when setting up a synthetic division problem
- Starting with an exponent one smaller than that in the divisor when writing the quotient in a synthetic division problem

Dividing with Monomial Divisors

366–375 *Divide each numerator by the monomial.*

366. $\frac{3x^2+6}{3}$

367. $\frac{12y^3+8y}{4}$

368. $\frac{x^2-2x}{x}$

369. $\frac{4x^2-5x}{x}$

370. $\frac{6y^4-2y^2-8y}{2y}$

371. $\frac{3z+6z^2-12z^3}{3z}$

372. $\frac{9x^2y+12xy^2}{3xy}$

373. $\frac{40abc-60ab}{20ab}$

374. $\frac{12x^2y^3-18xy^4+24y^5}{6y^3}$

375. $\frac{42x^3y^2-35x^2y^3+14xy^4}{7xy^2}$

Monomial Divisors and Remainders

376–385 *Divide each numerator by the monomial. Write any remainders as fractions.*

376. $\frac{9y^2+3y-2}{3}$

377. $\frac{9y^2+3y-2}{3y}$

378. $\frac{50x^2+25x+1}{25}$

379. $\frac{50x^2+25x+1}{25x}$

380. $\frac{2x^4-4x^3+6x^2-4x+1}{x}$

381. $\frac{2x^4 - 4x^3 + 6x^2 - 4x + 1}{2x}$

382. $\frac{5xy^2 - 10x^2y^3 + 15x^3y^4 - 25}{5xy^2}$

383. $\frac{ax^2 + bx + c}{x}$

384. $\frac{ax^2 + bx + c}{a}$

385. $\frac{ax^2 + bx + c}{ax}$

Using Long Division to Divide with Binomials

386–395 *Divide each numerator by the binomial, using long division. Write any remainders as fractions.*

386. $\frac{x^2 - 3x - 2}{x - 1}$

387. $\frac{2x^2 + 4x + 1}{x - 2}$

388. $\frac{5x^2 + 3x - 7}{x + 3}$

389. $\frac{2x^2 - 4x + 1}{x + 1}$

390. $\frac{x^3 + 2x^2 - x - 2}{x - 1}$

391. $\frac{2x^3 + x^2 - 18x - 9}{x + 3}$

392. $\frac{3x^3 + x^2 - 2x + 1}{x - 4}$

393. $\frac{x^4 - 3x^3 + x - 2}{x + 1}$

394. $\frac{2x^4 + 3x^2 - x + 1}{x - 3}$

395. $\frac{2x^3 + 1}{x - 1}$

Dividing with Binomials Using Synthetic Division

396–405 *Divide each numerator by the binomial, using synthetic division. Write any remainders as fractions.*

396. $\frac{x^2-3x-2}{x-1}$

397. $\frac{2x^2+4x+1}{x-2}$

398. $\frac{5x^2+3x-7}{x+3}$

399. $\frac{2x^2-4x+1}{x+1}$

400. $\frac{x^3+2x^2-x-2}{x-1}$

401. $\frac{2x^3+x^2-18x-9}{x+3}$

402. $\frac{3x^3+x^2-2x+1}{x-4}$

403. $\frac{x^4-3x^3+x-2}{x+1}$

404. $\frac{2x^4+3x^2-x+1}{x-3}$

405. $\frac{2x^3+1}{x-1}$

Dividing with Higher Power Divisors

406–415 *Divide each numerator by the denominator, using long division. Write any remainders as fractions.*

406. $\frac{x^4-1}{x^2-1}$

407. $\frac{y^4-81}{y^2-9}$

408. $\frac{x^3-3x^2+3x-1}{x^2-2x+1}$

409. $\frac{x^4+4x^3+6x^2+4x+1}{x+1}$

410. $\frac{x^4-3x^3+2x^2+3x-1}{x^2+2x-1}$

411. $\frac{x^4+2x^3+2x^2+4x+1}{x^2+2}$

412. $\frac{x^4+x^3-2x^2+14x+1}{x^2-4}$

413. $\frac{x^4+3x^3+5x^2+12x+4}{x^2+3x+1}$

414. $\frac{x^6+3x^5+x^4+x^2+3}{x^4+1}$

415. $\frac{x^6-1}{x^5-1}$

Chapter 9

Factoring Basics

Factoring algebraic expressions is one of the most important techniques you'll practice. Not much else can be done in terms of solving equations, graphing functions and conics, and working on applications if you can't pull out a common factor and simplify an expression. Factoring changes an expression of two or more terms into one big product, which is really just one term. Having everything multiplied together allows for finding common factors in two or more expressions and reducing fractions. It also allows for the application of the multiplication property of zero. Factoring is crucial, essential, and basic to algebra.

The Problems You'll Work On

In this chapter, you work through factoring basics in the following ways:

- Determining what divides a number by using the rules of divisibility
- Creating prime factorizations of numbers
- Finding a numerical GCF (greatest common factor)
- Factoring out a GCF containing numbers and variables
- Reducing fractions with monomial divisors
- Reducing fractions with polynomial divisors

What to Watch Out For

Here are a few things to keep in mind as you factor your way through this chapter:

- Making sure you apply divisibility rules correctly
- Writing a prime factorization with the correct exponents on the prime factors
- Checking that the terms remaining after dividing out a GCF don't still have a common factor
- Reducing only factors, not terms
- Writing fractional answers with correct grouping symbols to distinguish remaining factors

Finding Divisors Using Rules of Divisibility

416–421 *Use divisibility rules for numbers 2 through 11 to determine values that evenly divide the given number.*

416. 88

417. 1,010

418. 3,492

419. 4,257

420. 1,940

421. 3,003

Writing Prime Factorizations

422–429 *Write the prime factorization of each number.*

422. 28

423. 45

424. 150

425. 108

426. 512

427. 500

428. 1,936

429. 2,700

Factoring Out a GCF

430–443 *Factor each using the GCF.*

430. $24x^4 - 30y^8$

431. $44z^5 + 60a - 8$

432. $300abc + 420xyz$

433. $121x^4 - 165z$

434. $24x^2y^3 - 48x^3y^2$

435. $36a^3b - 24a^2b^2 - 40ab^3$

436. $9z^{-4} + 15z^{-3} - 27z^{-1}$

437. $20y^{3/4} - 25y^{1/4}$

438. $16a^{1/2}b^{3/4}c^{4/5} - 48a^{3/2}b^{7/4}c^{9/5}$

439. $8x^2(5x-1) + 6x^3(5x-1)$

440. $36x^{-3}y^4 + 20x^{-5}y^2$

441. $125x^{-3}y^{-4} + 500x^{-5}y^{-2}$

442. $x(3x-1)^2 + 2x^2(3x-1)$

443. $4x^3(x-4)^4 - 6x^4(x-4)^3$

Reducing Fractions with a Common GCF

444–455 *Reduce the fractions by dividing with the GCF of the numerator and denominator.*

444. $\frac{4x(x-1)^2-16x^2(x-1)}{(x-1)^4}$

445. $\frac{12x^2(x-3)^{-4}-8x^3(x-3)^{-3}}{4(x-3)^{-6}}$

446. $\frac{7!a^5b^{-1}}{4!a^{-3}b^{-2}}$

447. $\frac{100!a^{3/2}b^{1/2}}{99!a^{1/2}b^{-1/2}}$

448. $\frac{14a^2b-35a}{28ab^3}$

449. $\frac{36x^2y^4-54xy^6}{60x^3y^2}$

450. $\frac{16w^3(w+1)^2-8w^4(w+1)^3}{20w^5(w+1)^3-12w^2(w+1)^4}$

451. $\frac{x^2+x}{(x+1)(x-1)}$

452. $\frac{x(x+1)-6(x+1)}{x^2-36}$

453. $\frac{6x^2-19x-7}{2x(x+1)-7(x+1)}$

454. $\frac{x^2(x+4)-4x(x+4)}{x^3+8x^2+16x}$

455. $\frac{(x-3)(x+3)(x-2)(x^2+2x+4)}{x^5(x-2)-3x^4(x-2)}$

Chapter 10

Factoring Binomials

A *binomial* is an expression with two terms. The terms can be separated by addition or subtraction. You have four possibilities for factoring binomials: (1) factor out a greatest common factor, (2) factor as the difference of perfect squares, (3) factor as the difference of perfect cubes, and (4) factor as the sum of perfect cubes. If one of these methods doesn't work, the binomial doesn't factor when using real numbers.

The Problems You'll Work On

The problems in this chapter focus on the following:

- Factoring when the two terms are the difference of perfect squares (both the numbers and the variables must be perfect squares)
- Factoring when the two terms are the difference of perfect cubes (both the numbers and the variables must be perfect cubes)
- Factoring when the two terms are the sum of perfect cubes (both the numbers and the variables must be perfect cubes)
- Using more than one factorization technique in a problem

What to Watch Out For

When working through the steps necessary for factoring binomials, pay careful attention to the following:

- Recognizing when a number is a perfect square so you can apply the factorization technique
- Knowing enough of the perfect cubes to recognize them in binomials
- Using the correct sign between the first and second terms of the trinomial when factoring sums and differences of cubes
- Trying to factor the sum of perfect squares and mistaking it for the technique used with cubes
- Using the correct exponents when factoring higher-power squares and cubes and dividing the exponents, not taking their root

Factoring the Difference of Perfect Squares

456–465 *Factor each binomial using the pattern for the difference of squares.*

456. $x^2 - 36$

457. $9y^2 - 100$

458. $81a^2 - y^2$

459. $4x^2 - 49z^2$

460. $64x^2y^2 - 25z^2w^4$

461. $36a^4b^6 - 121$

462. $121x^{1/2} - 144y^{1/4}$

463. $25x^{-2} - 9y^{-4}$

464. $16 - x^2y^{-1/4}$

465. $z^{-4/9} - 49w^{1/2}$

Factoring the Sum or Difference of Two Perfect Cubes

466–475 *Factor each as the sum or difference of perfect cubes.*

466. $x^3 + 8$

467. $x^3 + 343$

468. $a^3 - 216z^3$

469. $1 - y^3$

470. $125z^3 + 343$

471. $8a^3 + 27b^3$

472. $729x^3 - 1000y^6$

473. $512x^9 - 125y^{27}$

474. $27x^{1/3} - 1$

475. $8y^{-6} + 343z^{-1}$

Factoring More Than Once

476–495 *Completely factor each binomial.*

476. $3x^4y^3 - 75x^2y^3$

477. $6x^4y^2 - 96x^2y^4$

478. $36z^2 - 3600w^2$

479. $100x^3 - 900x$

480. $32y^4 + 4y$

481. $4x^4y^2 + 32xy^2$

482. $625x^4 - 1$

483. $16x^4 - 81y^8$

484. $x^{-4} - x^{-7}$

485. $y^{-8} - y^{-12}$

486. $216a^3b^3 + 216c^6$

487. $125a^2b^4 - 500c^6$

488. $a^3b^6 - 8b^6$

489. $81x^2y^3 + 3x^2$

490. $32x^4y - 4xy$

491. $9a^6b^5 + 72z^3b^2$

492. $2x^6 - 162x^2$

493. $x^6 - 1$

494. $y^{12} - 64$

495. $10a^4x^2 - 1000b^8x^2$

Chapter 11

Factoring Quadratic Trinomials

You can factor trinomials with the form $ax^2 + bx + c$ in one of two ways: (1) factor out a GCF, or (2) find two binomials whose product is that trinomial. When finding the two binomials whose product is a particular trinomial, you work from the factors of the constant term and the factors of the coefficient of the lead term to create a sum or difference that matches the coefficient of the middle term. This technique can be expanded to trinomials that have the same general format but with exponents that are multiples of the basic trinomial.

The Problems You'll Work On

Here are the types of things you work on in this chapter:

- Factoring out a GCF (greatest common factor)
- Creating the product of two binomials, both with variable coefficients of 1
- Creating the product of two binomials, one or both with variable coefficients not equal to 1
- Applying the techniques to quadratic-like trinomials
- Using more than one factorization method in a problem

What to Watch Out For

Be aware of the following when factoring quadratic trinomials:

- Assigning the correct sign to each term, especially when a factor or term is negative
- Positioning the signs correctly in the product of binomials so a difference has the correct sign after cross-multiplying
- Finding the correct factors of coefficients and constants when you have several to choose from in the problem
- Recognizing when a factor in a problem can be factored again

Factoring Out the GCF of a Trinomial

496–499 *Factor out the GCF of each.*

496. $12x^4y^2 - 6x^3y^3 + 21x^2y^4$

497. $70a^2b^3c + 63a^3b^2c^2 - 21a^4bc^3$

498. $3(x-4)^3 + 6x(x-4)^2 - 9x^2(x-4)$

499. $60x^5y - 48x^6y^2 + 36x^2y^3$

Factoring Trinomials into the Products of Binomials

500–511 *Factor each trinomial into the product of two binomials.*

500. $x^2 - 8x - 20$

501. $x^2 + 10x + 9$

502. $y^2 - 6y - 16$

503. $z^2 + 2z - 48$

504. $2x^2 + x - 6$

505. $3x^2 + 5x - 12$

506. $9z^2 + 24z + 16$

507. $16x^2 - 40x + 25$

508. $w^2 - 63w - 64$

509. $4x^2 + 15x - 25$

510. $40x^2 - 3x - 54$

511. $16x^2 - 14x - 15$

Factoring Quadratic-Like Expressions

512–519 *Factor the quadratic-like expressions into the product of two binomials.*

512. $x^{10} - 5x^5 + 4$

513. $y^6 - 4y^3 - 21$

514. $y^{16} - 25$

515. $25a^4 - 49b^{10}$

516. $x^{-8} - 3x^{-4} - 18$

517. $x^{-6} + 5x^{-3} + 4$

518. $5x^{1/3} - 11x^{1/6} + 2$

519. $6x^{2/5} - x^{1/5} - 12$

Factoring Completely Using More Than One Technique

520–535 *Completely factor each trinomial.*

520. $5z^2 + 30z + 45$

521. $18x^3 + 12x^2 + 2x$

522. $4y^3 - 8y^2 - 12y$

523. $6x^6 + x^5 - x^4$

524. $x^5+8x^4+16x^3$

525. $96y-48y^2+6y^3$

526. w^4-13w^2+36

527. x^6-9x^3+8

528. $5x^2(x+3)^3+15x(x+3)^3-50(x+3)^3$

529. $4x^2(x+1)-6x(x+1)-4(x+1)$

530. $a^2(x^2-81)+13a(x^2-81)+22(x^2-81)$

531. $4y^2(x^3-1)+4y(x^3-1)-3(x^3-1)$

532. $300y^{1/4}+70y^{1/8}-150$

533. $6y-6y^{1/2}-12$

534. $3x^{-3}-19x^{-2}+20x^{-1}$

535. $12x^{-2}+5x^{-3}-2x^{-4}$

Chapter 12

Other Factoring Techniques

The process of factoring binomials and quadratic trinomials is pretty much scripted with the various choices available for each format. When you start factoring expressions with more than three terms, you need different techniques to create the factorization — or to recognize that factors may not even exist.

The Problems You'll Work On

In this chapter on factoring polynomials, you deal with the following situations:

- Factoring four, six, and eight (or more) terms, using grouping
- Recognizing and using uneven grouping to create differences of squares
- Starting with grouping and then finding differences of squares in the factors
- Starting with grouping and then finding sums or differences of cubes in the factors
- Recognizing that factoring out the GCF (greatest common factor) first makes factoring easier

What to Watch Out For

Here are a few things to keep in mind while you work on the factoring:

- Making sure factors created from grouping are exactly the same
- Factoring negative terms correctly
- Recognizing when a problem can be factored again and knowing when to stop

Factoring by Grouping

536–543 *Factor by grouping.*

536. $bc - 3b + 2c - 6$

537. $x^2 - abx + xyz - abyz$

538. $2x^3 - 3x^2 + 2x - 3$

539. $2xz^2 + 8x - 3z^2 - 12$

540. $n^{3/2} + 2n - 4n^{1/2} - 8$

541. $y^{5/2} - 3y^2 + 2y^{1/2} - 6$

542. $4x - 12 + xy - 3y - xz + 3z$

543. $kx + 4x + ky + 4y + kz + 4z$

Combining Other Factoring Techniques with Grouping

544–547 *Factor each completely, beginning with grouping.*

544. $x^2y^2 + 3x^2 - xy^2 - 3x - 12y^2 - 36$

545. $2x^4 - 4x^2 + 3x^3 - 6x + x^2 - 2$

546. $m^2n + 3m^2 - 25n - 75$

547. $4x^3 + 16x^2 - 25x - 100$

Using Multiple Factoring Methods

548–565 *Completely factor each expression.*

548. $4x^3 - 196x$

549. $6x^5 - 48x^2$

550. $y^5 - 4y^3 - 27y^2 + 108$

551. $x^5 - 13x^3 + 36x$

552. $16x^4 + 23x^2 - 75$

553. $4x^6 - 4x^2$

554. $z^6 - 729$

555. $y^8 - 1$

556. $64b^5 - 64b^3 + b^2 - 1$

557. $27z^5 - 243z^3 - 8z^2 + 72$

558. $z^8 - 17z^4 + 16$

559. $x^5 - 2x^4 + x^3$

560. $x^4 - 8x^2 + 16$

561. $y^{-3} - 27y^{-6}$

562. $\left(x^2-1\right)^2(3x+4)^2 + (3x+4)^3\left(x^2-1\right)$

563. $\left(y^3+8\right)^4\left(y^2-9\right) - \left(y^2-9\right)^2\left(y^3+8\right)^3$

564. $(z+1)^{1/2}\left(z^3-1\right)^2 - \left(z^3-1\right)(z+1)^{3/2}$

565. $4x^6 - 25x^4 + 500x^3 - 3125x$

Chapter 13

Solving Linear Equations

Linear equations are of the form $ax + b = c$, where x is some variable, and a, b, and c are real numbers. To solve a linear equation, you perform a series of *opposites:* If a number is added to the term containing x, you subtract that number from both sides of the equation; if a number is subtracted from the term containing the variable, you add; if a number multiplies the variable, you divide; and if a number divides the variable, you multiply. Just be sure that whatever you do to one side of the equation, you also do to the other side. Think of the equation as two expressions pivoting on either side of a balance scale — you need to keep the sides evenly balanced.

The Problems You'll Work On

Here are a few things you do in this chapter to find answers:

- Using the addition property to solve linear equations
- Using multiplication and division — inverses of one another — to solve linear equations
- Combining operations — doing more than one operation for a solution
- Rewriting equations by eliminating grouping symbols
- Getting rid of fractions and fractional expressions before solving linear equations

What to Watch Out For

As you get in a groove solving linear equations, be sure you're doing the following:

- Performing the same operation on each side of the equation to keep it balanced
- Being careful to use the inverse/opposite number when adding and subtracting
- Never multiplying or dividing by 0
- Distributing correctly to remove grouping symbols, especially with negative numbers

Solving Linear Equations Using Basic Operations

566–583 *Solve the linear equations for the value of the variable.*

566. $x + 7 = 15$

567. $y - 3 = 10$

568. $z + 14 = 2$

569. $x - 3 = -4$

570. $5x + 3 = 3x - 1$

571. $3x + 9 = 5x - 7$

572. $4y + 9 + 5y - 6 = 4y + 7 + 3y - 2$

573. $5z + 1 - 3z + 5 = 12 - 4z$

574. $-8x = 24$

575. $-4x = -2$

576. $4y - 9y = 20$

577. $-12 = 3z$

578. $3x - 4 = 11$

579. $5x + 7 = 2$

580. $4x - 3 = 7x + 9$

581. $4x + 7 = 3x - 8$

582. $8y + 14 - 3y = 5 + 3y$

583. $9x + 4 - 5x = 3 + 3x$

Dealing with Fractions in Linear Equations

584–605 *Solve for the value of the variable by eliminating the fraction.*

584. $\frac{z}{3} = 9$

585. $\frac{w}{6} = -12$

586. $\frac{w}{-4} = -8$

587. $\frac{x}{-5} = 10$

588. $6 - \frac{y}{2} = 11$

589. $\frac{x}{3} + 7 = 5$

590. $\frac{z}{4} - 3 = z + 9$

591. $4 + \frac{w}{2} = 13 - w$

592. $\frac{5}{4}x - 4 = \frac{9}{4}x + 8$

593. $\frac{5}{2}x + 1 = 2 + \frac{15}{2}x$

594. $\frac{3x+1}{4} = x + 2$

595. $w - 2 = \frac{3w-2}{4}$

596. $\frac{x-3}{6} - 1 = 2$

597. $\frac{2z+1}{3}+5=2$

598. $\frac{4x}{3}-\frac{3x}{5}=11$

599. $\frac{x}{5}+6=3+\frac{x}{8}$

600. $\frac{2(y+3)}{3}-1=\frac{5(y-2)}{7}+2$

601. $\frac{4(x-2)}{7}+3=\frac{x+3}{6}+5$

602. $\frac{x}{5}+\frac{x}{4}+\frac{x}{2}=19$

603. $\frac{y}{4}+\frac{y}{5}-\frac{y}{6}=17$

604. $\frac{12}{y}+\frac{15}{y}=9$

605. $\frac{8}{x}-\frac{3}{x}=1$

Chapter 14

Taking on Quadratic Equations

A quadratic equation has the form $ax^2 + bx + c = 0$. The equation can have exactly two solutions, only one solution (a double root), or no solutions among the real numbers. Where no real solution occurs, imaginary numbers are brought into the picture. Quadratic equations are solved most easily when the expression that's set to 0 factors, but the quadratic formula is also a nice means to finding solutions.

The Problems You'll Work On

In this chapter, you work with quadratic equations in the following ways:

- Applying the square root rule
- Solving equations by using factoring and the multiplication property of 0
- Using the quadratic formula and simplifying radicals when possible
- Solving quadratic equations by completing the square
- Introducing imaginary solutions
- Simplifying complex solutions with or without radicals

What to Watch Out For

Don't get too caught up in your work and neglect the following:

- Applying the square root rule only when you have $ax^2 = c$
- Using the correct signs when applying the multiplication property of 0
- Watching the order of operations when simplifying the work in the quadratic formula
- Simplifying the fraction correctly in the quadratic formula
- Pulling out the square root of –1 when determining imaginary roots

Applying the Square Root Rule to Quadratic Equations

606–613 *Solve each quadratic equation using the square root rule.*

606. $x^2 = 25$

607. $x^2 = 121$

608. $3y^2 = 27$

609. $5z^2 = 80$

610. $n^2 - 100 = 0$

611. $m^2 - 1 = 0$

612. $4x^2 - 9 = 0$

613. $24x^2 - 150 = 0$

Solving Quadratic Equations Using Factoring

614–629 *Solve the quadratic equations using factoring.*

614. $x^2 - 2x - 15 = 0$

615. $y^2 + 15y + 44 = 0$

616. $2x^2 + x - 6 = 0$

617. $3x^2 - 8x + 5 = 0$

618. $y^2 - 3y = 0$

619. $z^2 = 7z$

620. $2x^2 + x = 0$

621. $3y^2 = 2y$

622. $8x^2 - 6x - 9 = 0$

623. $10x^2 + 29x + 10 = 0$

624. $16x^2 + 4x - 2 = 0$

625. $6x^2 - 9x - 15 = 0$

626. $\frac{1}{2}x^2 - x - \frac{3}{2} = 0$

627. $x^2 - \frac{11}{6}x + \frac{2}{3} = 0$

628. $y^2 = \frac{3}{2}y$

629. $\frac{1}{4}x^2 = 4x$

Applying the Quadratic Formula to Quadratic Equations

630–641 *Solve each quadratic equation using the quadratic formula.*

630. $x^2 + 3x - 4 = 0$

631. $x^2 - 8x + 12 = 0$

632. $2x^2 + x - 6 = 0$

633. $10x^2 + 13x + 4 = 0$

634. $x^2 - 3x - 1 = 0$

635. $x^2 + 5x + 2 = 0$

636. $2x^2 - x - 5 = 0$

637. $2x^2 - 4x - 5 = 0$

638. $3x^2 + 6x + 1 = 0$

639. $x^2 - 7x - 17 = 0$

640. $2x^2 + 8x + 3 = 0$

641. $x^2 - 12x + 9 = 0$

Completing the Square to Solve Quadratic Equations

642–645 *Solve each quadratic equation by "completing the square."*

642. $x^2 + 2x - 24 = 0$

643. $2x^2 + 11x - 40 = 0$

644. $x^2 - 4x + 2 = 0$

645. $x^2 - 12x - 9 = 0$

Writing Complex Numbers in the Standard a + bi Form

646–653 *Rewrite each as a complex number in the form a + bi.*

646. $\sqrt{-9}$

647. $\sqrt{-25}$

648. $4-\sqrt{-36}$

649. $-5+\sqrt{-49}$

650. $-2+\sqrt{-12}$

651. $3-\sqrt{-27}$

652. $-2+\sqrt{-200}$

653. $4-\sqrt{-75}$

Finding Complex Solutions Using the Quadratic Formula

654–655 *Use the quadratic formula to solve the equations. Write your answers as complex numbers.*

654. $x^2+4x+8=0$

655. $x^2+x+25=0$

Chapter 15

Solving Polynomials with Powers Three and Higher

A *polynomial* is a smooth curve that goes on and on forever, using input variables going from negative infinity to positive infinity. To *solve* a polynomial means to set the equation equal to 0 and determine which, if any, numbers create a true statement. Any numbers satisfying this equation give you important information: They tell you where the graph of the polynomial crosses or touches the x-axis.

The Problems You'll Work On

Solving polynomials in this chapter requires the following techniques:

- Counting the number of possible real roots/zeros, using Descartes's Rule of Signs
- Making a list of the possible rational roots/zeros, using the Rational Root Theorem
- Putting Descartes's Rule of Signs and the Rational Root Theorem together to find roots
- Applying the Factor Theorem
- Solving polynomial equations by factoring
- Applying synthetic division

What to Watch Out For

As you probably know, you can come up with a different answer to a math problem by simply confusing or forgetting one step; here are some things to watch out for:

- Confusing *real roots* with *rational roots;* rational roots are real, but real roots aren't necessarily rational
- Being sure to list all the possible divisors of a number, not missing multiples
- Remembering to change the sign of the numerical part of the divisor when using synthetic division
- Taking roots with *multiplicity* of more than one into account when looking for factors

Applying Descartes's Rule of Signs to Count Real Roots

656–659 *Count the possible number of positive and negative real roots of the equation.*

656. $x^4 - 3x^3 + 2x^2 - 4x - 9 = 0$

657. $x^5 - x^3 + 4x + 1 = 0$

658. $5x^4 - 3x^3 + 6x - 2 = 0$

659. $x^6 + x^4 - x^3 + 6x^2 - x + 9 = 0$

Applying the Rational Root Theorem to List Roots

660–663 *List all the possible rational roots for each polynomial equation.*

660. $x^5 + x^4 - 4x^3 - 2x^2 - 4x + 8 = 0$

661. $5x^4 - 3x^2 + 6x - 6 = 0$

662. $2x^5 - 5x^4 + 2x^3 - 3x^2 + 4 = 0$

663. $6x^4 - 3x^3 + 2x^2 + 5x + 3 = 0$

Determining Whether Numbers Are Roots

664–667 *Check to see which of the given values are roots of the equation.*

664. Given $x^3 - 3x^2 + 2x + 24 = 0$, check to see whether 2, –2, 3, or 4 is a root.

665. Given $x^4 - 5x^3 + 3x^2 + 8x + 3 = 0$, check to see whether 1, –1, 3, or –3 is a root.

666. Given $x^5 - 4x^4 - 3x^3 + 4x + 2 = 0$, determine whether 1, –1, 2, or –2 is a root.

667. Given $x^6 - x^5 + x^3 - 2x + 1 = 0$, determine whether 1 or –1 are roots.

Solving for the Roots of Polynomials

668–685 *Solve for all real roots.*

668. $x^3 + 3x^2 - 4x - 12 = 0$

669. $x^3 - x^2 - 25x + 25 = 0$

670. $x^3 + 4x^2 + x - 6 = 0$

671. $x^3 - x^2 - 26x - 24 = 0$

672. $x^4 - 81 = 0$

673. $x^6 - 64 = 0$

674. $x^3 + 7x^2 + 8x - 16 = 0$

675. $x^3 - 9x^2 + 24x - 20 = 0$

676. $x^4 - 37x^2 + 36 = 0$

677. $x^4 - 73x^2 + 576 = 0$

678. $x^4 - 4x^3 - 3x^2 + 10x + 8 = 0$

679. $x^5 - x^4 - 22x^3 - 44x^2 - 24x = 0$

680. $4x^3 - 9x^2 - 4x + 9 = 0$

681. $4x^3 + 12x^2 - 9x - 27 = 0$

682. $2x^5 + 5x^4 - 5x^3 - 5x^2 + 3x = 0$

683. $3x^4 - 5x^3 - 77x^2 + 125x + 50 = 0$

684. $8x^4 - 30x^3 - 51x^2 + 263x - 210 = 0$

685. $5x^5 - 6x^4 - 14x^3 + 28x^2 - 15x + 2 = 0$

Chapter 16

Reining in Radical and Absolute Value Equations

A radical equation contains at least one term that's a square root, cube root, or some other root. When solving radical equations, you apply a method that's effective but comes with a built-in error possibility; you may find (and need to recognize) *extraneous* solutions. You need to rewrite absolute value equations to solve them. The solutions of the *rewrites* are then the solutions of the original equation.

The Problems You'll Work On

Here's just a sampling of the radical things you work on in this chapter:

- Rewriting equations with only two terms and squaring both sides to solve
- Squaring both sides of an equation when starting with three terms
- Dealing with more than one radical term
- Catching extraneous solutions
- Graphing absolute value statements for clarity
- Solving absolute value equations after writing corresponding equations
- Checking for nonsense answers

What to Watch Out For

Here are a few things that may rock your boat, so be on the lookout:

- Squaring binomials correctly and not forgetting the middle term
- Factoring statements containing radicals correctly
- Checking for extraneous solutions by using the *original* equation, not the version changed in the process
- Catching impossible situations in initial absolute value equations

Solving Basic Radical Equations

686–689 *Solve each radical equation by squaring both sides.*

686. $\sqrt{x+2}=4$

687. $\sqrt{2-y}=3$

688. $\sqrt{z^2+7}=4$

689. $\sqrt{x^2-11}=5$

Checking for Extraneous Roots

690–697 *Solve the radical equations by squaring both sides; check for extraneous solutions.*

690. $\sqrt{x+3}+x=9$

691. $\sqrt{2x+9}+x=13$

692. $\sqrt{x-5}+7=x$

693. $\sqrt{2x-1}+2=x$

694. $\sqrt{x+9}-x=3$

695. $\sqrt{2x+10}-x=5$

696. $\sqrt{x-3}=x-3$

697. $\sqrt{x-7}=7-x$

Squaring Both Sides of Equations Twice

698–701 *Solve each radical equation by squaring both sides of the equation twice.*

698. $\sqrt{x+1}+1=\sqrt{2x+3}$

699. $\sqrt{3x-3}+2=\sqrt{5x+5}$

700. $4\sqrt{x+5}+\sqrt{x+8}=6$

701. $3\sqrt{x-1}-\sqrt{x+6}=5$

Solving Radicals with Roots Other Than Square Roots

702–703 *Solve the radical equations.*

702. $\sqrt[3]{x-4}+2=3$

703. $\sqrt[5]{3-y}-2=1$

Solving Absolute Value Equations

704–713 *Solve each absolute value equation by writing the two corresponding linear equations and solving.*

704. $|x+3|=8$

705. $|y-4|=3$

706. $|5z+3|=2$

707. $|3-2x|=4$

708. $|4w-1|-6=9$

709. $8+|2-w|=10$

710. $5|3x+1|=10$

711. $3|x+4|-2=7$

712. $|-3x|=4$

713. $|-2x-3|=15$

Handling Absolute Value Equations with No Solution

714–715 *Solve the absolute value equations, and check the answers carefully.*

714. $|3x - 2| + 4 = 1$

715. $3 - |4 - 5x| = 7$

Chapter 17

Making Inequalities More Fair

An inequality is a statement involving more than one expression and/or number. When two expressions are set greater than or less than one another, you want to determine for what numbers the statement is true. Inequalities can also involve several statements, one greater than the next, greater than the next, and so on. Solving these statements involves treating each section exactly the same and using the rules for dealing with inequalities.

The Problems You'll Work On

In short, here's what you'll be doing in this chapter:

- Using the rules special to inequality statements
- Writing solutions in both inequality notation and *interval* notation
- Solving linear inequalities
- Taking compound inequalities section by section
- Using a number line for nonlinear inequalities
- Dealing carefully with rational inequalities
- Rewriting and solving absolute value inequalities
- Finding the intersection of solutions when solving complex inequalities

What to Watch Out For

As you zip through the problems in this chapter, keep the following in mind:

- Reversing the signs when using multiplication or division of negative numbers
- Never multiplying or dividing by 0
- Using a parenthesis when writing interval notation involving infinity
- Treating the numerator and denominator as factors when solving rational inequalities

Performing Operations on Inequalities

716–719 *Perform the indicated operation on the inequalities.*

716. Starting with $7 > 3$, add -2 to each side, and then multiply each side by -4.

717. Starting with $-4 < 1$, multiply each side by -2, and then subtract 3 from each side.

718. Starting with $-6 \le 6$, divide each side by -3, and then add 3 to each side of the equation.

719. Starting with $0 \ge -4$, add 3 to each side of the equation and then multiply each side by -1.

Writing Inequalities Using Interval Notation

720–723 *Change the inequality notation to interval notation.*

720. $-3 \le x < 2$

721. $0 \le x \le 4$

722. $x > -3$

723. $x \le 7$

Changing Interval Notation to Inequality Notation

724–727 *Change the interval notation to inequality notation.*

724. $[-6, \infty)$

725. $(-\infty, -2)$

726. $[-4, 7)$

727. $(2, 3)$

Solving Linear Inequalities

728–733 *Solve each linear inequality for the values of the variable.*

728. $2x - 5 < 3$

729. $3x - 2 \geq 4x + 3$

730. $-3(x + 7) \leq 2x + 9$

731. $\frac{x-3}{-2} > x$

732. $\frac{x}{3} + \frac{x}{4} > 7$

733. $\frac{3}{4} - \frac{x}{5} < \frac{2x}{3} - \frac{7}{60}$

Taking on Compound Inequalities

734–737 *Solve each compound inequality.*

734. $-5 \leq 3x + 1 < 7$

735. $-4 < 6 - 5x < 11$

736. $2 \leq \frac{x-1}{4} \leq 7$

737. $-15 < -3(3 - 2x) < -9$

Solving Quadratic Inequalities

738–745 *Solve each quadratic inequality using a number line.*

738. $(x - 3)(x + 4) < 0$

739. $(2x + 5)(x + 8) \geq 0$

740. $x^2 - 8x - 9 \leq 0$

741. $x^2 - 4x - 21 > 0$

742. $48 - x^2 > -2x$

743. $36 - x^2 \leq 0$

744. $5x^2 < 15x$

745. $x^2 + 4x + 4 \geq 0$

Finding Solutions of Nonlinear Inequalities

746–753 *Solve each nonlinear inequality using a number line.*

746. $x(x + 3)(x - 2) > 0$

747. $(x + 1)^2(x + 5)(x - 7) \leq 0$

748. $x^3 + x^2 - 36x - 36 \geq 0$

749. $x^3 - 2x^2 + x < 0$

750. $\frac{x+1}{x+2} < 0$

751. $\frac{x^2-25}{x+3} > 0$

752. $\frac{x}{x-4} \leq 0$

753. $\frac{x^2+4x-5}{x^2-9} \geq 0$

Rewriting and Solving Absolute Value Inequalities

754–761 *Solve the absolute value inequalities by rewriting the statements.*

754. $|3x+2| \geq 7$

755. $|4-x| < 6$

756. $5|2x+1| \leq 20$

757. $\frac{1}{2}|5x+3| \geq 11$

758. $|x+4|-5>3$

759. $|5-2x|+4 \leq 7$

760. $2|x-5|-4 \geq -2$

761. $\frac{1}{6}|3x-1|+3<5$

Delving into Complex Inequalities

762–765 *Solve the complex inequalities.*

762. $-4 < 3x + 2 \leq 2x + 3$

763. $1 \leq 2x-5 \leq \frac{1}{3}x+10$

764. $-5 < 4x - 1 < 6x + 7$

765. $x + 1 \leq 3x + 5 < 8$

Chapter 18

Using Established Formulas

A formula is a rule that describes a situation that happens consistently or exists without variation. One of the first formulas that people learn is that for the area of a rectangle — just multiply the length by the width. The trick to using formulas is to understand what the different symbols represent and then to be able to apply the mathematical rules correctly.

The Problems You'll Work On

The majority of the problems in this chapter involve simply determining which formula to use, where to use it, and applying the following techniques:

- Figuring the interest by using the simple interest or compound interest formula
- Determining the height of an object after a certain amount of time
- Computing how far you've traveled given rate and time
- Calculating the sum of the measures of the angles in a polygon
- Finding the average or weighted average of items
- Summing a series of numbers
- Figuring out the value of a term in a sequence of numbers

What to Watch Out For

Whether you struggle remembering formulas or are a formula whiz, be sure you don't overlook the following:

- Assigning the correct value to the different variables in a formula
- Changing units, if necessary, to have consistency in the formula's input values
- Performing the operations correctly by using the order of operations

Getting Interested in Interest Problems

766–769 *Solve each using the simple interest formula,* $I = Prt$.

766. How much interest is earned if you invest $20,000 at 2.5% for ten years?

767. If you're lending a friend $4,000 at 4% simple interest for two years, what is the total amount you'll be paid back?

768. You're buying a television from an appliance store for $3,600. They're charging you 11% simple interest for three years. How much will your monthly payments be?

769. You have $10,000 to invest. You want to put it in an account that earns $4\frac{1}{4}\%$ simple interest. How long will it take for your investment to total $13,400 (so you can buy that boat)?

Heating It Up with Temperature Problems

770–773 *Solve the problems using the temperature formulas:* $°F = \frac{9}{5}°C + 32$

$$°C = \frac{5}{9}(°F - 32)$$

770. What temperature in degrees Fahrenheit corresponds to 37 degrees Celsius?

771. What temperature in degrees Celsius corresponds to 59 degrees Fahrenheit?

772. What temperature in degrees Celsius corresponds to 212 degrees Fahrenheit?

773. What temperature in degrees Fahrenheit corresponds to –40 degrees Celsius?

Adding up Natural Numbers

774–777 *Solve the problems using* $S_n = \frac{n(n+1)}{2}$ *for the sum of the first n natural numbers.*

774. What is the sum of the numbers from 1 through 50?

775. If the sum of the first n numbers is 5,050, then what are the numbers?

776. What is the sum of the natural numbers 40 through 60?

777. A theater has seating that begins with 36 seats in the front row and increases by one seat each row through the 30th row back. How many seats are in the theater?

Going the Distance with the Distance Formula

778–785 *Solve the problems using the distance, rate, and time formula, $d = rt$.*

778. How far do you travel if you're driving at 55 mph for six hours?

779. How fast were you driving if you traveled 450 miles in 7 hours, 30 minutes?

780. How long does it take to travel 1,050 miles if you're averaging 60 miles per hour?

781. What was your average speed if you left home at 8:00 a.m., drove 150 miles, stopped for an hour, drove another 200 miles, and arrived at your destination at 4:00 p.m.?

782. Hank left home at 8:00 a.m. traveling at 50 mph. Helen found his briefcase on the table and left at 8:30 trying to catch up with Hank. She was traveling at 60 mph. What time did Helen catch up with Hank?

783. A bus left Chicago at 6:00 a.m. traveling due east at 40 mph. A second bus left Chicago at 7:00 traveling due west at 55 mph. At what time are the two buses 800 miles apart?

784. Claire and Charlie decided to walk around the lake. Claire started in a clockwise direction walking 4 mph, and Charlie started at the same time in a counterclockwise direction walking 5 mph. After half an hour, Charlie took a 15-minute break. If the lake is 10 miles around, how long did it take for them to meet?

785. Bill and Will drove from Peoria to their home in Missouri in separate vehicles. It took Bill one hour longer than Will, because Bill drives an average of 10 mph slower than Will. If it took Bill five hours, then how far was the trip in miles?

Getting the Inside Scoop with Sums of Interior Angles

786–789 *Solve the problems using the formula for the sum of the measures of all the interior angles of a polygon with n sides: $A = 180(n - 2)$.*

786. What is the sum of the measures of all the angles in a hexagon (six-sided)?

787. What is the sum of the measures of all the angles in a decagon?

788. If the sum of the measures of the angles in a polygon is 1,080°, then what is the polygon?

789. If the sum of the measures of the angles in a *regular* (all sides equal) polygon is 1,800 degrees, then what is the measure of just one of the angles in the polygon?

Averaging Out the Numbers

790–793 *Solve the problems using the formula for the average of n numbers, $A = \frac{x_1 + x_2 + x_3 + \cdots + x_n}{n}$.*

790. Stephan got scores of 81, 67, 93, and 99 on his exams. What is his average?

791. Stephanie got three scores of 9, two scores of 10, four scores of 7, and one score of 8 on her quizzes. What is her average quiz score?

792. Joel has test scores of 85, 87, 93, and 100 on his first exams. What does he need on the last exam to have an average of at least 90?

793. Tomas has an average of 91 on his first three exams. What does he have to earn on the last exam to bring the average up to at least a 93?

Summing the Squares of Numbers

794–795 *Solve the problems using the formula for the sum of n squares, $S = \frac{n(n+1)(2n+1)}{6}$.*

794. What is the sum of the numbers 1, 4, 9, 16, 25, and 36?

795. What is the sum of the squares of the numbers from 1 through 15?

Finding the Terms of a Sequence

796–799 *Solve the problems using the formula for the nth term in an arithmetic sequence, $a_n = a_1 + (n-1)d$.*

796. What is the tenth term in an arithmetic sequence whose first term is 1, and where the difference between the terms is 4?

797. What is the 100th term in an arithmetic sequence whose first term is –6, and where the difference between the terms is 2?

798. What is the 40th term in the arithmetic sequence that starts: 3, 7, 11, 15, . . . ?

799. What is the 20th term in the arithmetic sequence that starts: 100, 97, 94, 91, . . . ?

Adding the Terms in an Arithmetic Sequence

800–801 *Solve the problems using the formula for the sum of the first n terms in an arithmetic sequence, $S_n = \frac{n}{2}(a_1 + a_n)$.*

800. What is the sum of the even numbers from 40 through 60?

801. What is the sum of the first 40 numbers in the arithmetic sequence starting: 2, 5, 8, 11, . . . ?

Using the Formula for the Sum of Cubes

802–803 *Solve the problems using the formula for the sum of the first n cubes, $S_n = \frac{n^2(n+1)^2}{4}$.*

802. What is the sum of the numbers 1, 8, 27, 64, 125, 216, and 343?

803. What is the sum of the cubes of the numbers from 5 through 20?

Compounding the Problems Involving Compound Interest

804–807 *Solve the problems using the compound interest formula* $A = P\left(1+\frac{r}{n}\right)^{nt}$, *where A is the total accumulated amount, P is the principal (deposit), r is the interest rate written as a decimal, n is the number of times per year the money is compounded, and t is the number of years.*

804. What is the total amount of money that you accumulate after ten years if you deposit $5,000 in an account earning 3% interest compounded quarterly?

805. How much interest do you earn on $40,000 deposited for five years in an account earning $5\frac{1}{4}$% compounded monthly?

806. What is the total amount of money you accumulate in an account earning 6% interest compounded daily if you deposit $20,000 and leave it in the account for 15 years?

807. A member of Columbus's crew deposited $1 in the Bank of the West Indies in 1492. The deposit was earning 2% interest compounded quarterly. You were contacted in 2012, because you are a descendant of this crew member, and the account has been abandoned. They're going to send you a check for the amount in the account. How much money will you get?

Chapter 19

Using Formulas in Geometric Story Problems

Story problems (practical applications) appear in all types and levels of mathematics. A really nice feature of geometric story problems is that you can almost always find a formula to apply.

The Problems You'll Work On

From the beginning and through the middle and end, the geometric story problems in this chapter require the following skills:

- Finding the perimeter and area of rectangles
- Tackling trapezoids with their unequal sides
- Squaring up with properties of squares
- Recognizing hexagons and their usefulness in traffic control
- Applying the Pythagorean theorem and Pythagorean triples
- Computing the volume of a right rectangular prism
- Determining the surface area and volume of cylinders
- Dealing with the difference between radian and degree measures for angles
- Using Heron's formula for the area of a triangle

What to Watch Out For

Don't get ahead of yourself when working these problems. Watch out for the following:

- Recognizing which formula to use for which figure or application
- Matching up the correct numbers and units with the symbols in a formula
- Performing the formula's operations correctly

Working around the Perimeter of Rectangles

808–813 *Solve the problems using the formula for the perimeter of a rectangle: $P = 2(l + w)$.*

808. What is the perimeter of a rectangular yard measuring 6 yards wide and 8 yards long?

809. If the perimeter of a rectangular plot is 400 feet, and the width is 50 feet, then what is the length of the plot?

810. A rectangular room has a length that's three times the width. The perimeter is 480 feet. What is the length?

811. A rectangular pool has a length that's 4 feet greater than the width. It has a perimeter of 200 feet. What is the length of the pool?

812. A rectangle has a width that's 30 feet more than one-third the length. If the perimeter is 420 feet, then what is the length of the rectangle?

813. You have 600 feet of fencing and need to create a rectangular pen for your llama. You want the length and width to be in a ratio of 3:2. What length and width will work?

Using the Area Formula for a Trapezoid

814–815 *Solve the problems using the formula for the area of a trapezoid: $A = \frac{1}{2}h(b_1 + b_2)$.*

814. What is the area of a trapezoid if the parallel bases measure 5 feet and 8 feet, and if the perpendicular distance between those bases is 4 feet?

815. The area of a trapezoid is 170 square feet. What is the length of the second base if the first base measures 16 feet and the height is 5 feet?

Tackling Area and Perimeter of a Square

816–817 *Solve the problems using the formulas for the perimeter and area of a square: $P = 4s$ and $A = s^2$.*

816. If the area of a square is 64 square feet, then what is its perimeter?

817. If the perimeter of a square is 64 inches, then what is the area of the square?

Solving Problems Using the Perimeter of a Triangle

818–819 *Solve the problems using the fact that the perimeter of a triangle is equal to the sum of the measures of its sides.*

818. A triangle with a perimeter of 42 inches has one side twice the length of the shortest side and the third side 6 inches greater than the shortest side. What are the lengths of the three sides?

819. If you have a fenced-in triangular garden where one side is 8 feet, the second side is three times that length, and the third side is 8 feet shorter than the second side, then what would the lengths of the sides of an equilateral triangle be, if you made it from the current fencing?

Working with the Area and Perimeter of a Rectangle

820–821 *Solve the problems using the formulas for the perimeter and area of a rectangle:* $P = 2(l + w)$, $A = lw$.

820. A rectangle's length is 2 feet less than twice its width, and its area is 180 square feet. What is the rectangle's width?

821. A rectangle's width is 4 inches greater than half its length, and its perimeter is 248 inches. What is the rectangle's area?

Using the Formula for the Area of a Hexagon

822–823 *Solve the problems using the formula for the area of a regular hexagon:* $A = \frac{3\sqrt{3}}{2}x^2$, *where* x *is the length of a side.*

822. The perimeter of a regular hexagon is 360 cm. What is its area?

823. The area of a regular hexagon is $216\sqrt{3}$ square feet. What is the hexagon's perimeter?

Getting Pythagoras Involved in the Area of a Triangle

824–827 *Solve the problems using the formula for the area of a triangle,* $A = \frac{1}{2}bh$ *and the Pythagorean theorem,* $a^2 + b^2 = c^2$.

824. A triangle has an area of 60 square centimeters and a height of 10 centimeters. What is the length of the base of the triangle?

825. A triangle has an area of 2 square feet and a base measuring 36 inches. What is the height of the triangle? (Remember: 1 sq ft = 12 in × 12 in = 144 sq in)

826. A right triangle has a hypotenuse measuring 5 feet and a leg measuring 3 feet. What is its area?

827. The area of a right triangle is 30 square feet. If the measure of one of the legs is two more than twice the other leg, then what is the measure of the hypotenuse?

Making Use of the Volume of a Box

828–829 *Solve the problems using the formula for the volume of a right rectangular prism (box), $V = lwh$.*

828. The volume of a box is 48 cubic inches. If the length is four times the width, and the height is 3 inches, then what is the length of the box?

829. A cube has a volume of 64 cubic centimeters. What is its height?

Working with the Volume of a Cylinder

830–831 *Solve the problems using the formula for the volume of a right circular cylinder (can): $V = \pi r^2 h$*

830. What is the height of a cylinder whose volume is 54 cubic inches and whose radius is 3 inches?

831. What is the radius of a cylinder whose volume is 192π cubic feet and whose height is 12 feet?

Changing Radians to Degrees and Degrees to Radians

832–837 *Solve the problems using the angle measure proportion: $\frac{\theta^D}{180} = \frac{\theta^R}{\pi}$.*

832. What degree measure corresponds to $\frac{\pi}{4}$ radians?

833. What radian measure corresponds to 150°?

834. What radian measure corresponds to 300°?

835. What degree measure corresponds to $\frac{11\pi}{12}$ radians?

836. What radian measure corresponds to –30°?

837. What degree measure corresponds to 2 radians?

Determining the Height of an Object

838–841 *Solve the problems using the formula for the height, in feet, of a tennis ball shot from a launcher after t seconds:* $h = -16t^2 + 48t$.

838. How high is the tennis ball after 2 seconds?

839. How high is the tennis ball after 2.5 seconds?

840. After how many seconds is the height 20 feet?

841. After how many seconds does the ball hit the ground?

Working with Heron's Formula for the Area of a Triangle

842–845 *Solve the problems using Heron's formula for the area of a triangle:* $A = \sqrt{s(s-a)(s-b)(s-c)}$, *where a, b, and c are the lengths of the sides and s is the semi-perimeter.*

842. What is the area of a triangle whose sides measure 5, 12, and 13 inches?

843. What is the area of a triangle whose sides measure 6, 8, and 10 feet?

844. The area of a triangle is 6 square yards. If two of the sides measure 3 and 5 yards, respectively, then what is the measure of the third side of the triangle?

845. The area of an *equilateral* (all sides the same length) triangle is $25\sqrt{3}$ square inches. What is the perimeter of the triangle?

Chapter 20

Tackling Traditional Story Problems

Traditional story problems sometimes have a less-than-desirable reputation. You probably won't ever need to know how old a person is if he's twice as old as another person was 16 years ago. But the structure of these story problems and the discipline of working out the logic and mathematics go far to improve mathematical *agility*. The problems in this chapter are pretty traditional and classic in the mathematics classroom.

The Problems You'll Work On

The traditional story problems you'll work on in this chapter include the following scenarios:

- Determining a person's age with respect to others and the passage of time
- Figuring out how much work a person is doing and how long it will take to complete the project
- Computing how much of one liquid to add to another to get the desired mixture
- Counting pennies, nickels, and dimes without ever touching the money
- Regrouping and commuting values for ease and accuracy
- Determining one or more numbers in a list of consecutive integers

What to Watch Out For

Here are some ways to ensure that you're correctly solving the problems in this chapter:

- Creating an equation or expression that represents what's going on in the problem
- Always letting variables represent numbers, not people or things
- Taking advantage of drawing pictures to help create the correct equation
- Solving the equations correctly, using the rules of equations
- Checking your answers to see whether they fit the situation and make sense

Tackling Age Problems

846–853 *Solve the "age" problems.*

846. Jon is twice as old as Jim. Five years ago, Jon was three times as old as Jim at that time. How old are they now?

847. Grace is three times as old as Greta. In six years, Grace will be twice as old as Greta at that time. How old are they now?

848. Amanda is three times as old as Stefanie. In eight years, Amanda will be four years more than twice Stefanie's age then. How old are they?

849. Hank is two years older than Hal. Twelve years ago, Hank's age was four years less than twice Hal's age at that time. How old are they now?

850. Betty is four years older than Bart. In three years, the sum of their ages will be 46. How old are they?

851. Les is ten years older than Maura. Five years ago, the sum of their ages was 20. How old are they now?

852. Joe is twice as old as Moe, and Louie is two years older than Joe. In two years, the sum of their ages will be 33. How old are they now?

853. Barb is three years older than Karen. Mary is twice Barb's age. Five years ago, the sum of their ages was 34. How old are they now?

Working with Consecutive Integers

854–867 *Solve the following consecutive integer problems.*

854. The sum of three consecutive integers is 90. What are the integers?

855. The sum of five consecutive integers is 5. What are the integers?

856. The sum of three consecutive even integers is 138. What is the largest of the integers?

857. The sum of four consecutive even integers is 412. What is the smallest of the integers?

858. The sum of four consecutive odd integers is 176. What is the largest of the integers?

859. The sum of five consecutive odd integers is 755. What is the middle integer in the list?

860. The sum of three consecutive multiples of 4 is 120. What are the multiples of 4?

861. The sum of four consecutive multiples of 6 is 108. What is the largest of those numbers?

862. The sum of five consecutive multiples of 7 is 525. What is the middle number in the list?

863. The sum of the first and last numbers in a list of five even integers is 108. What is the middle number?

864. The sum of the first and last numbers in a list of six multiples of 4 is 12. What is the last number?

865. The sum of the first and last numbers in a list of five multiples of three is 108. What is the first number?

866. The product of two consecutive positive even integers is 142 more than their sum. What are the numbers?

867. The product of the first and last numbers in a list of three consecutive multiples of 4 is 16 less than their sum. What are the three numbers?

Getting the Job Done on Work Problems

868–875 *Solve the work problems letting x represent the time it takes to do the job working together.*

868. Shirley can weed the garden in 4 hours, and John can weed it in 6 hours. How many hours will it take to weed the garden if they work together?

869. Ken can paint the room in 5 hours, but Paula is much more careful and would need 20 hours. How long would it take them to paint the room if they worked together?

870. Madeline can clean the house in 4 hours, but it would take Katie 8 hours to do the job. How many hours would it take if they worked together?

871. Larry, Moe, and Curly were given the job of cleaning out the garage. Larry likes to get things done quickly and could do the whole job himself in 3 hours. Moe spends a lot of time looking at things and remembering, so it would take him twice as long as Larry, or 6 hours. Curly cannot stay on task, so he'd need 9 hours to clean the garage by himself. If they can work together without getting into trouble, how many hours would it take them to clean the job working together?

872. Three TAs are assigned the task of grading the test papers from all the calculus sections. Dan could grade all the papers in 4 hours, Don would need 8 hours, and Duane couldn't do it in less than 18 hours. How many hours will it take if they all work together?

873. Fred and Ted power wash the deck together in 1 hour and 20 minutes. It would take Fred 4 hours to do it himself. How long would it take Ted to do the job if he did it by himself?

874. It takes Jake and Blake 2 and a half hours to wash and dry all the banquet dishes when they work together. Doing the whole job by himself, it takes Jake 3 hours. How long would Blake need if he had to do the job himself?

875. When working together, it takes Dasher, Dancer, and Prancer 4 hours to decorate the Christmas tree. Working alone, Dasher would need 12 hours. Left to his own devices, Dancer could decorate the tree by himself in 9 hours. How long would it take Prancer to do the job by himself?

Counting on Quality and Quantity Problems

876–879 *Solve the quality/quantity problems involving solutions.*

876. How many quarts of 70% apple juice do you have to add to 10 quarts of 20% apple juice to produce a mixture that's 50% apple juice?

877. How many ounces of 60% solution do you have to add to 6 ounces of 10% solution to create a mixture of 50% solution?

878. How many ounces of chocolate syrup do you need to add to 8 ounces of milk to make a chocolate milk mix that's 20% chocolate?

879. How many ounces of antifreeze do you need to drain from your 16-quart radiator and replace with pure antifreeze to bring the current 50% solution up to a 70% solution?

Solving Money Problems

880–885 *Solve the quality/quantity problems involving money.*

880. Janie has $50 in dimes and quarters. If she has 60 more quarters than dimes, then how many quarters does she have?

881. Stan has $17 in nickels and quarters. If he has exactly 100 coins, then how many of them are quarters?

882. Elliott has $4 in nickels, dimes, and quarters. If he has twice as many nickels as quarters, and four more dimes than quarters, then how much money does he have in nickels?

883. Ryan has 78 coins totaling $8.10. He has three times as many dimes as quarters, and the rest are nickels. How many of each coin does he have?

884. Lynn is giving Roger $50 in newly minted quarters, half-dollars and silver dollars. There are five more quarters than half-dollars and 110 coins total. How much of the money is in quarters?

885. Janet has $10, $5, $2, and $1 bills in her piggy bank. She has four times as many $5 bills as $10 bills, six more $2 than $5, and ten fewer than five times as many $1 bills as $2 bills. If she has $90 in her piggy bank, then how many of the bills are $2 bills?

Chapter 21

Graphing Basics

Graphing in algebra amounts to plotting points and, often, connecting them. The points are placed by using the *Cartesian coordinates,* so named for Rene Descartes, a prolific mathematician who dabbled in many areas and made many contributions. The points are assigned their positions by distances from a central point, called the *origin.* The *ordered pairs* that name points, (x, y), always have the horizontal movement listed first and the vertical movement listed second.

The Problems You'll Work On

In this chapter, you'll graph and plot and work with Cartesian coordinates in the following ways:

- Plotting points correctly on the coordinate plane
- Recognizing point positions in terms of their quadrant or position on an axis
- Finding the intersection of two lines
- Computing slopes from points or determining slopes from equations of lines
- Graphing lines by using points and slopes
- Graphing lines by using more than one point

What to Watch Out For

The following points are important to keep in mind as you work through this chapter:

- Plotting points on the correct axis; $(0, y)$ is on the y-axis and $(x, 0)$ is on the x-axis.
- Using the slope formula correctly by keeping the order of the coordinates the same
- Remembering that slope is change-in-y divided by change-in-x
- Counting off slope correctly when graphing lines

Plotting Points on the Coordinate System

886–889 *Identify the graphed point.*

886. Which is the graph of (–2, 3)?

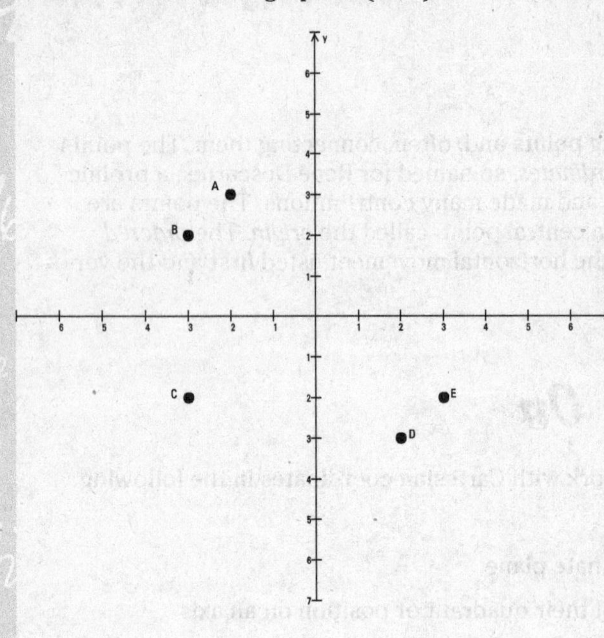

887. Which is the graph of (4, –1)?

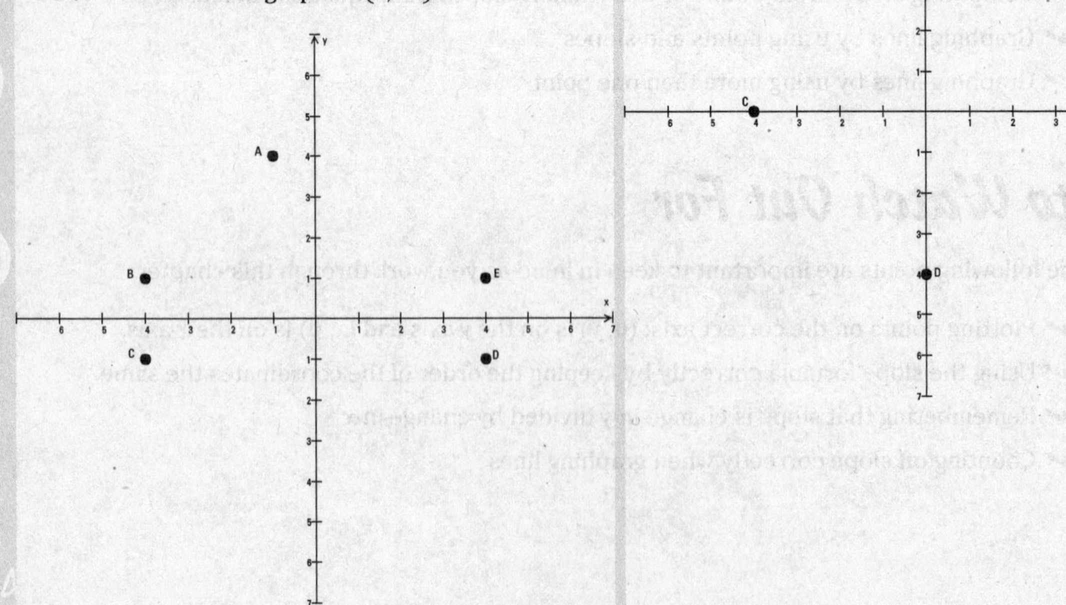

888. Which is the graph of (0, 2)?

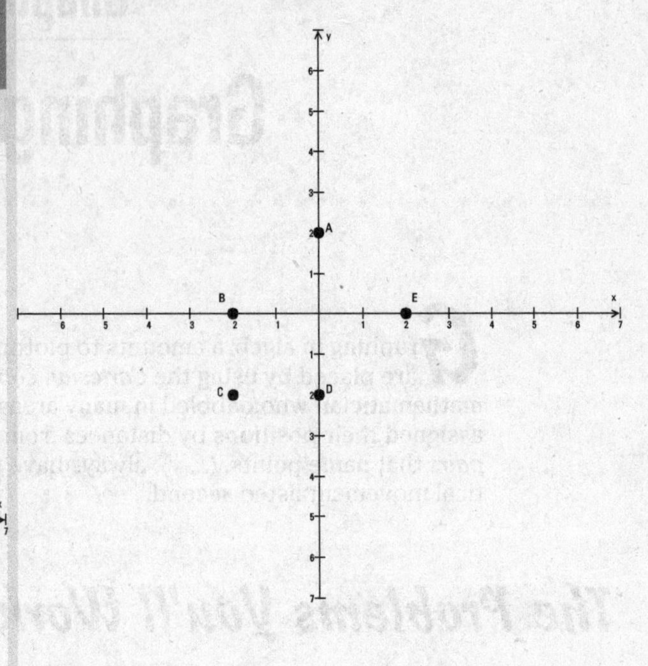

889. Which is the graph of (–4, 0)?

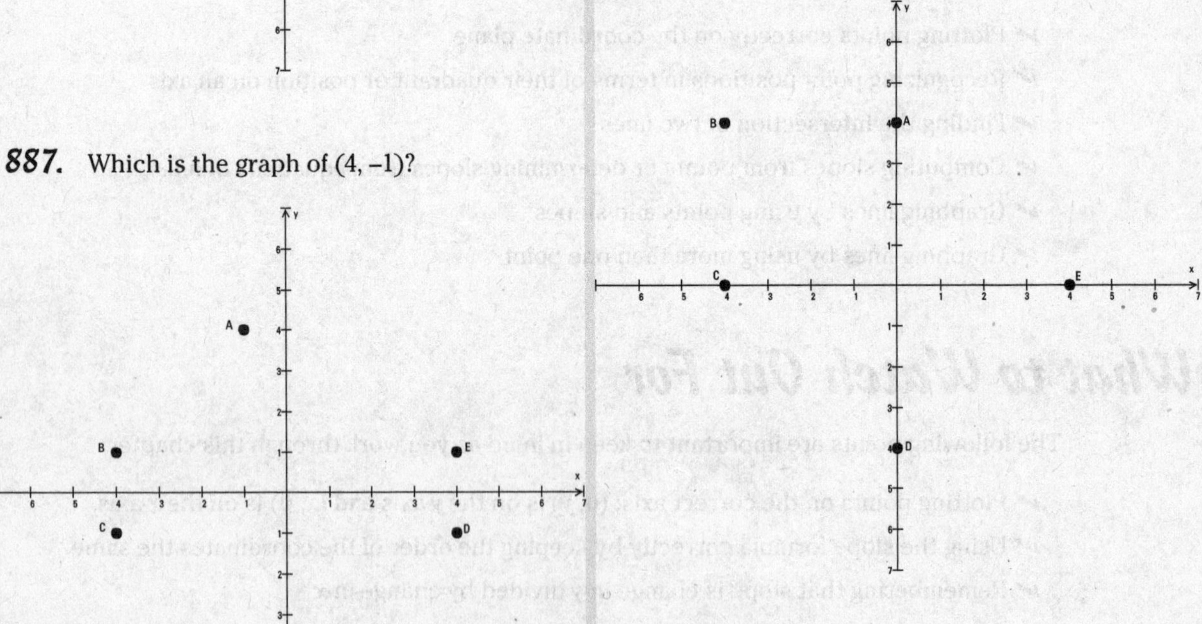

Determining the Quadrant of a Point

890–893 *Name the quadrant or axis where you find the point.*

890.

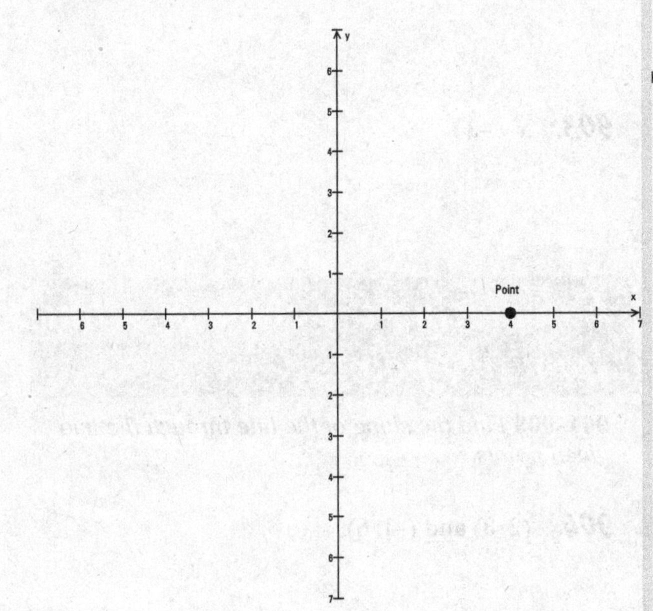

891.

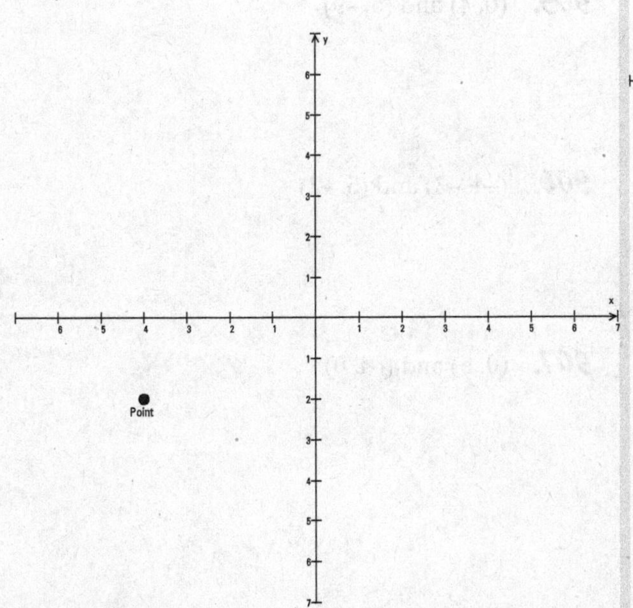

892.

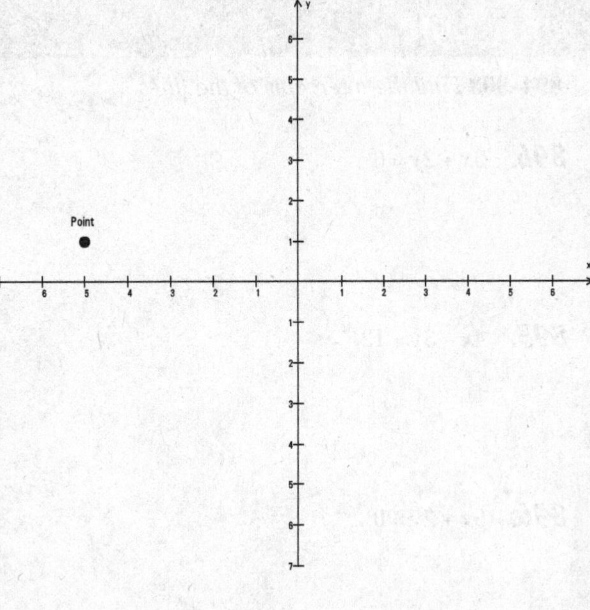

893.

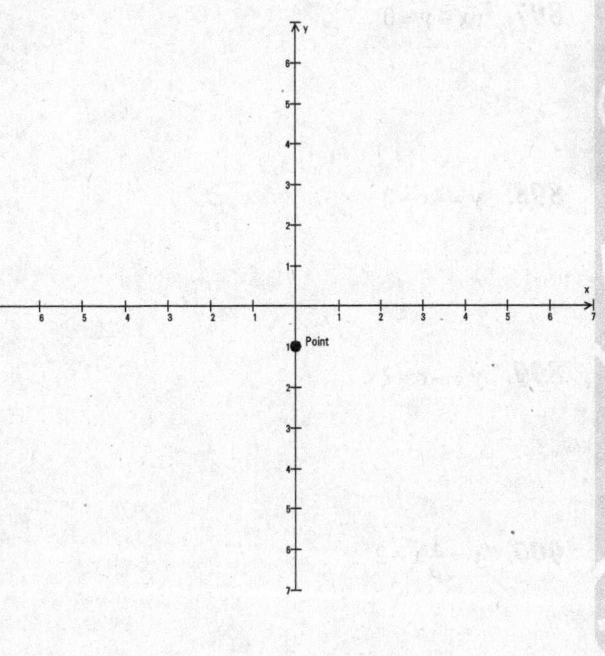

Finding the Intercepts of a Line

894–903 *Find the intercepts of the lines.*

894. $3x + 2y = 6$

895. $4x - 3y = 12$

896. $5x + 2y = 0$

897. $6x - y = 0$

898. $y = 4x - 3$

899. $y = -x + 2$

900. $y = \frac{1}{3}x + 2$

901. $y = -\frac{3}{4}x - 12$

902. $y = 8$

903. $x = -3$

Calculating the slope of a line from two points

904–909 *Find the slope of the line through the two given points.*

904. (2, 3) and (–1, 6)

905. (0, 4) and (5, –9)

906. (–4, –3) and (5, –2)

907. (0, 5) and (–4, 0)

908. (6, 5) and (–3, 5)

909. (–4, 2) and (–4, –4)

Determining a Line's Slope from Its Equation

910–915 *Find the slope of the line given its equation.*

910. $y = -4x + 3$

911. $y = 2x - 1$

912. $3x + 6y = 11$

913. $4x - 3y = 7$

914. $y = -6$

915. $x = 3$

Sketching the Graph of a Line from Its Equation

916–921 *Sketch a graph of the line using the slope-intercept form, and determine a point that the line passes through.*

916. $y = 3x - 1$

917. $y = -2x + 3$

918. $y = -\frac{4}{3}x - 2$

919. $y = \frac{1}{4}x - 3$

920. $y = 2$

921. $y = -4$

Sketching Lines Using Two Points

922–925 *Sketch the graph of a line using the two points.*

922. (–2, 2) and (1, –3)

923. (3, 0) and (–1, –1)

924. (–2, 3) and (5, 3)

925. (0, –2) and (–4, 0)

Chapter 22

Using the Algebra of Lines

The equation of a line can take on one of many forms. The more popular are the *slope-intercept* form, $y = mx + b$, and the *standard* form, $Ax + By = C$. Each has its advantages and uses in the different applications for lines. Formulas involving lines are often based on one of these forms.

The Problems You'll Work On

Lining up equations and applications in this chapter requires the following techniques:

- Writing the equation of a line given slope and a point
- Writing the equation of a line given two points
- Determining the slopes of lines parallel or perpendicular to a particular line
- Writing the equations of lines parallel or perpendicular to a particular line through a particular point
- Finding the distance between points
- Finding the midpoint of a segment between two points

What to Watch Out For

Although working with line formulas is the main skill in this chapter, don't forget the following:

- Calculating the slope correctly when given two points
- Remembering to use the *negative* reciprocal when determining slopes of perpendicular lines
- Calculating squares and square roots correctly and using order of operations correctly in the distance formula
- Adding signed numbers correctly when determining midpoints

Writing Equations of Lines Using the Slope and a Point

926–931 *Write the equation of the line given the slope and a point on the line.*

926. $m = 2$, (–3, 4)

927. $m = -\frac{1}{2}$, (1, –1)

928. $m = -3$, (4, 0)

929. $m = \frac{6}{7}$, (3, 5)

930. $m = 0$, (4, 3)

931. m is undefined, (–3, 5)

Writing the Equation of a Line Using Two Points

932–937 *Write the equation of the line that passes through the two points.*

932. (4, –1) and (6, –5)

933. (3, 3) and (–5, –7)

934. (1, 6) and (3, 6)

935. (4, –3) and (0, 2)

936. (–4, 5) and (–4, –5)

937. (0, 0) and (3, –8)

Dealing with Slopes of Parallel and Perpendicular Lines

938–943 *Determine the slopes of the lines parallel and perpendicular to the given line.*

938. $y = 4x - 3$

939. $y = -\frac{5}{3}x + 7$

940. $2x - 3y = 7$

941. $x - 4y = 8$

942. $x = 5$

943. $y = -6$

Finding Equations of Lines Parallel or Perpendicular to One Given

944–947 *Find the equations of lines parallel and perpendicular to the line through the given point.*

944. $y = -2x + 1$ through (0, 3)

945. $y = \frac{4}{3}x - 3$ through (–3, –7)

946. $4x - y = 3$ through (0, 0)

947. $6x + 3y = 7$ through (–1, 1)

Computing the Distance Between Points

948–957 *Find the distance between the two points.*

948. (3, 4) and (–2, –8)

949. (1, –3) and (–5, 5)

950. (–4, –3) and (0, 0)

951. (5, –2) and (–2, 22)

952. (3, 3) and (–2, –2)

953. (–4, 1) and (6, 9)

954. (–3, 7) and (0, –2)

955. (0, 4) and (4, 0)

956. (–3, 6) and (–3, 8)

957. (–5, 4) and (5, –4)

Determining the Midpoint

958–965 *Find the midpoint of the two points.*

958. (4, 7) and (2, –5)

959. (–3, 6) and (–5, –4)

960. (1, 6) and (–3, –2)

961. (–6, –6) and (–8, 0)

962. (4, –3) and (5, 9)

963. (–3, 8) and (–3, 10)

964. (4, 0) and (0, 0)

965. $\left(\frac{5}{2},\frac{3}{2}\right)$ and $\left(-\frac{7}{2},-\frac{9}{2}\right)$

Chapter 23

Other Graphing Topics

Not all graphing involves lines, although lines play an important role in graphs (as either axes or asymptotes or reflection structures). You can quickly graph circles if you know the center and radius. A parabola has a unique *U* or cup shape. And once you have the basic shape of a particular graph, you can quickly create variations on that graph by using a few transformational rules.

The Problems You'll Work On

Here are the graphing techniques you'll use in this chapter:

- Finding the intersection of two lines by graphing
- Graphing circles
- Graphing parabolas
- Using transformations in graphing

What to Watch Out For

Keeping all the graphing rules and tips in mind can be tricky; here are a few things to specifically watch out for in this chapter:

- Checking the discovered point of intersection of two lines by substituting the point into the equations of the lines
- Remembering to compute the square root of the constant in the equation of a circle when using the radius in the graph
- Watching the direction of the parabola based on the sign of the lead coefficient
- Remembering that translations to the left and right are the opposite of the sign in the equation

Graphing to Find the Intersection of Two Lines

966–973 *Find the intersection of the lines by graphing.*

966. $\begin{cases} y = x + 4 \\ y = -x + 6 \end{cases}$

967. $\begin{cases} y = 2x - 3 \\ y = -2x + 1 \end{cases}$

968. $\begin{cases} y = x - 5 \\ y = 2x - 7 \end{cases}$

969. $\begin{cases} y = 3x + 1 \\ y = x - 1 \end{cases}$

970. $\begin{cases} y = 4 \\ x = 3 \end{cases}$

971. $\begin{cases} y = -2 \\ x = 0 \end{cases}$

972. $\begin{cases} x + 3y = 4 \\ 3x + 4y = 2 \end{cases}$

973. $\begin{cases} 2x - y = 7 \\ x + y = 2 \end{cases}$

Sketching the Graph of a Circle

974–979 *Sketch a graph of each circle, indicating the center and the radius.*

974. $x^2 + y^2 = 4$

975. $x^2 + y^2 = \frac{1}{9}$

976. $(x - 2)^2 + (y + 1)^2 = 9$

977. $(x + 1)^2 + y^2 = 16$

978. $(x - 4)^2 + (y - 4)^2 = 1$

979. $x^2 + (y - 2)^2 = 25$

Creating the Graph of a Parabola

980–989 *Sketch the graph of the parabola* $y = a(x-h)^2 + k$

980. Vertex: (–1, –2); passes through (0, 0)

981. Vertex: (3, 4); passes through (1, 1)

982. Vertex: (0, 2); passes through (1, 0)

983. Vertex: (3, –1); passes through (–1, 3)

984. $y = (x-3)^2 + 1$

985. $y = (x+1)^2 - 2$

986. $y = 6 - (x-2)^2$

987. $y = -1 - (x+1)^2$

988. $y - 3 = (x+3)^2$

989. $y + 2 = (x+4)^2$

Transforming the Graph of a Figure

990–1001 *Perform the indicated transformation when graphing a new graph from a basic graph.*

990. Using the basic graph for $y = |x|$, sketch the graph of $y = |x| + 3$.

991. Using the basic graph for $y = |x|$, sketch the graph of $y = |x| - 1$.

992. Using the basic graph for $y = |x|$, sketch the graph of $y = |x-2|$.

993. Using the basic graph for $y = |x|$, sketch the graph of $y = -|x+4|$.

994. Using the basic graph for $y = x^3$, sketch the graph of $y = x^3 + 2$.

995. Using the basic graph for $y = x^3$, sketch the graph of $y = -x^3 - 1$.

996. Using the basic graph for $y = x^3$, sketch the graph of $y = (x - 4)^3$.

997. Using the basic graph for $y = x^3$, sketch the graph of $y = -(x + 1)^3$.

998. Using the basic graph for $y = (x - 1)^2$, sketch the graph of $y = -(x - 1)^2 + 1$.

999. Using the basic graph for $y = (x - 1)^2$, sketch the graph of $y = 2(x - 1)^2 - 4$.

1000. Using the basic graph for $y = (x + 3)^2$, sketch the graph of $y = -(x + 3)^2$.

1001. Using the basic graph for $y = (x + 3)^2$, sketch the graph of $y = -3(x + 3)^2 - 2$.

Part II
The Answers

To access the Cheat Sheet created specifically for this book, go to www.dummies.com/cheatsheet/1001algebra1

In this part . . .

You get answers and explanations for all 1001 problems. As you're going over your work, you may realize that you need a little more instruction. Fortunately, the *For Dummies* series offers several excellent resources. The following titles, all written by your humble author, are available at your favorite bookstore or in ebook format:

- *Algebra I For Dummies*
- *Algebra I Essentials For Dummies*
- *Algebra I Workbook For Dummies*
- *Math Word Problems For Dummies*

After you've mastered algebra I and you're ready to step up to algebra II, you'll find all the help you'll need in these titles, which I also wrote:

- *Algebra II For Dummies*
- *1001 Algebra II Practice Problems For Dummies.*

Visit `Dummies.com` for more information.

Chapter 24

Answers

1. **–4, –3, –1, 0, 4**

The number –4 has a greater absolute value than –3, so it's farther to the left on the number line. And –3 has a greater absolute value than –1.

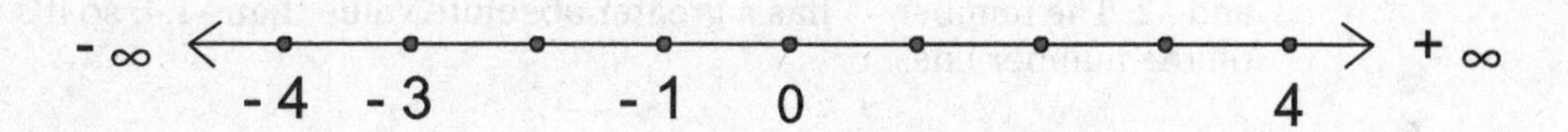

2. **–3, –2, 0, 1, 3**

The number –3 has a greater absolute value than –2, so it's farther to the left on the number line.

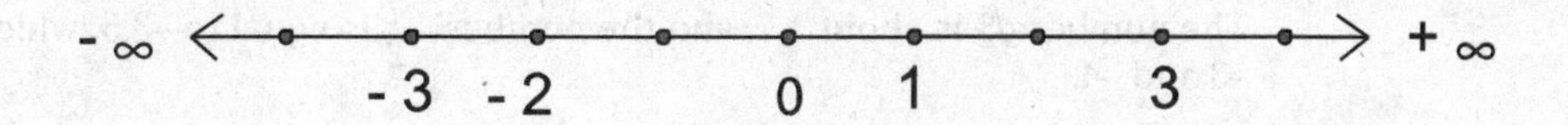

3. $-5, -\frac{7}{3}, -1, \frac{3}{7}, 2$

The number –5 has a greater absolute value than –1, so it's farther to the left on the number line. The number $-\frac{7}{3}$ is between –2 and –3. And the number $\frac{3}{7}$ is between 0 and 1.

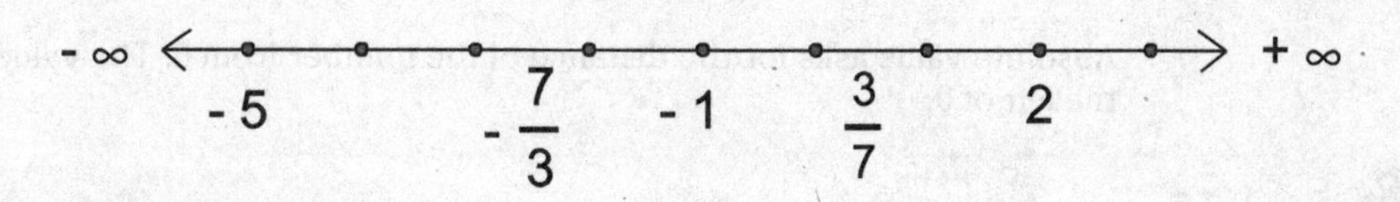

4. $-4, -2, -\frac{6}{5}, 0, \frac{5}{6}$

The number –4 has a greater absolute value than –2, so it's farther to the left on the number line. And the number $-\frac{6}{5}$ is between –1 and –2.

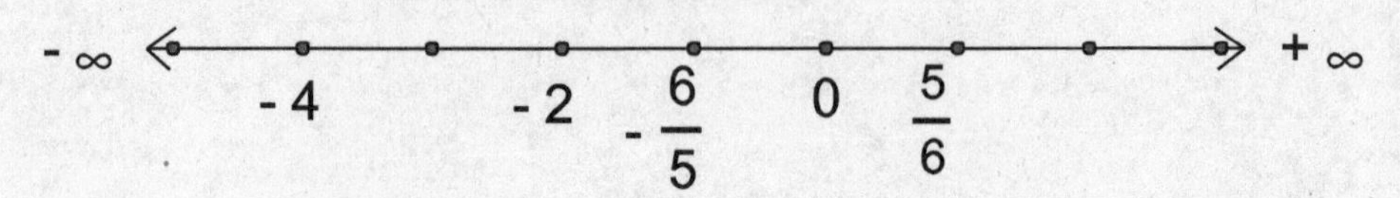

5. $-4, -\sqrt{2}, 0, \sqrt{3}, 3$

The number $\sqrt{3}$ is about 1.7, and the number $-\sqrt{2}$ is about –1.4, which is between –1 and –2. The number –4 has a greater absolute value than –1.4, so it's farther to the left on the number line.

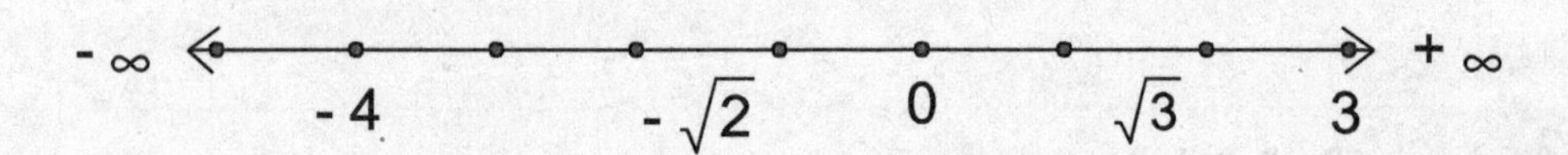

6. $-\frac{7}{2}, -3, 0, \sqrt{3}, 2, 4$

The number $\sqrt{3}$ is about 1.7, and the number $-\frac{7}{2}$ is equal to –3.5, which is between –3 and –4.

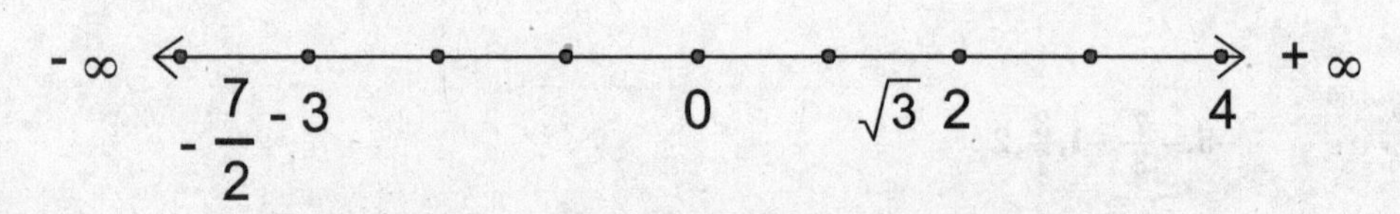

7. **4**

Absolute value asks for the distance of the number from 0. The value of –4 is 4 units to the left of 0.

8. **7.6**

Absolute value asks for the distance of the number from 0. The value of –7.6 is 7.6 units to the left of 0.

9. -2

First find the absolute value of the –2 within the absolute value symbols.

$$|-2| = 2$$

Then find the opposite of that answer.

$$-(2) = -2$$

10. $-\frac{2}{3}$

First find the absolute value of the $-\frac{2}{3}$ within the absolute value symbols.

$$\left|-\frac{2}{3}\right| = \frac{2}{3}$$

Then find the opposite of that answer.

$$-\left(\frac{2}{3}\right) = -\frac{2}{3}$$

11. -6

The signs are the same, so you find the sum of the absolute values of the numbers, 4 + 2 = 6, and then attach their negative sign. The answer is –6.

12. -2

The signs are different, so first you find the difference between the absolute values of the numbers, 4 – 2 = 2, and then you attach the sign of the number with the greater absolute value, the –4. The answer is –2.

13. 2

The signs are different, so first you find the difference between the absolute values of the numbers, 4 – 2 = 2, and then you attach the sign of the number with the greater absolute value, the +4. The answer is +2.

14. -2

The signs are different, so first you find the difference between the absolute values of the numbers, 5 – 3 = 2, and then you attach the sign of the number with the greater absolute value, the –5. The answer is –2.

15. 0

The signs are different, so you find the difference between the absolute values of the numbers, 6 – 6 = 0. The number 0 has no sign, so the answer is 0.

16. 5

The signs are different, so first you find the difference between the absolute values of the numbers, 7 – 2 = 5, and then you attach the sign of the number with the greater absolute value, the +7. The answer is 5.

17. –1

First, add the two negative numbers together, –4 + –2 = –6.

Now the problem looks like this: 5 + (–6).

Add the two numbers with the different signs by finding the difference between their absolute values, 6 – 5 = 1. Attach the sign of the number with the greater absolute value, the –6. The answer is –1.

18. 2

Add the two negative numbers together to get –4. Then add the two positive numbers together to get 6.

The sum of –4 and 6 is +2, because you find the difference between the absolute values of the numbers and attach the sign of the number with the greater absolute value.

19. 2

Add the first two numbers together by finding the difference between their absolute values, 1. Next, add the last two numbers together by finding the difference between their absolute values, 1.

Finally, add the two results together: 1 + 1 = 2.

20. –11

Add the four negative numbers together by finding the sum of their absolute values: 4 + 5 + 6 + 7 = 22. Attached a negative sign to the sum: –22.

Now add the two positive numbers together: 7 + 4 = 11.

The sum of the two results, –22 + 11, is found by finding the difference between the absolute values, 22 – 11 = 11. The sign of the final answer is negative, because the –22 has the greater absolute value. So the answer is – 11.

21. –10

Create an equivalent addition problem by changing the subtraction sign to addition and the sign of the 6 to –6. Then do the addition problem: –4 + (–6).

The signs are the same, so you find the sum of the absolute values of the numbers and then apply the negative sign. The answer is –10.

22. 15

Create an equivalent addition problem by changing the subtraction sign to addition and the sign of the –8 to +8. Then do the addition problem:

7 + (+8).

The signs are the same, so you find the sum of the absolute values of the numbers.

23. 3

This is just a simple subtraction problem involving two positive numbers. Subtract as usual.

24. –5

Create an equivalent addition problem by changing the subtraction sign to addition and the sign of the –4 to +4. Then do the addition problem:

–9 + (+4).

The signs are different, so you find the difference of the absolute values of the numbers and then apply the sign of the larger value. The answer is –5.

25. –14

Create an equivalent addition problem by changing the subtraction sign to addition and the sign of the 7 to –7. Then do the addition problem:

–7 + (–7).

The signs are the same, so you find the sum of the absolute values of the numbers and then apply the negative sign. The answer is –14.

26. 0

Create an equivalent addition problem by changing the subtraction sign to addition and the sign of the –7 to +7. Then do the addition problem:

–7 + (+7).

The signs are different, so you find the difference of the absolute values of the numbers, 7 – 7 = 0.

27. 5

Create an equivalent addition problem by changing the subtract to add and the sign of the –2 to +2. Then do the addition problem: 3 + (+2).

The signs are the same, so you find the sum of the absolute values of the numbers.

28. –1

First find the opposite of the –2 in the brackets, giving you +2. Then, create an equivalent addition problem by changing the subtraction sign to addition and the sign of the 3 to –3. Then do the addition problem:

2 + (–3).

The signs are different, so you find the difference of the absolute values of the numbers and apply the sign of the larger value. The answer is –1.

29. 8

First, find the opposite of the –4 in the brackets, giving you +4. Then create an equivalent addition problem by changing the subtraction sign to addition and the sign of the –4 in the parentheses to +4. Then do the addition problem:

4 + (+4).

The signs are the same, so you find the sum of the absolute values of the numbers.

30. 5

Create an equivalent addition problem by changing the subtraction sign to addition and the sign of the –5 in the parentheses to +5. Then do the addition problem:

0 + (+5).

The number 0 is the additive identity, so adding it to +5 does not change the value of the +5.

31. –6

The product of 2 and 3 is 6.

There's one negative sign (odd number), so the product is negative.

32. 20

The product of 4 and 5 is 20.

There are two negative signs (even number), so the product is positive.

33. –30

The product of 5 and 6 is 30.

There's one negative sign (odd number), so the product is negative.

34. –3

The product of 3 and 1 is 3.

There's one negative sign (odd number), so the product is negative.

35. 49

The product of 7 and 7 is 49.

There are two negative signs (even number), so the product is positive.

36. –64

The product of 8 and 8 is 64.

There's one negative sign (odd number), so the product is negative.

37. 10

The product of 6 and $\frac{5}{3}$ is $\frac{{}^{2}\cancel{6}}{1}\cdot\frac{5}{\cancel{3}_1}=\frac{10}{1}$.

There are two negative signs (even number), so the product is positive.

38. –15

The product of 20 and $\frac{3}{4}$ is $\frac{{}^{5}\cancel{20}}{1}\cdot\frac{3}{\cancel{4}_1}=\frac{15}{1}$.

There is one negative sign (odd number), so the product is negative.

39. 0

The product of any number and 0 is 0.

40. 1

The product of $1\cdot1\cdot1\cdot1$ is 1.

There are four negative signs (even number), so the product is positive.

41. –3

When you divide 6 by 2, you get 3.

There is one negative sign (odd number), so the quotient is negative.

42. 2

When you divide 8 by 4, you get 2.

There are two negative signs (even number), so the quotient is positive.

43. –4

When you divide 12 by 3, you get 4.

There is one negative sign (odd number), so the quotient is negative.

44. 4

When you divide 60 by 15, you get 4.

There are two negative signs (even number), so the quotient is positive.

45. 0

When you divide 0 by any number (other than 0), the answer is 0.

46. –5

When you divide 5 by 1, you get 5.

There is one negative sign (odd number), so the quotient is negative.

47. 2

First, find the product of the two numbers in the denominator.

$$2(-4) = -8$$

Now divide –16 by –8.
When you divide 16 by 8, you get 2.

There are two negative signs (even number), so the quotient is positive.

48. –1

First, find the product in the numerator and in the denominator.

In the numerator, $2(-6)(-1) = 12$, because there are an even number of negative signs.

In the denominator, $4(-3) = -12$, because there's only one (odd number) negative sign.
Now divide 12 by –12, and you get –1, because there's one negative sign.

A good way to check the sign of your answer is to just count the number of negative signs in the original problem. You see three negative signs (odd number), so the final result should be negative.

49. –4

First, find the product in the numerator and in the denominator.

In the numerator, $-4(-3)(-2)(-1) = 24$, because there are an even number of negative signs.

In the denominator, $6(-1)(-1)(-1) = -6$, because there are an odd number of negative signs.

Now divide 24 by –6, and you get –4, because there's one negative sign.

A good way to check the sign of your answer is to just count the number of negative signs in the original problem. You see seven negative signs, so the final result should be negative.

50. 1

One way to solve this would be to find the products of the numerator and denominator, separately and then divide. Both the numerator and denominator have products of 36, so, when you divide, you get 1. There are an even number of negative signs so the answer is positive.

Another way to handle the problem is to divide out (reduce) all the equal factors in the numerator and denominator. The two 2's and two 3's all divide out, leaving you with 1.

51. 4

First, perform the subtraction inside the parentheses.

$$6 - (5 - 3) = 6 - (2)$$

Now subtract.

$$6 - 2 = 4$$

52. –4

First, perform the subtraction inside the parentheses.

$$(4 - 3) - 5 = (1) - 5$$

Now subtract.

$$1 - 5 = -4$$

53. 20

First, perform the subtraction inside the parentheses.

$$5[6 + (3 - 5)] = 5[6 + (-2)]$$

Add the two numbers inside the brackets.

$$= 5[4]$$

Now multiply.

$$= 20$$

54. 0

First, perform the subtraction inside the parentheses.

$$8\{3 - [4 + (5 - 6)]\} = 8\{3 - [4 + (-1)]\}$$

Next, add the numbers inside the brackets.

$$= 8\{3 - [3]\}$$

Add the two numbers inside the braces.

$$= 8\{0\}$$

The product of 0 and anything is 0.

55. **2**

First, subtract the two numbers in the numerator.

$$\frac{8-4}{2}=\frac{4}{2}$$

Then divide to get 2.

56. **–6**

First, subtract the two numbers in the denominator.

$$\frac{12}{9-11}=\frac{12}{-2}$$

Then, when dividing to get –6, the quotient is negative, because there is one negative sign.

57. **1**

First, subtract the two numbers in the parentheses.

$$\frac{3-(6-2)}{7-8}=\frac{3-(4)}{7-8}$$

Next, subtract the two numbers in the numerator, and then the two numbers in the denominator.

$$\frac{3-4}{7-8}=\frac{-1}{-1}$$

Dividing a number by itself, you get 1.

58. **1**

First, do the two subtractions inside the parentheses.

$$\frac{(5-6)-(9-3)}{3-10}=\frac{(-1)-(6)}{3-10}$$

Now subtract the two numbers in the numerator, and then the two numbers in the denominator.

$$\frac{-1-6}{3-10}=\frac{-7}{-7}$$

Dividing a number by itself, you get 1.

59. **2**

First, find the difference of the numbers under the radical.

$$\sqrt{12-8}=\sqrt{4}$$

The square root of 4 is 2.

60. 4

First, subtract the numbers in the numerator, and then subtract the numbers under the radical sign.

$$\frac{11-3}{\sqrt{9-5}} = \frac{8}{\sqrt{4}}$$

Next, find the square root of 4.

$$\frac{8}{\sqrt{4}} = \frac{8}{2}$$

The final division gives you 4.

61. –1

First, subtract the numbers in the numerator under the radical sign, and then subtract the numbers in the denominator.

$$\frac{\sqrt{11-10}}{3-4} = \frac{\sqrt{1}}{-1}$$

The square root of 1 is 1.

$$\frac{\sqrt{1}}{-1} = \frac{1}{-1}$$

Dividing 1 by –1 gives you –1, because there's one negative sign.

62. –6

First, find the difference between the two numbers inside the absolute value symbol.

$$|5-6|-7=|-1|-7$$

Now evaluate the result inside the absolute value symbol.

$$|-1|-7=1-7$$

Subtracting, you get –6.

63. 2

First, find the difference between the two numbers inside the absolute value symbol.

$$5-|4-7|=5-|-3|$$

Now evaluate the result inside the absolute value symbol.

$$5-|-3|=5-3$$

Subtracting, you get 2.

64. **2**

First, subtract the numbers under the radical, and then subtract the numbers inside the absolute value symbol.

$$\frac{\sqrt{20-4}}{|9-11|}=\frac{\sqrt{16}}{|-2|}$$

The square root of 16 is 4, and the absolute value of –2 is 2.

$$\frac{\sqrt{16}}{|-2|}=\frac{4}{2}$$

Dividing, you get 2.

65. **$14-2y$**

Multiply the 2 times each term in the parentheses. Be sure to carry the negative sign along with the y.

$$2(7)+2(-y)=14-2y$$

66. **$-6x-24$**

Multiply the –6 times each term in the parentheses.

$$-6(x)+-6(4)=-6x-24.$$

67. **$-3x+1$**

Multiply the –3 times each term in the parentheses.

$$-3(x)-3\left(-\frac{1}{3}\right)=-3x+1$$

68. **$6-12y$**

Multiply each term in the parentheses by $\frac{3}{4}$. Be sure to carry the subtraction sign along with the second term.

$$\frac{3}{\not{4}_1}\left(\not{8}^{2}\right)+\frac{3}{\not{4}_1}\left(-\not{16}^{4}y\right)=6-12y$$

69. **$xy-6x$**

Multiply each term in the parentheses by x. Be sure to carry the subtraction sign along with the second term.

$$x(y)+x(-6)=xy-6x$$

70. $-4x^2 + 8xy - 12x$

Multiply each term in the parentheses by $-4x$.

$$-4x(x) + -4x(-2y) + -4x(3)$$
$$= -4x^2 + 8xy - 12x$$

71. $2 - 8 + 9 = 3$

Multiply each fraction in the parentheses by 12.

$$\not{12}^2\left(\frac{1}{\not{6}_1}\right)+\not{12}^4\left(-\frac{2}{\not{3}_1}\right)+\not{12}^3\left(\frac{3}{\not{4}_1}\right)$$
$$=2-8+9=3$$

72. $-20x + 2$

Multiply each term in the parentheses by -5.

$$-5(4x)-\not{5}\left(-\frac{2}{\not{5}}\right)=-20x+2$$

73. 90

Regroup the 47 with the -47; their sum is 0.

$$(47 + (-47)) + 90 = 0 + 90 = 90$$

74. -6

Regroup the 23 with the -23; their sum is 0.

$$-6 + (23 - 23) = -6 + 0 = -6$$

75. 29

Regroup the two fractions, which are reciprocals of one another. Their product is 1.

$$\left(\frac{3}{5}\cdot\frac{5}{3}\right)29=(1)29=29$$

76. 811

Regroup the $\frac{1}{15}$ and the 15 together. They're reciprocals, and their product is 1.

$$811\left(\frac{1}{15}\cdot 15\right)=811(1)=811$$

77. 20

Regroup the 19 and –19 together. Their sum is 0.

$$16 + (19 + (-19)) + 4 = 16 + 0 + 4 = 20$$

78. 77

Regroup the –53.2 and 53.2 together. Their sum is 0.

$$77 + (-53.2 + 53.2) = 77 + 0 = 77$$

79. 47

Reverse the order of the 47 and 16. Then add the –16 and 16, which are additive inverses.

$$-16 + 16 + 47 = (-16 + 16) + 47 = 0 + 47 = 47.$$

80. $\frac{4}{3}$

Reverse the order of the first two fractions. Then the two additive inverses can be combined to give you 0.

$$\frac{4}{3}+\frac{3}{11}+\left(-\frac{3}{11}\right)=\frac{4}{3}+\left(\frac{3}{11}+\left(-\frac{3}{11}\right)\right)$$
$$=\frac{4}{3}+0=\frac{4}{3}$$

81. 673

Reverse the last two numbers. Then you have additive inverses that can be combined to give you 0.

$$432 - 432 + 673 = 0 + 673 = 673$$

82. $-\frac{12}{13}$

Reverse the second two fractions. The product of a fraction and its reciprocal is 1.

$$\frac{31}{17}\left(\frac{17}{31}\right)\left(-\frac{12}{13}\right)=\left(\frac{\cancel{31}}{\cancel{17}}\cdot\frac{\cancel{17}}{\cancel{31}}\right)\left(-\frac{12}{13}\right)$$
$$=1\left(-\frac{12}{13}\right)=-\frac{12}{13}$$

83. $\frac{50}{11}$

Reverse the two fractions. Then multiply the $\frac{1}{9}$ times 18 before multiplying the result times $\frac{25}{11}$.

$$\frac{25}{11}\cdot\frac{1}{9}\cdot 18=\frac{25}{11}\left(\frac{1}{9}\cdot 18\right)=\frac{25}{11}(2)=\frac{50}{11}$$

84. 4

Reverse the –3 and 4, and then reverse the 23 and 3.

$$4 + -3 + 23 + 3 - 23$$
$$= 4 + -3 + 3 + 23 - 23$$

Associate the two pairs of additive inverses and simplify.

$$= 4 + (-3 + 3) + (23 - 23) = 4 + 0 + 0 = 4$$

85. $-4x$

First, distribute the 4 over the terms in the parentheses.

$$-32 + 4(8 - x) = -32 + 32 - 4x$$

The first two terms have a sum of 0.

$$0 - 4x = -4x$$

86. $-5x$

First, distribute the –5 over the two terms in the parentheses.

$$-5(x - 2) - 10 = -5x + 10 - 10$$

The last two terms have a sum of 0.

$$-5x + 0 = -5x$$

87. 48

First, distribute the 48 over the two terms in the parentheses.

$$\cancel{48}^{8}\left(\frac{x}{\cancel{6}_1}\right) + 48(1) - 8x = 8x + 48 - 8x$$

Now reverse the first and second terms.

$$48 + 8x - 8x$$

The second two terms have a sum of 0.

$$48 + 0 = 48$$

88. $-x$

First, distribute the $-\frac{1}{4}$ over the two terms in the parentheses.

$$-\frac{1}{\cancel{4}_1}\left(\cancel{4}^1 x\right) - \frac{1}{\cancel{4}_1}\left(-\cancel{20}^5\right) - 5 = -x + 5 - 5$$

The sum of the second two terms is 0.

$$-x + 0 = -x$$

Answers 1–100

89. y

First distribute the –2 over the terms in the first parentheses, and distribute the 3 over the terms in the second parentheses.

$$-2(3) + -2(y) + 3(y) + 3(2) = -6 - 2y + 3y + 6$$

Move the first term all the way to the right.

$$-2y + 3y + 6 - 6$$

The first two terms combine, and the last two terms have a sum of 0.

$$y + 0 = y$$

90. $-\frac{1}{4}$

First reverse the last two terms.

$$\frac{1}{8}\left(-\frac{8}{5}\right)\frac{5}{4}$$

Now write the middle fraction as the product of two fractions, one of them $\frac{8}{1}$, and write the last fraction as the product of two fractions, one of them $\frac{5}{1}$.

$$=\frac{1}{8}\left(\frac{8}{1}\cdot\left(-\frac{1}{5}\right)\right)\left(\frac{5}{1}\cdot\frac{1}{4}\right)$$

Grouping the pairs of reciprocals together,

$$\left(\frac{1}{8}\cdot\frac{8}{1}\right)\left(-\frac{1}{5}\cdot\frac{5}{1}\right)\frac{1}{4}=(1)(-1)\frac{1}{4}=-\frac{1}{4}$$

91. $1\frac{13}{30}$

Find a common denominator, and change each fraction to an equivalent fraction using that denominator.

Since the denominators are relatively small, find the common denominator by multiplying them together, $6\times5=30$.

$$\frac{3}{5}\cdot\frac{6}{6}+\frac{5}{6}\cdot\frac{5}{5}=\frac{18}{30}+\frac{25}{30}$$

Find the sum of the numerators; then change the answer to a mixed number.

$$\frac{18}{30}+\frac{25}{30}=\frac{43}{30}=1\frac{13}{30}$$

92. $\frac{32}{75}$

Find a common denominator, and change each fraction to an equivalent fraction using that denominator.

You can find the least common denominator by checking multiples of the larger denominator, 25. When you get to $25 \times 6 = 150$, you have a number divisible by 15, also.

$$\frac{24}{25} \cdot \frac{6}{6} - \frac{8}{15} \cdot \frac{10}{10} = \frac{144}{150} - \frac{80}{150}$$

Find the difference between the numerators; then reduce the fraction.

$$\frac{144}{150} - \frac{80}{150} = \frac{64}{150} = \frac{32}{75}$$

93. $11\frac{5}{24}$

Find a common denominator for the two fractions, and change each fraction to an equivalent fraction using that denominator.

You can find the least common denominator by checking multiples of the larger denominator, 8. When you get to $8 \times 3 = 24$, you have a number divisible by 6, also.

$$4\frac{3}{8} \cdot \frac{3}{3} + 6\frac{5}{6} \cdot \frac{4}{4} = 4\frac{9}{24} + 6\frac{20}{24}$$

Add the whole numbers together, then the fractions.

$$4\frac{9}{24} + 6\frac{20}{24} = 10\frac{29}{24}$$

The fractional part of the answer is improper. Change the $\frac{29}{24}$ to $1\frac{5}{24}$, and then add the whole number part to the rest of the answer.

$$10\frac{29}{24} = 10 + 1\frac{5}{24} = 11\frac{5}{24}$$

94. $3\frac{7}{15}$

Find a common denominator for the two fractions, and change each fraction to an equivalent fraction using that denominator.

You can find the least common denominator by checking multiples of the larger denominator, 10. When you get to $10 \times 3 = 30$, you have a number divisible by 6, also.

$$6\frac{3}{10} \cdot \frac{3}{3} - 2\frac{5}{6} \cdot \frac{5}{5} = 6\frac{9}{30} - 2\frac{25}{30}$$

The numerator of the second fraction is larger than the numerator it's being subtracted from, so you borrow one, $\frac{30}{30}$, from the 6 in the first number and add it on to the fraction.

$$\begin{aligned} 6\frac{9}{30} - 2\frac{25}{30} &= 5 + \frac{30}{30} + \frac{9}{30} - 2\frac{25}{30} \\ &= 5\frac{39}{30} - 2\frac{25}{30} \end{aligned}$$

Now subtract the whole numbers and then the fractions by finding the difference between the numerators. Then reduce the fraction.

$$5\frac{39}{30} - 2\frac{25}{30} = 3\frac{14}{30} = 3\frac{7}{15}$$

95. $\frac{227}{840}$

First, find a common denominator for the four fractions, and change each fraction to an equivalent fraction using that denominator. The four denominators have no factors in common, so the least common denominator will be their product, 840.

$$\frac{4}{5}\cdot\frac{8\cdot3\cdot7}{8\cdot3\cdot7}+\frac{3}{8}\cdot\frac{5\cdot3\cdot7}{5\cdot3\cdot7}-\frac{1}{3}\cdot\frac{5\cdot8\cdot7}{5\cdot8\cdot7}-\frac{4}{7}\cdot\frac{5\cdot8\cdot3}{5\cdot8\cdot3}$$
$$=\frac{4}{5}\cdot\frac{168}{168}+\frac{3}{8}\cdot\frac{105}{105}-\frac{1}{3}\cdot\frac{280}{280}-\frac{4}{7}\cdot\frac{120}{120}$$
$$=\frac{672}{840}+\frac{315}{840}-\frac{280}{840}-\frac{480}{840}$$

Next, find the sum of the first two fractions. Then find the sum of the second two fractions, but place a subtraction sign in front of that result.

$$\frac{672}{840}+\frac{315}{840}-\frac{280}{840}-\frac{480}{840}$$
$$=\left(\frac{672}{840}+\frac{315}{840}\right)-\left(\frac{280}{840}+\frac{480}{840}\right)$$
$$=\frac{987}{840}-\frac{760}{840}$$

Subtract the two fractions.

$$\frac{987}{840}-\frac{760}{840}=\frac{227}{840}$$

96. $5\frac{5}{24}$

First, find a common denominator for the three fractions in the problem, and change each fraction to an equivalent fraction using that denominator. The fractions have common factors, so just try multiples of the largest denominator, the 12. The number 24 is the least common denominator.

$$3\frac{1}{8}\cdot\frac{3}{3}-1\frac{5}{12}\cdot\frac{2}{2}+4-\frac{1}{2}\cdot\frac{12}{12}$$
$$=3\frac{3}{24}-1\frac{10}{24}+4-\frac{12}{24}$$

Next, add the first mixed number and the 4 together. Then add the two negative numbers together and put a subtraction sign in front of the result.

$$3\frac{3}{24}-1\frac{10}{24}+4-\frac{12}{24}$$
$$=\left(3\frac{3}{24}+4\right)-\left(1\frac{10}{24}+\frac{12}{24}\right)$$
$$=7\frac{3}{24}-1\frac{22}{24}$$

The numerator in the second number is larger than that in the first, so borrow from the 7 to create an equivalent number.

$$7\frac{3}{24}-1\frac{22}{24}$$
$$=6+\frac{24}{24}+\frac{3}{24}-1\frac{22}{24}$$
$$=6\frac{27}{24}-1\frac{22}{24}$$

Now perform the subtraction.

$$6\frac{27}{24}-1\frac{22}{24}=5\frac{5}{24}$$

97. $\frac{13}{20}$

Before multiplying the two numerators and the two denominators, reduce the fractions — dividing one numerator and one denominator by the same number.

$$\frac{{}^{1}\cancel{16}}{25}\times\frac{65}{\cancel{64}_{4}}=\frac{1}{{}_{5}\cancel{25}}\times\frac{\cancel{65}^{13}}{4}=\frac{1}{5}\times\frac{13}{4}$$

Multiplying the two numerators and the two denominators:

$$\frac{1}{5}\times\frac{13}{4}=\frac{13}{20}$$

98. $-\frac{26}{31}$

Before multiplying the two numerators and the two denominators, reduce the fractions — dividing one numerator and one denominator by the same number.

$$-\frac{{}^{2}\cancel{70}}{93}\times\frac{39}{\cancel{35}_{1}}=-\frac{2}{{}_{31}\cancel{93}}\times\frac{\cancel{39}^{13}}{1}=-\frac{2}{31}\times\frac{13}{1}$$

Multiply the two numerators and the two denominators:

$$-\frac{2}{31}\times\frac{13}{1}=-\frac{26}{31}$$

99. $18\frac{2}{3}$

First, change each mixed number to an improper fraction.

$$3\frac{3}{7}\times5\frac{4}{9}=\frac{24}{7}\times\frac{49}{9}$$

Before multiplying the two numerators and the two denominators, reduce the fractions — dividing one numerator and one denominator by the same number.

$$\frac{{}^{8}\cancel{24}}{7}\times\frac{49}{\cancel{9}_{3}}=\frac{8}{{}_{1}\cancel{7}}\times\frac{\cancel{49}^{7}}{3}=\frac{8}{1}\times\frac{7}{3}$$

Multiply the two numerators and the two denominators.

$$\frac{8}{1}\times\frac{7}{3}=\frac{56}{3}$$

Now rewrite the answer as a mixed number.

$$\frac{56}{3}=18\frac{2}{3}$$

100. $34\frac{2}{7}$

First, change each mixed number to an improper fraction.

$$-3\frac{1}{3}\times2\frac{2}{5}\times\left(-4\frac{2}{7}\right)=-\frac{10}{3}\times\frac{12}{5}\times\left(-\frac{30}{7}\right)$$

Before multiplying the three numerators and the three denominators, reduce the fractions — dividing one numerator and one denominator by the same number.

$$-\frac{{}^{2}\cancel{10}}{3}\times\frac{12}{\cancel{5}_{1}}\times\left(-\frac{30}{7}\right)=-\frac{2}{{}_{1}\cancel{3}}\times\frac{\cancel{12}^{4}}{1}\times\left(-\frac{30}{7}\right)$$

$$=-\frac{2}{1}\times\frac{4}{1}\times\left(-\frac{30}{7}\right)$$

Multiply the numerators and the denominators.

$$-\frac{2}{1}\times\frac{4}{1}\times\left(-\frac{30}{7}\right)=\frac{240}{7}$$

Now rewrite the answer as a mixed number.

$$\frac{240}{7}=34\frac{2}{7}$$

101. $1\frac{11}{25}$

First, change the division problem to an equivalent multiplication problem using the reciprocal of the second fraction.

$$\frac{48}{35}\div\frac{20}{21}=\frac{48}{35}\times\frac{21}{20}$$

Now, before multiplying the numerators and the denominators, reduce the fractions — dividing one numerator and one denominator by the same number.

$$\frac{{}^{12}\cancel{48}}{35}\times\frac{21}{\cancel{20}_{5}}=\frac{12}{{}_{5}\cancel{35}}\times\frac{\cancel{21}^{3}}{5}=\frac{12}{5}\times\frac{3}{5}$$

Multiply the numerators and the denominators.

$$\frac{12}{5}\times\frac{3}{5}=\frac{36}{25}$$

Finally, rewrite the improper fraction as a mixed number.

$$\frac{36}{25}=1\frac{11}{25}$$

102. $1\frac{2}{5}$

First, rewrite the mixed numbers as improper fractions.

$$7\frac{1}{5}\div5\frac{1}{7}=\frac{36}{5}\div\frac{36}{7}$$

Next, change the division problem to an equivalent multiplication problem using the reciprocal of the second fraction.

$$\frac{36}{5}\div\frac{36}{7}=\frac{36}{5}\times\frac{7}{36}$$

Now, before multiplying the numerators and the denominators, reduce the fractions — dividing one numerator and one denominator by the same number.

$$\frac{\cancel{36}}{5}\times\frac{7}{\cancel{36}}=\frac{1}{5}\times\frac{7}{1}$$

Multiply the numerators and the denominators.

$$\frac{1}{5}\times\frac{7}{1}=\frac{7}{5}$$

Finally, rewrite the improper fraction as a mixed number.

$$\frac{7}{5}=1\frac{2}{5}$$

103. $2\frac{1}{7}$

First, rewrite the whole number and mixed number as improper fractions.

$$12\div5\frac{3}{5}=\frac{12}{1}\div\frac{28}{5}$$

Next, change the division problem to an equivalent multiplication problem using the reciprocal of the second fraction.

$$\frac{12}{1}\div\frac{28}{5}=\frac{12}{1}\times\frac{5}{28}$$

Now, before multiplying the numerators and the denominators, reduce the fractions — dividing one numerator and one denominator by the same number.

$$\frac{{}^{3}\cancel{12}}{1}\times\frac{5}{\cancel{28}_{7}}=\frac{3}{1}\times\frac{5}{7}$$

Multiply the numerators and the denominators.

$$\frac{3}{1}\times\frac{5}{7}=\frac{15}{7}$$

Finally, rewrite the improper fraction as a mixed number.

$$\frac{15}{7}=2\frac{1}{7}$$

104. $-10\frac{4}{5}$

First, rewrite the integer and mixed number as improper fractions.

$$-45\div4\frac{1}{6}=-\frac{45}{1}\div\frac{25}{6}$$

Next, change the division problem to an equivalent multiplication problem using the reciprocal of the second fraction.

$$-\frac{45}{1}\div\frac{25}{6}=-\frac{45}{1}\times\frac{6}{25}$$

Now, before multiplying the numerators and the denominators, reduce the fractions — dividing one numerator and one denominator by the same number.

$$-\frac{{}^{9}\cancel{45}}{1}\times\frac{6}{\cancel{25}_{5}}=-\frac{9}{1}\times\frac{6}{5}$$

Multiply the numerators and the denominators.

$$-\frac{9}{1}\times\frac{6}{5}=-\frac{54}{5}$$

Finally, rewrite the improper fraction as a mixed number.

$$-\frac{54}{5}=-10\frac{4}{5}$$

105. $\frac{3}{5}$

Multiply the numerator and denominator by 3. This is, essentially, multiplying by 1. Be sure to distribute the 3 over both terms in the denominator.

$$\frac{3}{3}\cdot\frac{1}{1+\frac{2}{3}}=\frac{3\cdot 1}{3\left(1+\frac{2}{3}\right)}$$

$$=\frac{3\cdot 1}{3\cdot 1+\cancel{3}\cdot\frac{2}{\cancel{3}}}=\frac{3}{3+2}$$

Now simplify the denominator.

$$\frac{3}{3+2}=\frac{3}{5}$$

Another technique for solving this type of problem is to add the two numbers in the denominator.

$$\frac{1}{1+\frac{2}{3}}=\frac{1}{1\frac{2}{3}}$$

Write the result as an improper fraction.

$$\frac{1}{1\frac{2}{3}}=\frac{1}{\left(\frac{5}{3}\right)}$$

Then multiply the numerator by the reciprocal of this new denominator.

$$\frac{1}{\left(\frac{5}{3}\right)}=1\cdot\left(\frac{3}{5}\right)=\frac{3}{5}$$

106. $\frac{3}{14}$

Multiply the numerator and denominator by 6. This is the least common denominator of the two fractions in the problem. Be sure to distribute the 6 over both terms in the denominator.

$$\frac{6}{6}\cdot\frac{\frac{1}{2}}{2+\frac{1}{3}}=\frac{6\cdot\frac{1}{2}}{6\left(2+\frac{1}{3}\right)}$$

$$=\frac{{}^{3}\cancel{6}\cdot\frac{1}{\cancel{2}}}{6\cdot 2+{}^{2}\cancel{6}\cdot\frac{1}{\cancel{3}}}=\frac{3}{12+2}$$

Now simplify the denominator.

$$\frac{3}{12+2}=\frac{3}{14}$$

Another technique for solving this type of problem is to add the two numbers in the denominator, write the result as an improper fraction, and then multiply the numerator by the reciprocal of this new denominator.

107. $7\frac{1}{2}$

Subtract the two fractions in the denominator by first creating common denominators.

$$\frac{\frac{3}{4}}{\frac{3}{5}-\frac{1}{2}}=\frac{\frac{3}{4}}{\frac{2}{2}\cdot\frac{3}{5}-\frac{5}{5}\cdot\frac{1}{2}}$$

$$=\frac{\frac{3}{4}}{\frac{6}{10}-\frac{5}{10}}=\frac{\frac{3}{4}}{\frac{1}{10}}$$

Now multiply the numerator by the reciprocal of the denominator.

$$\frac{\frac{3}{4}}{\frac{1}{10}}=\frac{3}{4}\times\frac{10}{1}$$

Reduce the fractions before multiplying and convert the final product to a mixed number.

$$=\frac{3}{\cancel{4}_2}\times\frac{\cancel{10}^5}{1}=\frac{15}{2}=7\frac{1}{2}$$

Another technique for solving this type of problem is to multiply the numerator and denominator by the least common denominator of all the fractions. In this case, it would be 20.

108. $\frac{27}{32}$

Subtract the two fractions in the denominator, and change the mixed number in the numerator to an improper fraction.

$$\frac{4\frac{1}{2}}{6-\frac{2}{3}}=\frac{\frac{9}{2}}{\frac{18}{3}-\frac{2}{3}}=\frac{\frac{9}{2}}{\frac{16}{3}}$$

Now multiply the numerator by the reciprocal of the denominator.

$$\frac{\frac{9}{2}}{\frac{16}{3}}=\frac{9}{2}\times\frac{3}{16}=\frac{27}{32}$$

Another technique for solving this type of problem is to multiply the numerator and denominator by the least common denominator of all the fractions. In this case, it would be 6.

Answers 101–200

109. $\frac{x^2}{x+1}$

Multiply the numerator and denominator by x. This is the denominator of the only fraction and is, essentially, multiplying by 1. Be sure to distribute the x over both terms in the denominator.

$$\frac{x}{x}\cdot\frac{x}{1+\frac{1}{x}}=\frac{x\cdot x}{x\left(1+\frac{1}{x}\right)}$$

$$=\frac{x^2}{x\cdot 1+\cancel{x}\cdot\frac{1}{\cancel{x}}}=\frac{x^2}{x+1}$$

Another technique for solving this type problem is to add the two terms in the denominator and then multiply the numerator by the reciprocal of this new denominator.

110. $\frac{-1}{x(x+3)}$

Subtract the two fractions in the numerator, and change the whole number in the denominator to an improper fraction.

$$\frac{\frac{1}{x+3}-\frac{1}{x}}{3}=\frac{\frac{x}{x}\cdot\frac{1}{x+3}-\frac{x+3}{x+3}\cdot\frac{1}{x}}{\frac{3}{1}}$$

$$=\frac{\frac{x}{x(x+3)}-\frac{x+3}{x(x+3)}}{\frac{3}{1}}$$

$$=\frac{\frac{x-(x+3)}{x(x+3)}}{\frac{3}{1}}$$

$$=\frac{\frac{x-x-3}{x(x+3)}}{\frac{3}{1}}$$

$$=\frac{\frac{-3}{x(x+3)}}{\frac{3}{1}}$$

Now multiply the numerator by the reciprocal of the denominator.

$$\frac{\frac{-3}{x(x+3)}}{\frac{3}{1}}=\frac{-\cancel{3}^1}{x(x+3)}\cdot\frac{1}{\cancel{3}_1}$$

$$=\frac{-1}{x(x+3)}$$

Another technique for solving this type problem is to multiply the numerator and denominator by the least common denominator of all the fractions. In this case, it would be $x(x+3)$.

111. 438.0401

When adding or subtracting decimal numbers, you need to line up the decimal points to align the digits having the same place value.

$$\begin{array}{r} 432.04 \\ +\ \ \ 6.0001 \\ \hline 438.0401 \end{array}$$

112. 10.277

When adding or subtracting decimal numbers, you need to line up the decimal points to align the digits having the same place value.

$$\begin{array}{r} 15.4 \\ -\ \ 5.123 \\ \hline \end{array}$$

Add 0's after the 4, and then subtract.

$$\begin{array}{r} 15.400 \\ -\ \ 5.123 \\ \hline 10.277 \end{array}$$

113. 1.043x

When adding like terms, you combine the coefficients of the variables. So you add 1 and 0.043 and make the sum be the new coefficient of the x.

When adding or subtracting decimal numbers, you need to line up the decimal points to align the digits having the same place value.

$$\begin{array}{r} 1. \\ +\ 0.043 \\ \hline 1.043 \end{array}$$

So the sum is 1.043x.

114. 0.588y

When subtracting like terms, you combine the coefficients of the variables. So you subtract 5.3 minus 4.712 and make the difference be the new coefficient of the y.

When adding or subtracting decimal numbers, you need to line up the decimal points to align the digits having the same place value.

$$\begin{array}{r} 5.300 \\ -\ 4.712 \\ \hline 0.588 \end{array}$$

So the difference is 0.588y.

115. **0.2408**

When multiplying decimal numbers, count the number of digits to the right of the decimal point in the multipliers. That will be the number of digits to the right of the decimal point in the answer. Then multiply the numbers together without considering their decimal points. Place the decimal in the final answer.

4.3 has one digit to the right of the decimal point.

0.056 has three digits to the right of the decimal.

The sum of those digits is four, which you need for the answer.

Multiplying 43×56 (the 0's in front of the 5 and 6 aren't necessary), you get 2408. You need four digits to the right of the decimal point, so that gives you 0.2408.

116. **–34.155**

When multiplying decimal numbers, count the number of digits to the right of the decimal point in the multipliers. That will be the number of digits to the right of the decimal point in the answer. Then multiply the numbers together without considering their decimal points. Place the decimal in the final answer.

6.21 has two digits to the right of the decimal point.

–5.5 has one digit to the right of the decimal.

The sum of those digits is three, which you need for the answer.

Multiplying 621×55, you get 34155. You need three digits to the right of the decimal point, so that gives you 34.155.

There's one negative sign in the problem, so the answer is negative.

117. **$0.0332x^2$**

In multiplication of algebraic expressions, the coefficients get multiplied together, and the variables get multiplied.

When multiplying decimal numbers, count the number of digits to the right of the decimal point in the multipliers. That will be the number of digits to the right of the decimal point in the answer. Then multiply the numbers together without considering their decimal points. Place the decimal in the final answer.

8.3 has one digit to the right of the decimal point.

0.004 has three digits to the right of the decimal.

The sum of those digits is four, which you need for the answer.

Multiplying 83×4, you get 332. You need four digits to the right of the decimal point, so add a 0 in front of the first 3, giving you 0.0332.

Now multiply the x variables to get x^2.

118. **$1.665y^3$**

In multiplication of algebraic expressions, the coefficients get multiplied together, and the variables get multiplied.

When multiplying decimal numbers, count the number of digits to the right of the decimal point. That will be the number of digits to the right of the decimal point in the answer. Then multiply the numbers together without considering their decimal points. Place the decimal in the final answer.

3.7, –4.5, and –0.1 each have just the one digit to the right of the decimal point.

The sum of those digits is three, which you need for the answer.

Multiplying $37 \times 45 \times 1$, you get 1665. You need three digits to the right of the decimal point, giving you 1.665.

Now multiply the y variables to get y^3.

There are two negative numbers in the problems, so the sign of the product is positive.

119. **730**

When dividing decimal numbers, you want to divide with a whole number. So move the decimal point in the divisor to the right to form a whole number, and then move the decimal point in the dividend that same number of places. Place a decimal point above the dividend's new decimal position for the quotient.

Dividing 36.5 by 0.05, move the decimal point two places to the right. You'll have to add a 0 after the 5 in the dividend.

$$0.05_\wedge \overline{)36.50_\wedge}$$

Becomes:

$$\begin{array}{r} 730. \\ 5\overline{)3650.} \end{array}$$

120. **0.13**

When dividing decimal numbers, you want to divide with a whole number. So move the decimal point in the divisor to the right to form a whole number, and then move the decimal point in the dividend that same number of places. Place a decimal point above the dividend's new decimal position for the quotient.

Dividing 0.143 by 1.1, move the decimal point one place to the right.

$$1.1_\wedge \overline{)0.1_\wedge 43}$$

Becomes:

$$\begin{array}{r} .13 \\ 11\overline{)1.43} \end{array}$$

121. ≈487.805

When dividing decimal numbers, you want to divide with a whole number. So move the decimal point in the divisor to the right to form a whole number, and then move the decimal point in the dividend that same number of places. Place a decimal point above the dividend's new decimal position for the quotient.

Dividing 6 by 0.0123, move the decimal point four places to the right.

$0.0123_{\wedge}\overline{)6.0000_{\wedge}}$

Becomes:

$$\begin{array}{r} 487.804878... \\ 123\overline{)60000.000000...} \end{array}$$

The problem doesn't divide evenly, so round to three decimal places, giving you approximately 487.805 for an answer.

122. ≈ –23.529

When dividing decimal numbers, you want to divide with a whole number. So move the decimal point in the divisor to the right to form a whole number, and then move the decimal point in the dividend that same number of places. Place a decimal point above the dividend's new decimal position for the quotient.

Dividing –72 by 3.06, move the decimal point two places to the right. You have to add two 0's to the end of the dividend.

$3.06_{\wedge}\overline{)-72.00_{\wedge}}$

Becomes:

$$\begin{array}{r} -23.529411... \\ 306\overline{)-7200.000000...} \end{array}$$

The problem doesn't divide evenly, so round to three decimal places, giving you approximately –23.529 for an answer.

123. ≈ 48.333

When dividing decimal numbers, you want to divide with a whole number. So move the decimal point in the divisor to the right to form a whole number, and then move the decimal point in the dividend that same number of places. Place a decimal point above the dividend's new decimal position for the quotient.

Dividing 1.45 by 0.03, move the decimal point two places to the right.

$0.03_{\wedge}\overline{)1.45_{\wedge}}$

Becomes:

$$\begin{array}{r} 48.333333... \\ 3\overline{)145.000000...} \end{array}$$

The problem doesn't divide evenly, so round to three decimal places, giving you approximately 48.333 for an answer.

The decimal is a non-terminating, repeating decimal which is equivalent to $\frac{1}{3}$, so the answer can also be written as $48\frac{1}{3}$.

124. $\approx$**1821.622**

When dividing decimal numbers, you want to divide with a whole number. So move the decimal point in the divisor to the right to form a whole number, and then move the decimal point in the dividend that same number of places. Place a decimal point above the dividend's new decimal position for the quotient.

Dividing 67.4 by 0.037, move the decimal point three places to the right. You have to add two 0's after the 4 in the dividend.

$0.037_{\wedge}\overline{)67.400_{\wedge}}$

Becomes:

$$\begin{array}{r} 1821.621621621... \\ 37\overline{)67400.000000000...} \end{array}$$

The problem doesn't divide evenly, so round to three decimal places, giving you approximately 1821.622 for an answer.

The decimal is a non-terminating, repeating decimal which is equivalent to $\frac{23}{37}$, so the answer can also be written as $1821\frac{23}{37}$.

125. **0.125**

Divide the numerator by the denominator. Add zeros to the dividend, and keep dividing until the decimal part of the answer either terminates or repeats.

$$\begin{array}{r} .125 \\ 8\overline{)1.000} \end{array}$$

This decimal terminates, so 0.125 is the exact equivalent value.

126. **0.28**

Divide the numerator by the denominator. Add zeros to the dividend, and keep dividing until the decimal part of the answer either terminates or repeats.

$$\begin{array}{r} .28 \\ 25\overline{)7.00} \end{array}$$

This decimal terminates, so 0.28 is the exact equivalent value.

127. **0.0792**

Divide the numerator by the denominator. Add zeros to the dividend, and keep dividing until the decimal part of the answer either terminates or repeats.

$$\begin{array}{r} .0792 \\ 1250\overline{)99.0000} \end{array}$$

This decimal terminates, so 0.0792 is the exact equivalent value.

128. **0.0006**

When the denominator is a power of 10, the decimal equivalent will have the value of the numerator with as many digits to the right of the decimal point as there are 0s in the denominator. In this case, you write 6 in the fourth place after the decimal, giving you the equivalent decimal of 0.0006.

129. **0.444444...**

Divide the numerator by the denominator. Add zeros to the dividend, and keep dividing until the decimal part of the answer either terminates or repeats.

$$9\overline{)4.0000000...}\quad .4444444...$$

This decimal repeats. The decimal equivalent can be written as 0.444444… or as $0.\overline{4}$ to show that the 4 repeats.

130. **0.727272...**

Divide the numerator by the denominator. Add zeros to the dividend, and keep dividing until the decimal part of the answer either terminates or repeats.

$$11\overline{)8.0000000...}\quad .7272727...$$

This decimal repeats. The decimal equivalent can be written as 0. 727272… or as $0.\overline{72}$ to show that the 72 repeats.

131. **0.810810...**

Divide the numerator by the denominator. Add zeros to the dividend, and keep dividing until the decimal part of the answer either terminates or repeats.

$$37\overline{)30.000000000...}\quad .810810810...$$

This decimal repeats. The decimal equivalent can be written as 0.810810… or as $0.\overline{810}$ to show that the 810 repeats.

132. **0.08910891...**

Divide the numerator by the denominator. Add zeros to the dividend, and keep dividing until the decimal part of the answer either terminates or repeats.

$$101\overline{)9.000000000000,..}\quad .089108910891...$$

This decimal repeats. The decimal equivalent can be written as 0.08910891… or as $0.\overline{0891}$ to show that the 0891 repeats.

133. $\frac{3}{4}$

When changing a terminating decimal to an equivalent fraction, write the digits to the right of the decimal point over a power of ten with the same number of 0's as the number of digits to the right of the decimal point. Reduce the fraction if possible.

For 0.75, write 75 over 100 and reduce the fraction.

$$\frac{75}{100} = \frac{3}{4}$$

134. $\frac{7}{8}$

When changing a terminating decimal to an equivalent fraction, write the digits to the right of the decimal point over a power of ten with the same number of 0's as the number of digits to the right of the decimal point. Reduce the fraction if possible.

For 0.875, write 875 over 1000 and reduce the fraction.

$$\frac{875}{1000} = \frac{7}{8}$$

135. $\frac{1}{1250}$

When changing a terminating decimal to an equivalent fraction, write the digits to the right of the decimal point over a power of ten with the same number of 0's as the number of digits to the right of the decimal point. Reduce the fraction if possible.

For 0.0008, write 8 over 10000 and reduce the fraction.

$$\frac{8}{10000} = \frac{1}{1250}$$

136. $\frac{61}{400}$

When changing a terminating decimal to an equivalent fraction, write the digits to the right of the decimal point over a power of ten with the same number of 0's as the number of digits to the right of the decimal point. Reduce the fraction if possible.

For 0.1525, write 1525 over 10000 and reduce the fraction.

$$\frac{1525}{10000} = \frac{61}{400}$$

137. $\frac{8}{9}$

When changing a repeating decimal (in which all the digits are part of the repeater) to an equivalent fraction, write the repeating digits over a number with as many 9's as digits that repeat. Reduce the fraction if possible.

For 0.888…, write 8 over 9. This fraction is already in lowest terms.

138. $\frac{7}{11}$

When changing a repeating decimal (in which all the digits are part of the repeater) to an equivalent fraction, write the repeating digits over a number with as many 9's as digits that repeat. Reduce the fraction if possible.

For 0.636363…, write 63 over 99; then reduce the fraction.

$$\frac{63}{99} = \frac{7}{11}$$

139. $\frac{29}{111}$

When changing a repeating decimal (in which all the digits are part of the repeater) to an equivalent fraction, write the repeating digits over a number with as many 9's as digits that repeat. Reduce the fraction if possible.

For 0.261261…, write 261 over 999; then reduce the fraction.

$$\frac{261}{999} = \frac{29}{111}$$

140. $\frac{2}{7}$

When changing a repeating decimal (in which all the digits are part of the repeater) to an equivalent fraction, write the repeating digits over a number with as many 9's as digits that repeat. Reduce the fraction if possible.

For 0.285714285714…, write 285714 over 999999; then reduce the fraction.

$$\frac{285714}{999999} = \frac{2}{7}$$

141. 243

Since the bases are the same, add the exponents.

$$3^2 \cdot 3^3 = 3^{2+3} = 3^5 = 243$$

142. 32

Since the bases are the same, add the exponents.

$$2^{-1} \cdot 2^6 = 2^{-1+6} = 2^5 = 32$$

143. 16

Since the bases are the same, add the exponents. The number 4 is written 4^1.

$$4 \cdot 4^2 \cdot 4^{-1} = 4^{1+2+(-1)} = 4^2 = 16$$

144. 125

Since the bases are the same, add the exponents. The number 5 is written 5^1.

$$5 \cdot 5^{-3} \cdot 5^5 = 5^{1+(-3)+5} = 5^3 = 125$$

145. 36

Since the bases are the same, subtract the exponents.

$$\frac{6^4}{6^2} = 6^{4-2} = 6^2 = 36$$

146. 8

Since the bases are the same, subtract the exponents.

The number 2 is written 2^1.

$$\frac{2^4}{2} = 2^{4-1} = 2^3 = 8$$

147. 27

Since the bases are the same, subtract the exponents.

$$\frac{3^2}{3^{-1}} = 3^{2-(-1)} = 3^{2+1} = 3^3 = 27$$

148. 125

Since the bases are the same, subtract the exponents.

The number 5 is written 5^1.

$$\frac{5}{5^{-2}} = 5^{1-(-2)} = 5^{1+2} = 5^3 = 125$$

149. $\frac{1}{64}$

Since the bases are the same, subtract the exponents.

$$\frac{4^{-1}}{4^2} = 4^{-1-2} = 4^{-3}$$

To write this without a negative exponent, move the base to the denominator.

$$4^{-3} = \frac{1}{4^3} = \frac{1}{64}$$

150. 3

Since the bases are the same, subtract the exponents.

$$\frac{3^{-2}}{3^{-3}} = 3^{-2-(-3)} = 3^{-2+3} = 3^1 = 3$$

151. 64

Multiply the exponents.

$$(2^2)^3 = 2^{(2)(3)} = 2^6 = 64$$

152. 81

Multiply the exponents.

$$(3^2)^2 = 3^{(2)(2)} = 3^4 = 81$$

153. 16

Multiply the exponents.

$$(4^4)^{1/2} = 4^{(4)\left(\frac{1}{2}\right)} = 4^2 = 16$$

154. 9

Multiply the exponents.

$$(3^{1/3})^6 = 3^{\left(\frac{1}{3}\right)(6)} = 3^2 = 9$$

155. 25

Multiply the exponents.

$$(5^{-2})^{-1} = 5^{(-2)(-1)} = 5^2 = 25$$

156. 64

Multiply the exponents.

$$(2^{-3})^{-2} = 2^{(-3)(-2)} = 2^6 = 64$$

157. $\frac{1}{27}$

Multiply the exponents.

$$(3^9)^{-1/3} = 3^{(9)\left(-\frac{1}{3}\right)} = 3^{-3}$$

To write this without a negative exponent, move the base to the denominator.

$$3^{-3} = \frac{1}{3^3} = \frac{1}{27}$$

158. 256

Multiply the exponents.

$$\left(4^{-2/5}\right)^{-10} = 4^{\left(-\frac{2}{5}\right)(-10)} = 4^4 = 256$$

159. $\frac{1}{25}$

Multiply the exponents.

$$\left(5^{-2/7}\right)^7 = 5^{\left(-\frac{2}{7}\right)(7)} = 5^{-2}$$

To write this without a negative exponent, move the base to the denominator.

$$5^{-2} = \frac{1}{5^2} = \frac{1}{25}$$

160. 6

Multiply the exponents.

$$\left(6^{-4/3}\right)^{-3/4} = 6^{\left(-\frac{4}{3}\right)\left(-\frac{3}{4}\right)} = 6^1 = 6$$

161. 2

First multiply the two factors in the numerator by adding the exponents.

$$\frac{2^2 \cdot 2^3}{2^4} = \frac{2^5}{2^4}$$

Now divide by subtracting the exponents.

$$\frac{2^5}{2^4} = 2^{5-4} = 2^1$$

162. $\frac{1}{9}$

First multiply the two factors in the numerator by adding the exponents.

$$\frac{3^5 \cdot 3^2}{3^9} = \frac{3^7}{3^9}$$

Now divide by subtracting the exponents.

$$\frac{3^7}{3^9} = 3^{7-9} = 3^{-2}$$

Rewrite without the negative exponent.

$$3^{-2} = \frac{1}{3^2} = \frac{1}{9}$$

163. 4

First raise the power in the parentheses to the power outside the parenthesis by multiplying exponents.

$$4^{-3}\left(4^2\right)^2 = 4^{-3}\left(4^4\right)$$

Now multiply by adding the exponents.

$$4^{-3}\left(4^4\right) = 4^1 = 4$$

164. $\frac{1}{5}$

First raise each of the powers in the parentheses to the power outside the parentheses by multiplying exponents.

$$\left(5^{-1}\right)^3\left(5^4\right)^{1/2} = \left(5^{-3}\right)\left(5^2\right)$$

Now multiply by adding the exponents.

$$\left(5^{-3}\right)\left(5^2\right) = 5^{-1}$$

Rewrite without the negative exponent.

$$5^{-1} = \frac{1}{5}$$

165. $\frac{1}{6}$

First raise each of the powers in the parentheses to the power outside the parentheses by multiplying exponents.

$$\left(6^{-3}\right)^{-1}\left(6^2\right)^{-2} = \left(6^3\right)\left(6^{-4}\right)$$

Now multiply by adding the exponents.

$$\left(6^3\right)\left(6^{-4}\right) = 6^{-1}$$

Rewrite without the negative exponent.

$$6^{-1} = \frac{1}{6}$$

166. $\frac{1}{64}$

First raise the power in the parentheses to the power outside the parentheses by multiplying exponents.

$$4^3 \div \left(4^8\right)^{3/4} = 4^3 \div \left(4^6\right)$$

Now divide by subtracting the exponents.

$$4^3 \div (4^6) = 4^{-3}$$

Rewrite without the negative exponent and simplify.

$$4^{-3} = \frac{1}{4^3} = \frac{1}{64}$$

167. **128**

First raise each of the powers in the parentheses to the power outside the parentheses by multiplying exponents.

$$(2^{-3})^{-4} \div (2^3)^2 (2^4)^{1/4} = 2^{12} \div (2^6)(2^1)$$

Now work from left to right — first dividing and then multiplying (subtracting and then adding exponents).

$$\begin{aligned} 2^{12} \div (2^6)(2^1) &= [2^{12} \div (2^6)](2^1) \\ &= [2^6](2^1) \\ &= 2^7 = 128 \end{aligned}$$

168. **16**

First raise each of the powers in the parentheses to the power outside the parentheses by multiplying exponents.

$$(4^2)^{-1/2}(2^3)^2 = (4^{-1})(2^6)$$

Now change the 4 to a power of 2.

$$= ([2^2]^{-1})(2^6)$$

Raise the power in the brackets to the power outside the brackets by multiplying exponents.

$$= (2^{-2})(2^6)$$

Now multiply by adding the exponents.

$$= 2^4 = 16$$

169. **9**

First raise each of the powers in the parentheses to the power outside the parentheses by multiplying exponents.

$$(6^{1/2})^4 \div (4^{-1/2})^{-2} = (6^2) \div (4^1)$$

Don't bother to change to the same bases — just evaluate the expression.

$$(6^2) \div (4^1) = 36 \div 4 = 9$$

170. 131,072

First raise each of the powers in the parentheses to the power outside the parentheses by multiplying exponents.

$$\left(2^2\right)^3\left(2^3\right)^2 \div \left(2^2\right)^2\left(2^3\right)^3 = \left(2^6\right)\left(2^6\right) \div \left(2^4\right)\left(2^9\right)$$

Now move from left to right, performing the operations.

$$= \left(2^{12}\right) \div \left(2^4\right)\left(2^9\right)$$
$$= 2^8\left(2^9\right)$$
$$= 2^{17}$$

Don't forget that, when operations are on the same level, such as multiplication and division, you move from left to right performing the operations.

171. 4

Change the 4 to a power of 2.

$$2^4 \cdot 4^{-1} = 2^4 \cdot \left(2^2\right)^{-1}$$

Raise the power to a power by multiplying the exponents.

$$= 2^4 \cdot 2^{-2}$$

Multiply by adding the exponents.

$$= 2^2 = 4$$

172. 81

Change the 27 to a power of 3.

$$3^{-2} \cdot 27^2 = 3^{-2} \cdot \left(3^3\right)^2$$

Raise the power to a power by multiplying the exponents.

$$= 3^{-2} \cdot 3^6$$

Multiply by adding the exponents.

$$= 3^4 = 81$$

173. 256

Change the 4 and 8 to powers of 2.

$$4^{1/3} \cdot 8^2 \cdot 2^{4/3} = \left(2^2\right)^{1/3} \cdot \left(2^3\right)^2 \cdot 2^{4/3}$$

Raise the powers to powers by multiplying the exponents.

$$= 2^{2/3} \cdot 2^6 \cdot 2^{4/3}$$

Multiply by adding the exponents.

$$= 2^{2/3+6+4/3} = 2^{6/3+6} = 2^8 = 256$$

174. $\frac{1}{5}$

Change the 25 to a power of 5.

$$5^3 \cdot 25^{-2} = 5^3 \cdot \left(5^2\right)^{-2}$$

Raise the power to a power by multiplying the exponents.

$$= 5^3 \cdot 5^{-4}$$

Multiply by adding the exponents.

$$= 5^{-1} = \frac{1}{5}$$

175. $\frac{1}{2048}$

Change the 4 to a power of 2, and find the cube root of 8 — which is 2.

$$\left(4^{-3}\right)^2\left(8^{1/3}\right) = \left(\left(2^2\right)^{-3}\right)^2 \cdot 2$$

Simplify the powers of 2 by multiplying all three.

$$= 2^{-12} \cdot 2$$

Multiply by adding the exponents.

$$= 2^{-12} \cdot 2^1 = 2^{-11} = \frac{1}{2^{11}} = \frac{1}{2048}$$

176. 243

Change the 9 and 27 to powers of 3. $\left(9^2\right)^{-1}(27)^3 = \left(\left(3^2\right)^2\right)^{-1} \cdot \left(3^3\right)^3$

Raise the powers to powers.

$$= 3^{-4} \cdot 3^9$$

Multiply by adding the exponents.

$$= 3^5 = 243$$

177. $\frac{2}{3}$

Write the 6 as the product of 2 and 3; then apply the power to each factor.

$$6^{1/3} \cdot 2^{2/3} \cdot 3^{-4/3} = (2 \cdot 3)^{1/3} \cdot 2^{2/3} \cdot 3^{-4/3}$$
$$= 2^{1/3} \cdot 3^{1/3} \cdot 2^{2/3} \cdot 3^{-4/3}$$

Group the factors of 2 and 3 together; then multiply the like bases by adding their exponents.

$$= 2^{1/3} \cdot 2^{2/3} \cdot 3^{1/3} \cdot 3^{-4/3}$$
$$= 2^{3/3} \cdot 3^{-1} = 2\left(\frac{1}{3}\right) = \frac{2}{3}$$

178. **3**

Write the 12 as the product of 3 and 4; then apply the power to each factor.

$$12^{-1}\cdot 3^2\cdot 4=(3\cdot 4)^{-1}\cdot 3^2\cdot 4$$
$$=3^{-1}\cdot 4^{-1}\cdot 3^2\cdot 4$$

Group the factors of 3 and 4 together; then multiply the like bases by adding their exponents.

$$=3^{-1}\cdot 3^2\cdot 4^{-1}\cdot 4^1$$
$$=3^1\cdot 4^0=3\cdot 1=3$$

179. $\frac{1}{4}$

Rewrite the 32 and 8 as powers of 2.

$$32^{1/2}\left(8^3\right)^{-1/2}=\left(2^5\right)^{1/2}\left(\left(2^3\right)^3\right)^{-1/2}$$

Raise the powers.

$$=2^{5/2}\cdot 2^{-9/2}$$

Now multiply by adding the exponents.

$$=2^{-4/2}=2^{-2}=\frac{1}{2^2}=\frac{1}{4}$$

180. $\frac{1}{7}$

Rewrite the 49 as a power of 7.

$$49^{-1/3}\left(7^2\right)^{-1/6}=\left(7^2\right)^{-1/3}\left(7^2\right)^{-1/6}$$

Raise the powers.

$$=7^{-2/3}\cdot 7^{-1/3}$$

Multiply by adding the exponents.

$$=7^{-1}=\frac{1}{7}$$

181. **8 × 106**

Multiply the coefficients together and the powers of 10 together.

$$\left(2\times 10^2\right)\left(4\times 10^4\right)=(2\times 4)\left(10^2\times 10^4\right)$$
$$=8\times 10^6$$

182. 5.1×10^2

Multiply the coefficients together and the powers of 10 together.

$$\begin{aligned}(3\times10^4)(1.7\times10^{-2}) &= (3\times1.7)(10^4\times10^{-2})\\ &= 5.1\times10^2\end{aligned}$$

183. 4.8×10^{11}

Multiply the coefficients together and the powers of 10 together.

$$\begin{aligned}(6\times10^3)(8\times10^7) &= (6\times8)(10^3\times10^7)\\ &= 48\times10^{10}\end{aligned}$$

For the answer to be in scientific notation, the coefficient has to be a number between 1 and 10, so rewrite it as such a number times a power of 10. Then multiply the powers of 10 again.

$$\begin{aligned}&= (4.8\times10^1)\times10^{10}\\ &= 4.8\times(10^1\times10^{10}) = 4.8\times10^{11}\end{aligned}$$

184. 4.5×10^{-6}

Multiply the coefficients together and the powers of 10 together.

$$\begin{aligned}(5\times10^{-3})(9\times10^{-4}) &= (5\times9)(10^{-3}\times10^{-4})\\ &= 45\times10^{-7}\end{aligned}$$

For the answer to be in scientific notation, the coefficient has to be a number between 1 and 10, so rewrite it as such a number times a power of 10. Then multiply the powers of 10 again.

$$\begin{aligned}&= (4.5\times10^1)\times10^{-7}\\ &= 4.5\times(10^1\times10^{-7}) = 4.5\times10^{-6}\end{aligned}$$

185. 3.328×10^1

Multiply the coefficients together and the powers of 10 together.

$$\begin{aligned}(6.4\times10^{10})(5.2\times10^{-10}) &= (6.4\times5.2)(10^{10}\times10^{-10})\\ &= 33.28\times10^0\end{aligned}$$

For the answer to be in scientific notation, the coefficient has to be a number between 1 and 10, so rewrite it as such a number times a power of 10. Then multiply the powers of 10 again.

$$\begin{aligned}&= (3.328\times10^1)\times10^0\\ &= 3.328\times(10^1\times10^0) = 3.328\times10^1\end{aligned}$$

186. 3×10^2

Divide the coefficients and the powers of 10 separately.

$$\begin{aligned}(9\times10^{-2})\div(3\times10^{-4}) &= (9\div3)\times(10^{-2}\div10^{-4})\\ &= 3\times10^2\end{aligned}$$

Remember, when you divide numbers with the same base, you subtract the exponents.

187. 5×10^5

Divide the coefficients and the powers of 10 separately.

$$\begin{aligned}(1.8\times10^{4})\div(3.6\times10^{-2}) &= (1.8\div3.6)\times(10^{4}\div10^{-2})\\ &= 0.5\times10^6\end{aligned}$$

Remember, when you divide numbers with the same base, you subtract the exponents.

Also, the coefficient has to be a number between 1 and 10, so rewrite it as such a number times a power of 10. Then multiply the powers of 10 again.

$$\begin{aligned}&= (5\times10^{-1})\times10^6\\ &= 5\times(10^{-1}\times10^6) = 5\times10^5\end{aligned}$$

188. 1.7×10^0

Divide the coefficients and the powers of 10 separately.

$$\begin{aligned}(5.1\times10^{-2})\div(3\times10^{-2}) &= (5.1\div3)\times(10^{-2}\div10^{-2})\\ &= 1.7\times10^0\end{aligned}$$

Remember, when you divide numbers with the same base, you subtract the exponents.

189. 9×10^{11}

Divide the coefficients and the powers of 10 separately.

$$\begin{aligned}(1.44\times10^{5})\div(1.6\times10^{-7}) &= (1.44\div1.6)\times(10^{5}\div10^{-7})\\ &= 0.9\times10^{12}\end{aligned}$$

Remember, when you divide numbers with the same base, you subtract the exponents.

Also, the coefficient has to be a number between 1 and 10, so rewrite it as such a number times a power of 10. Then multiply the powers of 10 again.

$$\begin{aligned}&= (9\times10^{-1})\times10^{12}\\ &= 9\times(10^{-1}\times10^{12}) = 9\times10^{11}\end{aligned}$$

190. 1.25×10^{-3}

Divide the coefficients and the powers of 10 separately.

$$\left(1\times10^{-17}\right)\div\left(8\times10^{-15}\right)=\left(1\div8\right)\times\left(10^{-17}\div10^{-15}\right)$$
$$=0.125\times10^{-2}$$

Remember, when you divide numbers with the same base, you subtract the exponents.

Also, the coefficient has to be a number between 1 and 10, so rewrite it as such a number times a power of 10. Then multiply the powers of 10 again.

$$=\left(1.25\times10^{-1}\right)\times10^{-2}$$
$$=1.25\times\left(10^{-1}\times10^{-2}\right)=1.25\times10^{-3}$$

191. $3\sqrt{2}$

The expression can be written $\sqrt{9\cdot2}$.

Simplifying,

$$\sqrt{9\cdot2}=\sqrt{9}\cdot\sqrt{2}=3\sqrt{2}$$

192. $6\sqrt{10}$

The expression can be written $\sqrt{36\cdot10}$.

Simplifying,

$$\sqrt{36\cdot10}=\sqrt{36}\cdot\sqrt{10}=6\sqrt{10}$$

193. $4\sqrt{2}$

The expression can be written $\sqrt{16\cdot2}$.

Simplifying,

$$\sqrt{16\cdot2}=\sqrt{16}\cdot\sqrt{2}=4\sqrt{2}$$

194. $8\sqrt{10}$

The expression can be written $\sqrt{64\cdot10}$.

Simplifying,

$$\sqrt{64\cdot10}=\sqrt{64}\cdot\sqrt{10}=8\sqrt{10}$$

195. $11\sqrt{2}$

First simplify the two radicals.

$$\sqrt{50}+\sqrt{72}$$
$$=\sqrt{25\cdot 2}+\sqrt{36\cdot 2}$$
$$=\sqrt{25}\cdot\sqrt{2}+\sqrt{36}\cdot\sqrt{2}$$
$$=5\sqrt{2}+6\sqrt{2}$$

Now add the radicals by adding the multipliers.

$$=11\sqrt{2}$$

196. $-\sqrt{5}$

First simplify the two radicals.

$$\sqrt{80}-\sqrt{125}$$
$$=\sqrt{16\cdot 5}-\sqrt{25\cdot 5}$$
$$=\sqrt{16}\cdot\sqrt{5}-\sqrt{25}\cdot\sqrt{5}$$
$$=4\sqrt{5}-5\sqrt{5}$$

Now add the radicals by subtracting the multipliers.

$$=-\sqrt{5}$$

197. $\sqrt{5}$

Multiply both the numerator and denominator by the radical in the denominator.

$$\frac{5}{\sqrt{5}}\cdot\frac{\sqrt{5}}{\sqrt{5}}=\frac{5\sqrt{5}}{\sqrt{25}}$$

Now simplify the radical in the denominator and reduce the fraction.

$$\frac{5\sqrt{5}}{\sqrt{25}}=\frac{\cancel{5}\sqrt{5}}{\cancel{5}}=\sqrt{5}$$

198. $2\sqrt{6}$

Multiply both the numerator and denominator by the radical in the denominator.

$$\frac{12}{\sqrt{6}}\cdot\frac{\sqrt{6}}{\sqrt{6}}=\frac{12\sqrt{6}}{\sqrt{36}}$$

Now simplify the radical in the denominator and reduce the fraction.

$$\frac{12\sqrt{6}}{\sqrt{36}}=\frac{\cancel{12}^{2}\sqrt{6}}{\cancel{6}}=2\sqrt{6}$$

199. $6\sqrt{3}$

Multiply both the numerator and denominator by the radical in the denominator.

$$\frac{18}{\sqrt{3}}\cdot\frac{\sqrt{3}}{\sqrt{3}}=\frac{18\sqrt{3}}{\sqrt{9}}$$

Now simplify the radical in the denominator and reduce the fraction.

$$\frac{18\sqrt{3}}{\sqrt{9}}=\frac{\cancel{18}^{6}\sqrt{3}}{\cancel{3}}=6\sqrt{3}$$

200. $\frac{\sqrt{2}}{2}$

Multiply both the numerator and denominator by the radical in the denominator.

$$\frac{\sqrt{20}}{\sqrt{40}}\cdot\frac{\sqrt{40}}{\sqrt{40}}=\frac{\sqrt{800}}{\sqrt{1600}}$$

Now simplify both radicals and reduce the fraction.

$$\frac{\sqrt{800}}{\sqrt{1600}}=\frac{\sqrt{400\cdot 2}}{40}$$
$$=\frac{\sqrt{400}\cdot\sqrt{2}}{40}=\frac{20\sqrt{2}}{40}$$
$$=\frac{\cancel{20}\sqrt{2}}{\cancel{40}^{2}}=\frac{\sqrt{2}}{2}$$

201. $\frac{\sqrt{5}}{5}$

Multiply both the numerator and denominator by the radical in the denominator.

$$\frac{\sqrt{12}}{\sqrt{60}}\cdot\frac{\sqrt{60}}{\sqrt{60}}=\frac{\sqrt{720}}{\sqrt{3600}}$$

Now simplify both radicals and reduce the fraction.

$$\frac{\sqrt{720}}{\sqrt{3600}}=\frac{\sqrt{144\cdot 5}}{60}$$
$$=\frac{\sqrt{144}\cdot\sqrt{5}}{60}=\frac{12\sqrt{5}}{60}$$
$$=\frac{\cancel{12}\sqrt{5}}{\cancel{60}^{5}}=\frac{\sqrt{5}}{5}$$

202. $\frac{2(8-\sqrt{6})}{29}$

Multiply both the numerator and denominator by the conjugate of the binomial in the denominator.

$$\frac{4}{8+\sqrt{6}}\cdot\frac{8-\sqrt{6}}{8-\sqrt{6}}=\frac{4(8-\sqrt{6})}{(8+\sqrt{6})(8-\sqrt{6})}$$

Now multiply in the denominator and simplify.

$$= \frac{4(8-\sqrt{6})}{64-\sqrt{36}}$$

$$= \frac{4(8-\sqrt{6})}{64-6}$$

$$= \frac{\cancel{4}^{2}(8-\sqrt{6})}{\cancel{58}^{29}}$$

$$= \frac{2(8-\sqrt{6})}{29}$$

203. $\frac{10(6-\sqrt{5})}{31}$

Multiply both the numerator and denominator by the conjugate of the binomial in the denominator.

$$\frac{10}{6+\sqrt{5}} \cdot \frac{6-\sqrt{5}}{6-\sqrt{5}} = \frac{10(6-\sqrt{5})}{(6+\sqrt{5})(6-\sqrt{5})}$$

Now multiply in the denominator and simplify.

$$= \frac{10(6-\sqrt{5})}{36-\sqrt{25}}$$

$$= \frac{10(6-\sqrt{5})}{36-5}$$

$$= \frac{10(6-\sqrt{5})}{31}$$

204. $\frac{7-\sqrt{7}}{3}$

Multiply both the numerator and denominator by the conjugate of the binomial in the denominator.

$$\frac{14}{7+\sqrt{7}} \cdot \frac{7-\sqrt{7}}{7-\sqrt{7}} = \frac{14(7-\sqrt{7})}{(7+\sqrt{7})(7-\sqrt{7})}$$

Now multiply in the denominator and simplify.

$$= \frac{14(7-\sqrt{7})}{49-\sqrt{49}}$$

$$= \frac{14(7-\sqrt{7})}{49-7}$$

$$= \frac{\cancel{14}(7-\sqrt{7})}{\cancel{42}^{3}}$$

$$= \frac{7-\sqrt{7}}{3}$$

205. $\frac{4(3+\sqrt{2})}{7}$

Multiply both the numerator and denominator by the conjugate of the binomial in the denominator.

$$\frac{4}{3-\sqrt{2}}\cdot\frac{3+\sqrt{2}}{3+\sqrt{2}}=\frac{4(3+\sqrt{2})}{(3-\sqrt{2})(3+\sqrt{2})}$$

Now multiply in the denominator and simplify.

$$=\frac{4(3+\sqrt{2})}{9-\sqrt{4}}$$

$$=\frac{4(3+\sqrt{2})}{9-2}$$

$$=\frac{4(3+\sqrt{2})}{7}$$

206. $\frac{3(5-\sqrt{10})}{5}$

Multiply both the numerator and denominator by the conjugate of the binomial in the denominator.

$$\frac{9}{5+\sqrt{10}}\cdot\frac{5-\sqrt{10}}{5-\sqrt{10}}=\frac{9(5-\sqrt{10})}{(5+\sqrt{10})(5-\sqrt{10})}$$

Now multiply in the denominator and simplify.

$$=\frac{9(5-\sqrt{10})}{25-\sqrt{100}}$$

$$=\frac{9(5-\sqrt{10})}{25-10}$$

$$=\frac{\cancel{9}^{3}(5-\sqrt{10})}{\cancel{15}^{5}}$$

$$=\frac{3(5-\sqrt{10})}{5}$$

207. $\frac{8-2\sqrt{6}+4\sqrt{3}-3\sqrt{2}}{10}$

Multiply both the numerator and denominator by the conjugate of the binomial in the denominator.

$$\frac{2+\sqrt{3}}{4+\sqrt{6}}\cdot\frac{4-\sqrt{6}}{4-\sqrt{6}}=\frac{(2+\sqrt{3})(4-\sqrt{6})}{(4+\sqrt{6})(4-\sqrt{6})}$$

Now multiply in the numerator and denominator and simplify.

$$=\frac{8-2\sqrt{6}+4\sqrt{3}-\sqrt{18}}{16-\sqrt{36}}$$

$$=\frac{8-2\sqrt{6}+4\sqrt{3}-\sqrt{9\cdot 2}}{16-6}$$

$$=\frac{8-2\sqrt{6}+4\sqrt{3}-3\sqrt{2}}{10}$$

That's it. Nothing combines, and nothing reduces.

208. $\frac{16-2\sqrt{6}-4\sqrt{10}+\sqrt{15}}{29}$

Multiply both the numerator and denominator by the conjugate of the binomial in the denominator.

$$\frac{4-\sqrt{10}}{8+\sqrt{6}}\cdot\frac{8-\sqrt{6}}{8-\sqrt{6}}=\frac{\left(4-\sqrt{10}\right)\left(8-\sqrt{6}\right)}{\left(8+\sqrt{6}\right)\left(8-\sqrt{6}\right)}$$

Now multiply in the numerator and denominator and simplify.

$$=\frac{32-4\sqrt{6}-8\sqrt{10}+\sqrt{60}}{64-\sqrt{36}}$$

$$=\frac{32-4\sqrt{6}-8\sqrt{10}+\sqrt{4\cdot 15}}{64-6}$$

$$=\frac{32-4\sqrt{6}-8\sqrt{10}+2\sqrt{15}}{58}$$

$$=\frac{\cancel{32}^{16}-\cancel{4}^{2}\sqrt{6}-\cancel{8}^{4}\sqrt{10}+\cancel{2}\sqrt{15}}{\cancel{58}^{29}}$$

$$=\frac{16-2\sqrt{6}-4\sqrt{10}+\sqrt{15}}{29}$$

209. $\frac{17-15\sqrt{5}}{76}$

Multiply both the numerator and denominator by the conjugate of the binomial in the denominator.

$$\frac{3-\sqrt{20}}{9-\sqrt{5}}\cdot\frac{9+\sqrt{5}}{9+\sqrt{5}}=\frac{\left(3-\sqrt{20}\right)\left(9+\sqrt{5}\right)}{\left(9-\sqrt{5}\right)\left(9+\sqrt{5}\right)}$$

Now multiply in the numerator and denominator and simplify.

$$=\frac{27+3\sqrt{5}-9\sqrt{20}-\sqrt{100}}{81-\sqrt{25}}$$

$$=\frac{27+3\sqrt{5}-9\sqrt{4\cdot 5}-10}{81-5}$$

$$=\frac{17+3\sqrt{5}-9\cdot 2\sqrt{5}}{76}$$

$$=\frac{17+3\sqrt{5}-18\sqrt{5}}{76}$$

$$=\frac{17-15\sqrt{5}}{76}$$

210. $\frac{16+6\sqrt{6}}{-5}$

Multiply both the numerator and denominator by the conjugate of the binomial in the denominator.

$$\frac{4+\sqrt{24}}{1-\sqrt{6}}\cdot\frac{1+\sqrt{6}}{1+\sqrt{6}}=\frac{\left(4+\sqrt{24}\right)\left(1+\sqrt{6}\right)}{\left(1-\sqrt{6}\right)\left(1+\sqrt{6}\right)}$$

Now multiply in the numerator and denominator and simplify.

$$=\frac{4+4\sqrt{6}+\sqrt{24}+\sqrt{144}}{1-\sqrt{36}}$$

$$=\frac{4+4\sqrt{6}+\sqrt{24}+12}{1-6}$$

$$=\frac{16+4\sqrt{6}+\sqrt{4\cdot 6}}{-5}$$

$$=\frac{16+4\sqrt{6}+2\sqrt{6}}{-5}$$

$$=\frac{16+6\sqrt{6}}{-5}$$

211. $a^{5/3}$

The cube root is written with an exponent of $1/3$.

$$\sqrt[3]{a^5}=\left(a^5\right)^{1/3}$$

Now multiply the exponents.

$$\left(a^5\right)^{1/3}=a^{5/3}$$

212. $4^{3/7}$

The seventh root is written with an exponent of $1/7$.

$$\sqrt[7]{4^3}=\left(4^3\right)^{1/7}$$

Now multiply the exponents.

$$\left(4^3\right)^{1/7}=4^{3/7}$$

213. a^5

The fifth root is written with an exponent of $1/5$.

$$\sqrt[5]{a^{25}}=\left(a^{25}\right)^{1/5}$$

Now multiply the exponents and simplify.

$$\left(a^{25}\right)^{1/5}=a^{25/5}=a^5$$

214. x^5

The third root is written with an exponent of $\frac{1}{3}$.

$$\sqrt[3]{x^{15}} = \left(x^{15}\right)^{1/3}$$

Now multiply the exponents and simplify.

$$\left(x^{15}\right)^{1/3} = x^{15/3} = x^5$$

215. $\frac{2}{3y^{1/2}}$

The square root is written with an exponent of $\frac{1}{2}$.

$$\frac{2}{3\sqrt{y}} = \frac{2}{3y^{1/2}}$$

216. $\frac{x}{yz^{1/3}}$

The cube root is written with an exponent of $\frac{1}{3}$.

$$\frac{x}{y\sqrt[3]{z}} = \frac{x}{yz^{1/3}}$$

217. 4

Find the root, and then raise the result to the power.

$$\begin{aligned} 8^{2/3} &= \left(8^{1/3}\right)^2 \\ &= \left(\sqrt[3]{8}\right)^2 \\ &= (2)^2 = 4 \end{aligned}$$

218. 8

Find the root, and then raise the result to the power.

$$\begin{aligned} 16^{3/4} &= \left(16^{1/4}\right)^3 \\ &= \left(\sqrt[4]{16}\right)^3 \\ &= (2)^3 = 8 \end{aligned}$$

219. 81

Find the root, and then raise the result to the power.

$$\begin{aligned} 27^{4/3} &= \left(27^{1/3}\right)^4 \\ &= \left(\sqrt[3]{27}\right)^4 \\ &= (3)^4 = 81 \end{aligned}$$

220. 243

Find the root, and then raise the result to the power.

$$\begin{aligned}9^{5/2} &= \left(9^{1/2}\right)^5 \\ &= \left(\sqrt{9}\right)^5 \\ &= (3)^5 = 243\end{aligned}$$

221. 32

Find the root, and then raise the result to the power.

$$\begin{aligned}64^{5/6} &= \left(64^{1/6}\right)^5 \\ &= \left(\sqrt[6]{64}\right)^5 \\ &= (2)^5 = 32\end{aligned}$$

222. 25

Find the root, and then raise the result to the power.

$$\begin{aligned}125^{2/3} &= \left(125^{1/3}\right)^2 \\ &= \left(\sqrt[3]{125}\right)^2 \\ &= (5)^2 = 25\end{aligned}$$

223. $\frac{1}{100}$

Find the root, and then raise the result to the power.

$$\begin{aligned}1000^{-2/3} &= \left(1000^{1/3}\right)^{-2} \\ &= \left(\sqrt[3]{1000}\right)^{-2} \\ &= (10)^{-2} = \frac{1}{10^2} = \frac{1}{100}\end{aligned}$$

224. $\frac{1}{2}$

Find the root, and then raise the result to the power.

$$\begin{aligned}32^{-1/5} &= \left(32^{1/5}\right)^{-1} \\ &= \left(\sqrt[5]{32}\right)^{-1} \\ &= (2)^{-1} = \frac{1}{2}\end{aligned}$$

225. $\frac{8}{27}$

First write the power of the fraction as a fraction with powers in the numerator and denominator.

$$\left(\frac{4}{9}\right)^{3/2} = \frac{4^{3/2}}{9^{3/2}}$$

In both numerator and denominator, find the root, and then raise the result to the power.

$$\frac{4^{3/2}}{9^{3/2}} = \left(\frac{4^{1/2}}{9^{1/2}}\right)^3$$
$$= \left(\frac{\sqrt{4}}{\sqrt{9}}\right)^3$$
$$= \left(\frac{2}{3}\right)^3 = \frac{8}{27}$$

226. $\frac{10}{3}$

First "flip" the fraction to make the exponent a positive number.

$$\left(\frac{9}{100}\right)^{-1/2} = \left(\frac{100}{9}\right)^{1/2}$$

Now write the power of the fraction as a fraction with powers in the numerator and denominator.

$$= \frac{100^{1/2}}{9^{1/2}}$$

In both numerator and denominator, find the root.

$$= \frac{\sqrt{100}}{\sqrt{9}} = \frac{10}{3}$$

227. 4

The number16 is a perfect fourth power, so change 16 to a power of 2, the radical to a fractional exponent, and multiply the powers.

$$\left(\sqrt[4]{16}\right)^2 = \left(\left(2^4\right)^{1/4}\right)^2$$
$$= \left(2^1\right)^2 = 2^2 = 4$$

228. 729

Since 81 isn't a perfect sixth power, change the radical to a fractional exponent and multiply the powers.

$$\left(\sqrt[6]{81}\right)^9 = \left(81^{1/6}\right)^9$$
$$= 81^{9/6} = 81^{3/2}$$

Now rewrite as a root and power and solve.

$$81^{3/2} = \left(81^{1/2}\right)^3$$
$$= \left(\sqrt{81}\right)^3$$
$$= 9^3 = 729$$

229. 512

Since 64 isn't a perfect fourth power, change the radical to a fractional exponent and multiply the powers.

$$\left(\sqrt[4]{64}\right)^6 = \left(64^{1/4}\right)^6$$
$$= 64^{6/4} = 64^{3/2}$$

Now rewrite as a root and power and solve.

$$64^{3/2} = \left(64^{1/2}\right)^3$$
$$= \left(\sqrt{64}\right)^3$$
$$= 8^3 = 512$$

230. $\frac{1}{100}$

Since 100 isn't a perfect cube, change the radical to a fractional exponent and multiply the powers.

$$\left(\sqrt[3]{100}\right)^{-3} = \left(100^{1/3}\right)^{-3}$$
$$= 100^{-1} = \frac{1}{100}$$

231. 4

First rewrite the numbers under the radicals as powers of 2.

$$\left(\sqrt[6]{256}\right)\left(\sqrt[3]{4}\right) = \left(\sqrt[6]{2^8}\right)\left(\sqrt[3]{2^2}\right)$$

Now write each radical as a fractional power.

$$= \left(2^8\right)^{1/6}\left(2^2\right)^{1/3}$$

Raise the powers to the powers and simplify.

$$= 2^{8/6} \cdot 2^{2/3} = 2^{4/3} \cdot 2^{2/3}$$

Multiply by adding the exponents and simplify.

$$2^{4/3} \cdot 2^{2/3} = 2^{6/3} = 2^2 = 4$$

232. 27

First rewrite the numbers under the radicals as powers of 3.

$$\left(\sqrt{27}\right)\left(\sqrt[3]{81}\right)\left(\sqrt[6]{3}\right) = \left(\sqrt{3^3}\right)\left(\sqrt[3]{3^4}\right)\left(\sqrt[6]{3^1}\right)$$

Now write each radical as a fractional power.

$$=\left(3^3\right)^{1/2}\left(3^4\right)^{1/3}\left(3^1\right)^{1/6}$$

Raise the powers to the powers.

$$=3^{3/2}\cdot 3^{4/3}\cdot 3^{1/6}$$

Now multiply by adding the exponents and simplify.

$$=3^{3/2+4/3+1/6}=3^{9/6+8/6+1/6}$$

$$=3^{18/6}=3^3=27$$

233. $\frac{1}{2}$

First rewrite the numbers under the radicals as powers of 2.

$$\frac{\sqrt[4]{64}}{\sqrt{32}}=\frac{\sqrt[4]{2^6}}{\sqrt{2^5}}$$

Now write each radical as a fractional power.

$$=\frac{\left(2^6\right)^{1/4}}{\left(2^5\right)^{1/2}}$$

Raise the powers to the powers and simplify.

$$=\frac{2^{6/4}}{2^{5/2}}=\frac{2^{3/2}}{2^{5/2}}$$

Now divide by subtracting the exponents and simplify.

$$2^{3/2-5/2}=2^{-2/2}=2^{-1}=\frac{1}{2}$$

234. $\frac{1}{3}$

First rewrite the numbers under the radicals as powers of 3.

$$\frac{\sqrt[6]{81}}{\sqrt[3]{243}}=\frac{\sqrt[6]{3^4}}{\sqrt[3]{3^5}}$$

Now write each radical as a fractional power.

$$=\frac{\left(3^4\right)^{1/6}}{\left(3^5\right)^{1/3}}$$

Raise the powers to the powers.

$$=\frac{3^{4/6}}{3^{5/3}}=\frac{3^{2/3}}{3^{5/3}}$$

Now divide by subtracting the exponents and simplify.

$$3^{2/3-5/3}=3^{-3/3}=3^{-1}=\frac{1}{3}$$

235. $216\sqrt[3]{36}$

The bases are the same, so add the exponents.

$$6^{10/3} \cdot 6^{1/3} = 6^{10/3+1/3} = 6^{11/3}$$

Now rewrite the power as the sum of a whole number and a fraction.

$$= 6^{9/3+2/3} = 6^{3+2/3}$$

Rewrite again as a product and simplify.

$$= 6^3 \cdot 6^{2/3} = 216 \cdot 6^{2/3}$$

As a radical expression, this is written as $216\sqrt[3]{6^2} = 216\sqrt[3]{36}$.

236. $625\sqrt{5}$

The bases are the same, so add the exponents.

$$5^{3/4} \cdot 5^{15/4} = 5^{3/4+15/4} = 5^{18/4} = 5^{9/2}$$

Now rewrite the power as the sum of a whole number and a fraction.

$$= 5^{8/2+1/2} = 5^{4+1/2}$$

Rewrite again as a product and simplify.

$$= 5^4 \cdot 5^{1/2} = 625 \cdot 5^{1/2}$$

As a radical expression, this is written as $625\sqrt{5}$.

237. $8\sqrt[7]{64}$

The bases are the same, so subtract the exponents.

$$\frac{8^{12/7}}{8^{3/7}} = 8^{12/7-3/7} = 8^{9/7}$$

Now rewrite the power as the sum of a whole number and a fraction.

$$= 8^{7/7+2/7} = 8^{1+2/7}$$

Rewrite again as a product and simplify.

$$= 8^1 \cdot 8^{2/7} = 8 \cdot 8^{2/7}$$

As a radical expression, this is written as $8\sqrt[7]{8^2} = 8\sqrt[7]{64}$.

238. $4\sqrt[4]{4}$

The bases are the same, so subtract the exponents.

$$\frac{4^{11/4}}{4^{3/2}} = 4^{11/4-3/2}$$

$$= 4^{11/4-6/4} = 4^{5/4}$$

Now rewrite the power as the sum of a whole number and a fraction.

$$= 4^{4/4+1/4} = 4^{1+1/4}$$

Rewrite again as a product and simplify.

$$= 4^1 \cdot 4^{1/4}$$

As a radical expression, this is written as $4\sqrt[4]{4}$.

239. 9

First rewrite the bases as powers of 3 and simplify the factors.

$$\frac{27^{5/6}}{9^{1/4}} = \frac{\left(3^3\right)^{5/6}}{\left(3^2\right)^{1/4}}$$

$$= \frac{3^{15/6}}{3^{2/4}} = \frac{3^{5/2}}{3^{1/2}}$$

The bases are now the same, so subtract the exponents.

$$3^{5/2-1/2} = 3^{4/2} = 3^2 = 9$$

240. 1

First rewrite the bases as powers of 2 and simplify the factors.

$$\frac{16^{3/4}}{32^{3/5}} = \frac{\left(2^4\right)^{3/4}}{\left(2^5\right)^{3/5}}$$

$$= \frac{2^{12/4}}{2^{15/5}} = \frac{2^3}{2^3}$$

The bases are now the same, so subtract the exponents.

$$= 2^{3-3} = 2^0 = 1$$

241. 2

First rewrite the bases as powers of 2 and simplify the factors.

$$\frac{8^{1/6} \cdot 2^{2/3}}{4^{1/12}} = \frac{\left(2^3\right)^{1/6} 2^{2/3}}{\left(2^2\right)^{1/12}}$$

$$= \frac{2^{3/6} \cdot 2^{2/3}}{2^{2/12}} = \frac{2^{1/2} \cdot 2^{2/3}}{2^{1/6}}$$

The bases are now the same, so add the exponents in the numerator.

$$= \frac{2^{1/2+2/3}}{2^{1/6}} = \frac{2^{3/6+4/6}}{2^{1/6}} = \frac{2^{7/6}}{2^{1/6}}$$

Divide by subtracting the exponents.

$$= 2^{7/6-1/6} = 2^{6/6} = 2^1 = 2$$

242. 25

First rewrite the bases as powers of 5 and simplify the factors.

$$\frac{25^{3/2}}{5^{1/2}\cdot 125^{1/6}} = \frac{\left(5^2\right)^{3/2}}{5^{1/2}\left(5^3\right)^{1/6}}$$

$$= \frac{5^{6/2}}{5^{1/2}5^{3/6}} = \frac{5^3}{5^{1/2}5^{1/2}}$$

The bases are now the same, so add the exponents in the denominator.

$$= \frac{5^3}{5^{1/2+1/2}} = \frac{5^3}{5^1}$$

Divide by subtracting the exponents.

$$= 5^{3-1} = 5^2 = 25$$

243. 4.4

First rewrite the radical as the product of an integer and a radical.

$$\sqrt{20} = \sqrt{4\cdot 5} = \sqrt{4}\sqrt{5} = 2\sqrt{5}$$

Now substitute the decimal approximation for the radical and multiply.

$$\approx 2(2.2) = 4.4$$

244. 17

First rewrite the radical as the product of an integer and a radical.

$$\sqrt{300} = \sqrt{100\cdot 3} = \sqrt{100}\sqrt{3} = 10\sqrt{3}$$

Now substitute the decimal approximation for the radical and multiply.

$$\approx 10(1.7) = 17$$

245. 8.5

First rewrite the radical as the product of an integer and a radical.

$$\sqrt{75} = \sqrt{25\cdot 3} = \sqrt{25}\sqrt{3} = 5\sqrt{3}$$

Now substitute the decimal approximation for the radical and multiply.

$$\approx 5(1.7) = 8.5$$

246. 16.8

First rewrite the radical as the product of an integer and a radical.

$$\sqrt{288} = \sqrt{144\cdot 2} = \sqrt{144}\sqrt{2} = 12\sqrt{2}$$

Now substitute the decimal approximation for the radical and multiply.

$$\approx 12(1.4) = 16.8$$

247. 7

First rewrite the radicals as the products of an integer and a radical. Then find the sum.

$$\begin{aligned}\sqrt{18}+\sqrt{8} &= \sqrt{9\cdot 2}+\sqrt{4\cdot 2}\\ &= \sqrt{9}\sqrt{2}+\sqrt{4}\sqrt{2}\\ &= 3\sqrt{2}+2\sqrt{2}=5\sqrt{2}\end{aligned}$$

Now substitute the decimal approximation for the radical and multiply.

$$\approx 5(1.4) = 7.0$$

248. 2.2

First rewrite the radicals as the products of an integer and a radical. Then find the difference.

$$\begin{aligned}\sqrt{125}-\sqrt{80} &= \sqrt{25\cdot 5}-\sqrt{16\cdot 5}\\ &= \sqrt{25}\sqrt{5}-\sqrt{16}\sqrt{5}\\ &= 5\sqrt{5}-4\sqrt{5}=1\sqrt{5}\end{aligned}$$

Now substitute the decimal approximation for the radical and multiply.

$$\approx 1(2.2) = 2.2$$

249. 2.38

First rewrite the radical as a product.

$$\sqrt{6}=\sqrt{2}\sqrt{3}$$

Now replace the radicals with the respective approximations.

$$\sqrt{2}\sqrt{3}\approx(1.4)(1.7)=2.38$$

250. 3.08

First rewrite the radical as a product.

$$\sqrt{10}=\sqrt{2}\sqrt{5}$$

Now replace the radicals with the respective approximations.

$$\sqrt{2}\sqrt{5}\approx(1.4)(2.2)=3.08$$

251. $10a$

Add the coefficients: $(4+6)a = 10a$

252. $8xy$

Add and subtract the coefficients:

$$(9 + 4 - 5)xy = 8xy$$

253. $3z + 4$

Rearrange the terms with like-terms together.

$$5z - 2z - 3 + 7$$

Perform the operations on the like-terms.

$$3z + 4$$

254. $-2y + 1$

Rearrange the terms with like-terms together.

$$6y - 8y + 4 - 3$$

Perform the operations on the like-terms.

$$-2y + 1$$

255. $11a - 4ab - 3$

Rearrange the terms with like-terms together.

$$7a + 4a + 2b - 2b + ab - 5ab - 3$$

Perform the operations on the like-terms.

$$11a - 4ab - 3$$

256. $7x^2 - 3x + 2$

Rearrange the terms with like-terms together.

$$3x^2 + 4x^2 + 2x - 5x - 1 + 3$$

Perform the operations on the like-terms.

$$7x^2 - 3x + 2$$

257. $9 - 3z - 8ab + 6b$

Rearrange the terms with like-terms together.

$$9 + 4 - 4 - 3z - 7ab - ab + 6b$$

Perform the operations on the like-terms.

$$9 - 3z - 8ab + 6b$$

258. $x + 10 - y - z^2$

Rearrange the terms with like-terms together.

$$x + 3 + 4 + 5 - 2 - y - z^2$$

The only like-terms are the constants.

Perform the operations on the like-terms.

$$x + 10 - y - z^2$$

259. $12x$

Multiply the two constants.

$$(4 \cdot 3)x = 12x$$

260. $-45y$

Multiply the two constants.

$$(-9 \cdot 5)y = -45y$$

261. $4x$

Reduce the fraction by dividing common factors.

$$\frac{\cancel{12}^4 x^2}{\cancel{3}^1 x} = \frac{4x^{\cancel{2}^1}}{\cancel{x}} = \frac{4x}{1} = 4x$$

262. -4

Reduce the fraction by dividing common factors.

$$\frac{\cancel{24}^4 y^2}{-\cancel{6}^1 y^2} = \frac{4\cancel{y^2}}{-\cancel{y^2}} = -4$$

263. $12x^2y^3$

Multiply the constants together. Add exponents when multiplying like variables.

$$(3 \cdot 4)(x \cdot x)(y \cdot y^2)$$

$$= 12x^2y^3$$

264. $-15y^3z^3$

Multiply the constants together. Add exponents when multiplying like variables.

$$(-5 \cdot 3)(y \cdot y^2)(z^2 \cdot z)$$

$$= -15y^3z^3$$

265. **$7a$**

Reduce the fraction by dividing common factors.

$$\frac{\cancel{42}^{7}a^{2}b}{\cancel{6}^{1}ab}=\frac{7a^{\cancel{2}^{1}}\cancel{b}}{\cancel{a}\cancel{b}}=\frac{7a}{1}=7a$$

266. **$2b$**

Reduce the fraction by dividing common factors.

$$\frac{\cancel{8}^{2}ab^{2}cd}{\cancel{4}abcd}=\frac{2\cancel{a}b^{\cancel{2}^{1}}\cancel{c}\cancel{d}}{\cancel{a}\cancel{b}\cancel{c}\cancel{d}}=\frac{2b}{1}=2b$$

267. **20**

Multiply first.

$= 2 + 18$

Then add.

$= 20$

268. **7**

Divide first.

$= 9 - 2$

Then subtract.

$= 7$

269. **–6**

Do the two multiplications first.

$= 6 + (-12)$

Then add.

$= -6$

270. **–4**

Do the two divisions first.

$= -2 - 2$

Then subtract.

$= -4$

271. **19**

Evaluate the two roots first.

$= 5(2) + 3(3)$

Next do the two multiplications.

$= 10 + 9$

Now add.

$= 19$

272. **–3**

Find the root first.

$= 7 - 2(5)$

Now multiply.

$= 7 - 10$

Finally, subtract.

$= -3$

273. **–1**

Do the multiplication and division.

$= 6 - 8 + 2 - 1$

Now subtract, add, and subtract — working from left to right.

$= -2 + 2 - 1$

$= 0 - 1$

$= -1$

274. **15**

Do the division and multiplication.

$= 4 + 12 - 1$

Now add and subtract, working from left to right.

$= 16 - 1$

$= 15$

275. **10**

First multiply.

$= 10 - 4 - 8 + 12$

Now subtract, subtract and add, working from left to right.

$= 6 - 8 + 12$

$= -2 + 12$

$= 10$

276. 32

First do the two divisions and two multiplications, in order from left to right.

$= 6 \div 3 + 3 \cdot 2 \cdot 5$

$= 2 + 3 \cdot 2 \cdot 5$

$= 2 + 6 \cdot 5$

$= 2 + 30$

Now add.

$= 32$

277. 12

First do the subtraction in the parentheses.

$= 4(3)$

Now multiply.

$= 12$

278. –5

First do the addition in the parentheses.

$= 5(-1)$

Now multiply.

$= -5$

279. 2

First add the terms in the numerator.

$= \frac{12}{6}$

Now divide.

$= 2$

280. 4

First add the two terms in the denominator.

$= \frac{40}{10}$

Now divide.

$= 4$

281. **20**

First add the two numbers under the radical.

$= 5\sqrt{16}$

Now find the square root.

$= 5(4)$

Finally, multiply.

$= 20$

282. **21**

First subtract the two numbers under the radical.

$= 7\sqrt{9}$

Now find the square root.

$= 7(3)$

Now multiply.

$= 21$

283. **7**

First subtract the two numbers in the parentheses.

$= 3 + 2(2)$

Next multiply.

$= 3 + 4$

Now add.

$= 7$

284. **–20**

First add the two numbers in the parentheses.

$= 8 - 7(4)$

Next multiply.

$= 8 - 28$

Now subtract.

$= -20$

285. **–12**

First add the two sets of numbers in the parentheses.

$= 4(7) - 8(5)$

Next do the two multiplications.

$= 28 - 40$

Finally, subtract.

$= -12$

286. **13**

First add and subtract in the denominators.

$= \frac{6}{6} + \frac{12}{1}$

Next do the two divisions.

$= 1 + 12$

Now add.

$= 13$

287. **12**

Replace the x in the expression with –2.

$3(-2)2$

Using the order of operations, first raise –2 to the power of 2.

$= 3(4)$

Now multiply.

$= 12$

288. **14**

Replace the x in the expression with –3.

$-5(-3) - 1$

Using the order of operations, first multiply.

$= 15 - 1$

Now subtract.

$= 14$

289. **–8**

Replace the x variables in the expression with 4.

$4(2 - 4)$

Using the order of operations, subtract in the parentheses.

$= 4(-2)$

Now multiply.

$= -8$

290. –4

Replace the x variables in the expression with –2.

$$\frac{-2-2}{-2+3}$$

Using the order of operations, first simplify in the numerator and denominator separately.

$$=\frac{-4}{1}$$

Now divide.

$$=-4$$

291. 14

Replace the l in the expression with 4 and the w with 3.

$$2(4+3)$$

Using the order of operations, first add the terms in the parentheses.

$$=2(7)$$

Now multiply.

$$=14$$

You'll use this formula to find the perimeter of a rectangle.

292. 18

Replace the b in the expression with 9 and the h with 4.

$$\frac{1}{2}\cdot 9\cdot 4$$

Even though the order of operations says to multiply from left to right, in order, you can take advantage of the associative property of multiplication and group the 9 and 4 together.

$$=\frac{1}{2}(9\cdot 4)$$

Multiply the numbers in the parentheses.

$$=\frac{1}{2}(36)$$

Now multiply.

$$=18$$

You get the same answer if you multiply the three numbers in order, but doing this grouping saves having to deal with fractions.

This formula is used to find the area of a triangle.

293. **34**

Replace the a_0 in the expression with 4, the n with 11, and the d with 3.

$4 + (11 - 1)(3)$

Using the order of operations, first subtract the terms in the parentheses.

$= 4 + (10)(3)$

Now multiply.

$= 4 + 30$

Finally, add.

$= 34$

This is a formula you can use to find the nth term in an arithmetic sequence.

294. **104**

Replace the C in the expression with 40.

$\frac{9}{5}(40)+32$

Using the order of operations, first multiply.

$= 72 + 32$

Now add.

$= 104$

This is the formula you use to change degrees Celsius (centigrade) to degrees Fahrenheit.

295. **2700**

Replace the A in the expression with 100, the r with 2, the n with 1, and the t with 3.

$100\left(1+\frac{2}{1}\right)^{(1)(3)}$

Using the order of operations, first do the division in the parentheses.

$=100(1+2)^{(1)(3)}$

Now add the terms in the parentheses.

$=100(3)^{(1)(3)}$

Now multiply the two factors in the exponent.

$= 100(3)^3$

Now raise the 3 to the 3rd power.

$=100(27)$

Finally, multiply.

$= 2700$

This is the compound interest formula, although "doubling" your money sounds pretty risky!

296. **6**

Replace the x in the expression with 6, the a with 4, the b with 3, and the c with 5.

$\sqrt{6(6-4)(6-3)(6-5)}$

Using the order of operations, first find all the differences in the parentheses.

$= \sqrt{6(2)(3)(1)}$

Now multiply the four factors under the radical.

$= \sqrt{36}$

Now find the square root.

$= 6$

This is Heron's formula for finding the area of a triangle.

297. **6**

Multiply $3 \cdot 2 \cdot 1 = 6$

298. **714**

First write:

$(6 \cdot 5 \cdot 4 \cdot 3 \cdot 2 \cdot 1) - (3 \cdot 2 \cdot 1)$

Multiply all the terms on either side of the subtraction symbol.

$= 720 - 6$

Now subtract.

$= 714$

299. **12**

First write:

$\frac{4 \cdot 3 \cdot 2 \cdot 1}{2 \cdot 1}$

Rather than multiplying and then dividing, first reduce the fraction.

$\frac{4 \cdot 3 \cdot \cancel{2} \cdot \cancel{1}}{\cancel{2} \cdot \cancel{1}} = \frac{12}{1} = 12$

300. 120

First write:

$$\frac{6\cdot5\cdot4\cdot3\cdot2\cdot1}{3\cdot2\cdot1}$$

Rather than multiplying and then dividing, first reduce the fraction.

$$\frac{6\cdot5\cdot4\cdot\cancel{3}\cdot\cancel{2}\cdot\cancel{1}}{\cancel{3}\cdot\cancel{2}\cdot\cancel{1}}=\frac{120}{1}=120$$

301. 11

Replace each x in the rule with 2.

$$f(2)=(2)^2+3(2)+1$$

First raise the power.

$$f(2)=4+3(2)+1$$

Now multiply.

$$f(2)=4+6+1$$

Finally, add.

$$f(2)=11$$

302. 6

Replace each x in the rule with –1.

$$g(-1)=9-3(-1)^2$$

First raise the power.

$$g(-1)=9-3(1)$$

Now multiply.

$$g(-1)=9-3$$

Finally, subtract

$$g(-1)=6$$

303. 3

Replace x in the rule with –4.

$$h(-4)=\sqrt{5-(-4)}$$

First subtract.

$$h(-4)=\sqrt{9}$$

Finally, find the root.

$$h(-4)=3$$

304. 3

Replace x in the rule with 10.

$$k(10)=\frac{10-4}{2}$$

First subtract the terms in the numerator.

$$k(10)=\frac{6}{2}$$

Now divide.

$$k(10) = 3$$

305. 16

Replace each x in the rule with 2.

$$n(2) = (2)^3 + 2(2)^2$$

Raise the two powers.

$$n(2) = 8 + 2(4)$$

Next multiply.

$$n(2) = 8 + 8$$

Now add.

$$n(2) = 16$$

306. 9

Replace each x in the rule with 3.

$$p(3)=\frac{(3+1)(3)^2}{4}$$

Now square the 3.

$$p(3)=\frac{(3+1)(9)}{4}$$

Next, add the two terms in the parentheses.

$$p(3)=\frac{(4)(9)}{4}$$

Multiply the two factors in the numerator.

$$p(3)=\frac{36}{4}$$

Finally, divide.

$$p(3) = 9$$

307. 30

First replace each x in the rule with 4.

$$q(4) = 4! + (4 - 1)!$$

Subtract the two terms in the parentheses.

$$q(4) = 4! + (3)!$$

Now compute the factorial operations.

$$q(4) = 4 \cdot 3 \cdot 2 \cdot 1 + 3 \cdot 2 \cdot 1$$
$$= 24 + 6$$

Finally, add.

$$q(4) = 30$$

308. 2

First replace each x in the rule with 8.

$$r(x) = \frac{8^2 - 4}{3(8) + 6}$$

Square the 8 in the numerator.

$$r(x) = \frac{64 - 4}{3(8) + 6}$$

Multiply the two factors in the denominator.

$$r(x) = \frac{64 - 4}{24 + 6}$$

Do the subtraction and addition.

$$r(x) = \frac{60}{30}$$

And now divide.

$$r(x) = 2$$

309. 8

Replace each x in the rule with –3.

$$t(-3) = \frac{5 - (-3)}{4 + (-3)}$$

Subtract and add.

$$t(-3) = \frac{8}{1}$$

And now divide.

$$t(-3) = 8$$

310. **5**

Replace the x in the rule with 4.

$$w(4)=\sqrt{2(4)^2-7}$$

Square the 4.

$$w(4)=\sqrt{2(16)-7}$$

Multiply.

$$w(4)=\sqrt{32-7}$$

And subtract.

$$w(4)=\sqrt{25}$$

Finally, find the root.

$$w(4) = 5$$

311. $\mathbf{6x + 12}$

Multiply:

$$3(2x) + 3(4)$$

$$= 6x + 12$$

312. $\mathbf{-20y + 24}$

Multiply:

$$-4(5y) - 4(-6)$$

$$= -20y + 24$$

313. $\mathbf{7x^2 - 14x + 21}$

Multiply:

$$7(x^2) + 7(-2x) + 7(3)$$

$$= 7x^2 - 14x + 21$$

314. $\mathbf{-8z + 4}$

Multiply:

$$-8(z)-8\left(-\frac{1}{2}\right)$$

$$=-8z+4$$

315. **35**

Multiply:

$$12(4)+12\left(\frac{1}{6}\right)+12\left(-\frac{5}{4}\right)$$
$$=48+2-15=35$$

316. **29**

Dividing:

$$\frac{18}{3}+\frac{60}{3}+\frac{9}{3}$$
$$=6+20+3=29$$

317. **–45**

Dividing:

$$\frac{-50}{-5}+\frac{75}{-5}+\frac{200}{-5}$$
$$=10-15-40=-45$$

318. **$1 + 2a - 3a^2$**

Dividing:

$$\frac{a^2}{a^2}+\frac{2a^3}{a^2}+\frac{-3a^4}{a^2}$$
$$=1+2a-3a^2$$

319. **$3x - 4y + 6z$**

Dividing:

$$\frac{6x}{2}+\frac{-8y}{2}+\frac{12z}{2}$$
$$=3x-4y+6z$$

320. **$5x^2 - 6xy + 8y^2$**

Dividing:

$$\frac{20x^2y}{4y}-\frac{24xy^2}{4y}+\frac{32y^3}{4y}$$
$$=5x^2-6xy+8y^2$$

321. $ax + x - 2a - 2$

$(a + 1)x + (a + 1)(-2)$

$= ax + x - 2a - 2$

322. $yz^2 - 4z^2 + 7y - 28$

$(y - 4)z^2 + (y - 4)(7)$

$= yz^2 - 4z^2 + 7y - 28$

323. $xy + 2y - 2x - 4$

$(x + 2)y + (x + 2)(-2)$

$= xy + 2y - 2x - 4$

324. $x^5 - 7x^3 - 8x^2 + 56$

$(x^2 - 7)x^3 + (x^2 - 7)(-8)$

$= x^5 - 7x^3 - 8x^2 + 56$

325. $x^4 - y^8$

$(x^2 + y^4)x^2 + (x^2 + y^4)(-y^4)$

$= x^4 + x^2y^4 - x^2y^4 - y^8$

$= x^4 - y^8$

326. $x^2 - x - 6$

First: $x \cdot x = x^2$

Outer: $x \cdot 2 = 2x$

Inner: $-3 \cdot x = -3x$

Last: $-3 \cdot 2 = -6$

$x^2 + 2x - 3x - 6 = x^2 - x - 6$

327. $y^2 + 10y + 24$

First: $y \cdot y = y^2$

Outer: $y \cdot 4 = 4y$

Inner: $6 \cdot y = 6y$

Last: $6 \cdot 4 = 24$

$y^2 + 4y + 6y + 24 = y^2 + 10y + 24$

328. $\mathbf{6x^2 - 13x + 6}$

First: $2x \cdot 3x = 6x^2$

Outer: $2x \cdot (-2) = -4x$

Inner: $-3 \cdot 3x = -9x$

Last: $-3 \cdot (-2) = 6$

$$6x^2 - 4x - 9x + 6 = 6x^2 - 13x + 6$$

329. $\mathbf{3z^2 - 20z + 32}$

First: $z \cdot 3z = 3z^2$

Outer: $z \cdot (-8) = -8z$

Inner: $-4 \cdot 3z = -12z$

Last: $-4 \cdot (-8) = 32$

$$3z^2 - 8z - 12z + 32 = 3z^2 - 20z + 32$$

330. $\mathbf{20x^2 + 2x - 6}$

First: $5x \cdot 4x = 20x^2$

Outer: $5x \cdot (-2) = -10x$

Inner: $3 \cdot 4x = 12x$

Last: $3 \cdot (-2) = -6$

$$20x^2 - 10x + 12x - 6 = 20x^2 + 2x - 6$$

331. $\mathbf{21y^2 - 16y - 16}$

First: $3y \cdot 7y = 21y^2$

Outer: $3y \cdot 4 = 12y$

Inner: $-4 \cdot 7y = -28y$

Last: $-4 \cdot 4 = -16$

$$21y^2 + 12y - 28y - 16 = 21y^2 - 16y - 16$$

332. $\mathbf{x^4 - 1}$

First: $x^2 \cdot x^2 = x^4$

Outer: $x^2 \cdot 1 = x^2$

Inner: $-1 \cdot x^2 = -x^2$

Last: $-1 \cdot 1 = -1$

$$x^4 + x^2 - x^2 - 1 = x^4 - 1$$

333. $6y^6 - y^3 - 2$

First: $2y^3 \cdot 3y^3 = 6y^6$

Outer: $2y^3 \cdot (-2) = -4y^3$

Inner: $1 \cdot 3y^3 = 3y^3$

Last: $1 \cdot (-2) = -2$

$$6y^6 - 4y^3 + 3y^3 - 2 = 6y^6 - y^3 - 2$$

334. $64x^2 - 49$

First: $8x \cdot 8x = 64x^2$

Outer: $8x \cdot 7 = 56x$

Inner: $-7 \cdot 8x = -56x$

Last: $-7 \cdot 7 = -49$

$$64x^2 + 56x - 56x - 49 = 64x^2 - 49$$

335. $4z^4 - 9$

First: $2z^2 \cdot 2z^2 = 4z^4$

Outer: $2z^2 \cdot (-3) = -6z^2$

Inner: $3 \cdot 2z^2 = 6z^2$

Last: $3 \cdot (-3) = -9$

$$4z^4 - 6z^2 + 6z^2 - 9 = 4z^4 - 9$$

336. $x^3 + x^2 - 5x + 3$

$$(x + 3)x^2 + (x + 3)(-2x) + (x + 3)(1)$$
$$= x^3 + 3x^2 - 2x^2 - 6x + x + 3$$
$$= x^3 + x^2 - 5x + 3$$

337. $y^3 + y^2 - 2y - 8$

$$(y - 2)y^2 + (y - 2)(3y) + (y - 2)(4)$$
$$= y^3 - 2y^2 + 3y^2 - 6y + 4y - 8$$
$$= y^3 + y^2 - 2y - 8$$

338. $\mathbf{2z^3 + 3z^2 + 15z + 7}$

$(2z + 1)z^2 + (2z + 1)z + (2z + 1)(7)$

$= 2z^3 + z^2 + 2z^2 + z + 14z + 7$

$= 2z^3 + 3z^2 + 15z + 7$

339. $\mathbf{8x^3 + 2x^2 - 2x - 3}$

$(4x - 3)(2x^2) + (4x - 3)(2x) + (4x - 3)(1)$

$= 8x^3 - 6x^2 + 8x^2 - 6x + 4x - 3$

$= 8x^3 + 2x^2 - 2x - 3$

340. $\mathbf{3y^3 + 14y^2 - 44y + 35}$

$(y + 7)(3y^2) + (y + 7)(-7y) + (y + 7)(5)$

$= 3y^3 + 21y^2 - 7y^2 - 49y + 5y + 35$

$= 3y^3 + 14y^2 - 44y + 35$

341. $\mathbf{x^2 + 10x + 25}$

Using the form $(a + b)^2 = a^2 + 2ab + b^2$,

a^2: x^2

$2ab$: $2(x)(5) = 10x$

b^2: 25

So the square is $x^2 + 10x + 25$.

342. $\mathbf{y^2 - 12y + 36}$

Using the form $(a + b)^2 = a^2 + 2ab + b^2$,

a^2: y^2

$2ab$: $2(y)(-6) = -12y$

b^2: 36

So the square is $y^2 - 12y + 36$.

343. $\mathbf{16z^2 + 24z + 9}$

Using the form $(a + b)^2 = a^2 + 2ab + b^2$,

a^2: $(4z)^2 = 16z^2$

$2ab$: $2(4z)(3) = 24z$

b^2: 9

So the square is $16z^2 + 24z + 9$.

344. $\mathbf{25x^2 - 20x + 4}$

Using the form $(a + b)^2 = a^2 + 2ab + b^2$,

a^2: $(5x)^2 = 25x^2$

$2ab$: $2(5x)(-2) = -20x$

b^2: 4

So the square is $25x^2 - 20x + 4$.

345. $\mathbf{64x^2 + 16xy + y^2}$

Using the form $(a + b)^2 = a^2 + 2ab + b^2$,

a^2: $(8x)^2 = 64x^2$

$2ab$: $2(8x)(y) = 16xy$

b^2: y^2

So the square is $64x^2 + 16xy + y^2$.

346. $\mathbf{x^3 + 6x^2 + 12x + 8}$

First, write the coefficients from the fourth row of Pascal's triangle.

1 3 3 1

Now, add the decreasing powers of x.

$1x^3 \quad 3x^2 \quad 3x^1 \quad 1x^0$

Next, add the increasing powers of 2.

$1x^3(2)^0 \quad 3x^2(2)^1 \quad 3x^1(2)^2 \quad 1x^0(2)^3$

Finally, simplify the terms and add them together.

$= x^3 + 6x^2 + 12x + 8$

347. $\mathbf{y^3 - 12y^2 + 48y - 64}$

First, write the coefficients from the fourth row of Pascal's triangle.

1 3 3 1

Now, add the decreasing powers of y.

$1y^3 \quad 3y^2 \quad 3y^1 \quad 1y^0$

Next, add the increasing powers of –4.

$1y^3(-4)^0 \quad 3y^2(-4)^1 \quad 3y^1(-4)^2 \quad 1y^0(-4)^3$

Finally, simplify the terms and add them together.

$= y^3 - 12y^2 + 48y - 64$

348. $27z^3 + 54z^2 + 36z + 8$

First, write the coefficients from the fourth row of Pascal's triangle.

1 3 3 1

Now, add the decreasing powers of $3z$.

$$1(3z)^3 \quad 3(3z)^2 \quad 3(3z)^1 \quad 1(3z)^0$$

Next, add the increasing powers of 2.

$$1(3z)^3(2)^0 \quad 3(3z)^2(2)^1 \quad 3(3z)^1(2)^2 \quad 1(3z)^0(2)^3$$

Finally, simplify the terms and add them together.

$$= 27z^3 + 54z^2 + 36z + 8$$

349. $8x^6 + 12x^4 + 6x^2 + 1$

First, write the coefficients from the fourth row of Pascal's triangle.

1 3 3 1

Now, add the decreasing powers of $2x^2$.

$$1(2x^2)^3 \quad 3(2x^2)^2 \quad 3(2x^2)^1 \quad 1(2x^2)^0$$

Next, add the increasing powers of 1.

$$1(2x^2)^3(1)^0 \quad 3(2x^2)^2(1)^1 \quad 3(2x^2)^1(1)^2 \quad 1(2x^2)^0(1)^3$$

Finally, simplify the terms and add them together.

$$= 8x^6 + 12x^4 + 6x^2 + 1$$

350. $a^6 - 3a^4b + 3a^2b^2 - b^3$

First, write the coefficients from the fourth row of Pascal's triangle.

1 3 3 1

Now, add the decreasing powers of a^2.

$$1(a^2)^3 \quad 3(a^2)^2 \quad 3(a^2)^1 \quad 1(a^2)^0$$

Next, add the increasing powers of $-b$.

$$1(a^2)^3(-b)^0 \quad 3(a^2)^2(-b)^1 \quad 3(a^2)^1(-b)^2 \quad 1(a^2)^0(-b)^3$$

Finally, simplify the terms and add them together.

$$= a^6 - 3a^4b + 3a^2b^2 - b^3$$

351. $x^4 + 12x^3 + 54x^2 + 108x + 81$

First, write the coefficients from the fifth row of Pascal's triangle.

1 4 6 4 1

Now, add the decreasing powers of x.

$1x^4 \quad 4x^3 \quad 6x^2 \quad 4x^1 \quad 1x^0$

Next, add the increasing powers of 3.

$1x^4(3)^0 \quad 4x^3(3)^1 \quad 6x^2(3)^2 \quad 4x^1(3)^3 \quad 1x^0(3)^4$

Finally, simplify the terms and add them together.

$= x^4 + 12x^3 + 54x^2 + 108x + 81$

352. $\mathbf{y^5 - 10y^4 + 40y^3 - 80y^2 + 80y - 32}$

First, write the coefficients from the sixth row of Pascal's triangle.

1 5 10 10 5 1

Now, add the decreasing powers of y.

$1y^5 \quad 5y^4 \quad 10y^3 \quad 10y^2 \quad 5y^1 \quad 1y^0$

Next, add the increasing powers of –2.

$1y^5(-2)^0 \quad 5y^4(-2)^1 \quad 10y^3(-2)^2 \quad 10y^2(-2)^3 \quad \cdots$

$\cdots \quad 5y^1(-2)^4 \quad 1y^0(-2)^5$

Finally, simplify the terms and add them together.

$= y^5 - 10y^4 + 40y^3 - 80y^2 + 80y - 32$

353. $\mathbf{z^6 + 6z^5 + 15z^4 + 20z^3 + 15z^2 + 6z + 1}$

First, write the coefficients from the seventh row of Pascal's triangle.

1 6 15 20 15 6 1

Now, add the decreasing powers of z.

$1z^6 \quad 6z^5 \quad 15z^4 \quad 20z^3 \quad 15z^2 \quad 6z^1 \quad 1z^0$

Next, add the increasing powers of 1.

$1z^6(1)^0 \quad 6z^5(1)^0 \quad 15z^4(1)^0 \quad 20z^3(1)^0 \cdots$

$\cdots \quad 15z^2(1)^0 \quad 6z^1(1)^0 \quad 1z^0(1)^0$

Finally, simplify the terms and add them together.

$= z^6 + 6z^5 + 15z^4 + 20z^3 + 15z^2 + 6z + 1$

354. $\mathbf{a^7 + 7a^6b^1 + 21a^5b^2 + 35a^4b^3 + 35a^3b^4 + 21a^2b^5 + 7ab^6 + b^7}$

First, write the coefficients from the eighth row of Pascal's triangle.

1 7 21 35 35 21 7 1

Now, add the decreasing powers of a.

$1a^7 \quad 7a^6 \quad 21a^5 \quad 35a^4 \quad 35a^3 \quad 21a^2 \quad 7a^1 \quad 1a^0$

Next, add the increasing powers of *b*.

$1a^7b^0 \quad 7a^6b^1 \quad 21a^5b^2 \quad 35a^4b^3 \quad 35a^3b^4 \quad \cdots$

$\cdots \quad 21a^2b^5 \quad 7a^1b^6 \quad 1a^0b^7$

Finally, simplify the terms and add them together.

$= a^7 + 7a^6b^1 + 21a^5b^2 + 35a^4b^3 + 35a^3b^4 + 21a^2b^5 + 7ab^6 + b^7$

355. $x^7 - 14x^6 + 84x^5 - 280x^4 + 560x^3 - 672x^2 + 448x - 128$

First, write the coefficients from the eighth row of Pascal's triangle.

1 7 21 35 35 21 7 1

Now, add the decreasing powers of *x*.

$1x^7 \quad 7x^6 \quad 21x^5 \quad 35x^4 \quad 35x^3 \quad 21x^2 \quad 7x^1 \quad 1x^0$

Next, add the increasing powers of –2.

$1x^7(-2)^0 \quad 7x^6(-2)^1 \quad 21x^5(-2)^2 \quad 35x^4(-2)^3 \quad 35x^3(-2)^4 \quad \cdots$

$\cdots \quad 21x^2(-2)^5 \quad 7x^1(-2)^6 \quad 1x^0(-2)^7$

Finally, simplify the terms and add them together.

$= x^7 - 14x^6 + 84x^5 - 280x^4 + 560x^3 - 672x^2 + 448x - 128$

356. $256z^4 + 256z^3 + 96z^2 + 16z + 1$

First, write the coefficients from the fifth row of Pascal's triangle.

1 4 6 4 1

Now, add the decreasing powers of 4*z*.

$1(4z)^4 \quad 4(4z)^3 \quad 6(4z)^2 \quad 4(4z)^1 \quad 1(4z)^0$

Next, add the increasing powers of 1.

$1(4z)^4(1)^0 \quad 4(4z)^3(1)^1 \quad 6(4z)^2(1)^2 \quad 4(4z)^1(1)^3 \quad 1(4z)^0(1)^4$

Finally, simplify the terms and add them together.

$= 256z^4 + 256z^3 + 96z^2 + 16z + 1$

357. $243y^5 - 810y^4 + 1080y^3 - 720y^2 + 240y - 32$

First, write the coefficients from the sixth row of Pascal's triangle.

1 5 10 10 5 1

Now, add the decreasing powers of 3*y*.

$1(3y)^5 \quad 5(3y)^4 \quad 10(3y)^3 \quad 10(3y)^2 \quad 5(3y)^1 \quad 1(3y)^0$

Next, add the increasing powers of –2.

$$1(3y)^5(-2)^0 \quad 5(3y)^4(-2)^1 \quad 10(3y)^3(-2)^2 \quad 10(3y)^2(-2)^3 \quad \cdots$$

$$\cdots \quad 5(3y)^1(-2)^4 \quad 1(3y)^0(-2)^5$$

Finally, simplify the terms and add them together.

$$= 243y^5 - 810y^4 + 1080y^3 - 720y^2 + 240y - 32$$

358. $\mathbf{64x^6 + 576x^5 + 2160x^4 + 4320x^3 + 4860x^2 + 2916x + 729}$

First, write the coefficients from the seventh row of Pascal's triangle.

1 6 15 20 15 6 1

Now, add the decreasing powers of $2x$.

$$1(2x)^6 \quad 6(2x)^5 \quad 15(2x)^4 \quad 20(2x)^3 \quad \cdots$$

$$\cdots \quad 15(2x)^2 \quad 6(2x)^1 \quad 1(2x)^0$$

Next, add the increasing powers of 3.

$$1(2x)^6(3)^0 \quad 6(2x)^5(3)^1 \quad 15(2x)^4(3)^2 \quad 20(2x)^3(3)^3 \quad \cdots$$

$$\cdots \quad 15(2x)^2(3)^4 \quad 6(2x)^1(3)^5 \quad 1(2x)^0(3)^6$$

Finally, simplify the terms and add them together.

$$= 64x^6 + 576x^5 + 2160x^4 + 4320x^3 + 4860x^2 + 2916x + 729$$

359. $\mathbf{81x^4 + 216x^3y + 216x^2y^2 + 96xy^3 + 16y^4}$

First, write the coefficients from the fifth row of Pascal's triangle.

1 4 6 4 1

Now, add the decreasing powers of $3x$.

$$1(3x)^4 \quad 4(3x)^3 \quad 6(3x)^2 \quad 4(3x)^1 \quad 1(3x)^0$$

Next, add the increasing powers of $2y$.

$$1(3x)^4(2y)^0 \quad 4(3x)^3(2y)^1 \quad 6(3x)^2(2y)^2 \quad 4(3x)^1(2y)^3 \quad 1(3x)^0(2y)^4$$

Finally, simplify the terms and add them together.

$$= 81x^4 + 216x^3y + 216x^2y^2 + 96xy^3 + 16y^4$$

360. $\mathbf{32z^5 - 240z^4w + 720z^3w^2 - 1080z^2w^3 + 810zw^4 - 243w^5}$

First, write the coefficients from the sixth row of Pascal's triangle.

1 5 10 10 5 1

Now, add the decreasing powers of $2z$.

$$1(2z)^5 \quad 5(2z)^4 \quad 10(2z)^3 \quad 10(2z)^2 \quad 5(2z)^1 \quad 1(2z)^0$$

Next, add the increasing powers of $-3w$.

$$1(2z)^5(-3w)^0 \quad 5(2z)^4(-3w)^1 \quad 10(2z)^3(-3w)^2 \quad \cdots$$

$$\cdots \quad 10(2z)^2(-3w)^3 \quad 5(2z)^1(-3w)^4 \quad 1(2z)^0(-3w)^5$$

Finally, simplify the terms and add them together.

$$= 32z^5 - 240z^4w + 720z^3w^2 - 1080z^2w^3 + 810zw^4 - 243w^5$$

361. $x^3 - 1$

$$(x-1)x^2 + (x-1)x + (x-1)(1)$$

$$= x^3 - x^2 + x^2 - x + x - 1$$

$$= x^3 - 1$$

362. $y^3 + 8$

$$(y+2)y^2 + (y+2)(-2y) + (y+2)(4)$$

$$= y^3 + 2y^2 - 2y^2 - 4y + 4y + 8$$

$$= y^3 + 8$$

363. $z^3 - 64$

$$(z-4)z^2 + (z-4)(4z) + (z-4)(16)$$

$$= z^3 - 4z^2 + 4z^2 - 16z + 16z - 64$$

$$= z^3 - 64$$

364. $27x^3 - 8$

$$(3x-2)(9x^2) + (3x-2)(6x) + (3x-2)(4)$$

$$= 27x^3 - 18x^2 + 18x^2 - 12x + 12x - 8$$

$$= 27x^3 - 8$$

365. $125z^3 + 8w^3$

$$(5z+2w)(25z^2) + (5z+2w)(-10zw) + (5z+2w)(4w^2)$$

$$= 125z^3 + 50wz^2 - 50wz^2 - 20w^2z + 20w^2z + 8w^3$$

$$= 125z^3 + 8w^3$$

366. $x^2 + 2$

Rewrite the fraction as two terms, each with 3 in the denominator.

$$\frac{3x^2}{3}+\frac{6}{3}$$

Reduce each fraction.

$$\frac{\not{3}x^2}{\not{3}}+\frac{{}^2\not{6}}{\not{3}}=x^2+2$$

Another way to do this problem, since the denominator divides each term evenly, is to factor out the GCF in the numerator and then divide.

$$\frac{3x^2+6}{3}=\frac{\not{3}(x^2+2)}{\not{3}}=x^2+2$$

367. $3y^3 + 2y$

Rewrite the fraction as two terms, each with 4 in the denominator.

$$\frac{12y^3}{4}+\frac{8y}{4}$$

Reduce each fraction.

$$\frac{{}^3\not{12}y^3}{\not{4}}+\frac{{}^2\not{8}y}{\not{4}}=3y^3+2y$$

Another way to do this problem, since the denominator divides each term evenly, is to factor out the GCF in the numerator and then divide.

$$\frac{12y^3+8y}{4}=\frac{\not{4}y(3y^2+2)}{\not{4}}$$
$$=y(3y^2+2)=3y^3+2y$$

368. $x - 2$

Rewrite the fraction as two terms, each with x in the denominator.

$$\frac{x^2}{x}-\frac{2x}{x}$$

Reduce each fraction.

$$\frac{x^{\not{2}^1}}{\not{x}}-\frac{2\not{x}}{\not{x}}=x-2$$

Another way to do this problem, since the denominator divides each term evenly, is to factor out the GCF in the numerator and then divide.

$$\frac{x^2-2x}{x}=\frac{\not{x}(x-2)}{\not{x}}=x-2$$

369. $4x - 5$

Rewrite the fraction as two terms, each with x in the denominator.

$$\frac{4x^2}{x} - \frac{5x}{x}$$

Reduce each fraction.

$$\frac{4x^{\not{2}^1}}{\not{x}} - \frac{5\not{x}}{\not{x}} = 4x - 5$$

Another way to do this problem, since the denominator divides each term evenly, is to factor out the GCF in the numerator and then divide.

$$\frac{4x^2 - 5x}{x} = \frac{\not{x}(4x-5)}{\not{x}} = 4x - 5$$

370. $3y^3 - y - 4$

Rewrite the fraction as three terms, each with $2y$ in the denominator.

$$\frac{6y^4}{2y} - \frac{2y^2}{2y} - \frac{8y}{2y}$$

Reduce each fraction.

$$\frac{{}^3\not{6}y^{\not{4}^3}}{\not{2}\not{y}} - \frac{\not{2}y^{\not{2}^1}}{\not{2}\not{y}} - \frac{{}^4\not{8}\not{y}}{\not{2}\not{y}} = 3y^3 - y - 4$$

Another way to do this problem, since the denominator divides each term evenly, is to factor out the GCF in the numerator and then divide.

$$\frac{6y^4 - 2y^2 - 8y}{2y} = \frac{\not{2}\not{y}(3y^3 - y - 4)}{\not{2}\not{y}}$$

$$= 3y^3 - y - 4$$

371. $1 + 2z - 4z^2$

Rewrite the fraction as three terms, each with $3z$ in the denominator.

$$\frac{3z}{3z} + \frac{6z^2}{3z} - \frac{12z^3}{3z}$$

Reduce each fraction.

$$\frac{\not{3}\not{z}}{\not{3}\not{z}} + \frac{{}^2\not{6}z^{\not{2}^1}}{\not{3}\not{z}} - \frac{{}^4\not{12}z^{\not{3}^2}}{\not{3}\not{z}} = 1 + 2z - 4z^2$$

Another way to do this problem, since the denominator divides each term evenly, is to factor out the GCF in the numerator and then divide.

$$\frac{3z + 6z^2 - 12z^3}{3z} = \frac{\not{3}\not{z}(1 + 2z - 4z^2)}{\not{3}\not{z}}$$

$$= 1 + 2z - 4z^2$$

372. $3x + 4y$

Rewrite the fraction as two terms, each with $3xy$ in the denominator.

$$\frac{9x^2y}{3xy}+\frac{12xy^2}{3xy}$$

Reduce each fraction.

$$\frac{{}^{3}\cancel{9}x^{\cancel{2}^1}\cancel{y}}{\cancel{3}\cancel{xy}}+\frac{{}^{4}\cancel{12}\cancel{x}y^{\cancel{2}^1}}{\cancel{3}\cancel{xy}}=3x+4y$$

Another way to do this problem, since the denominator divides each term evenly, is to factor out the GCF in the numerator and then divide.

$$\frac{9x^2y+12xy^2}{3xy}=\frac{\cancel{3xy}(3x+4y)}{\cancel{3xy}}=3x+4y$$

373. $2c - 3$

Rewrite the fraction as two terms, each with $20ab$ in the denominator.

$$\frac{40abc}{20ab}-\frac{60ab}{20ab}$$

Reduce each fraction.

$$\frac{{}^{2}\cancel{40}\cancel{ab}c}{\cancel{20}\cancel{ab}}-\frac{{}^{3}\cancel{60}\cancel{ab}}{\cancel{20}\cancel{ab}}=2c-3$$

Another way to do this problem, since the denominator divides each term evenly, is to factor out the GCF in the numerator and then divide.

$$\frac{40abc-60ab}{20ab}=\frac{\cancel{20ab}(2c-3)}{\cancel{20ab}}=2c-3$$

374. $2x^2 - 3xy + 4y^2$

Rewrite the fraction as three terms, each with $6y^3$ in the denominator.

$$\frac{12x^2y^3}{6y^3}-\frac{18xy^4}{6y^3}+\frac{24y^5}{6y^3}$$

Reduce each fraction.

$$\frac{{}^{2}\cancel{12}x^2\cancel{y^3}}{\cancel{6}\cancel{y^3}}-\frac{{}^{3}\cancel{18}xy^{\cancel{4}^1}}{\cancel{6}\cancel{y^3}}+\frac{{}^{4}\cancel{24}y^{\cancel{5}^2}}{\cancel{6}\cancel{y^3}}=2x^2-3xy+4y^2$$

Another way to do this problem, since the denominator divides each term evenly, is to factor out the GCF in the numerator and then divide.

$$\frac{12x^2y^3-18xy^4+24y^5}{6y^3}=\frac{\cancel{6y^3}(2x^2-3xy+4y^2)}{\cancel{6y^3}}$$

$$=2x^2-3xy+4y^2$$

375. $6x^2 - 5xy + 2y^2$

Rewrite the fraction as three terms, each with $7xy^2$ in the denominator.

$$\frac{42x^3y^2}{7xy^2} - \frac{35x^2y^3}{7xy^2} + \frac{14xy^4}{7xy^2}$$

Reduce each fraction.

$$\frac{{}^{6}\cancel{42}x^{\cancel{3}^2}\cancel{y^2}}{\cancel{7}\cancel{x}\cancel{y^2}} - \frac{{}^{5}\cancel{35}x^{\cancel{2}^1}y^{\cancel{3}^1}}{\cancel{7}\cancel{x}\cancel{y^2}} + \frac{{}^{2}\cancel{14}\cancel{x}y^{\cancel{4}^2}}{\cancel{7}\cancel{x}\cancel{y^2}} = 6x^2 - 5xy + 2y^2$$

Another way to do this problem, since the denominator divides each term evenly, is to factor out the GCF in the numerator and then divide.

$$\frac{42x^3y^2 - 35x^2y^3 + 14xy^4}{7xy^2} = \frac{\cancel{7xy^2}\left(6x^2 - 5xy + 2y^2\right)}{\cancel{7xy^2}}$$
$$= 6x^2 - 5xy + 2y^2$$

376. $3y^2 + y - \frac{2}{3}$

Rewrite the fraction as three terms, each with 3 in the denominator.

$$\frac{9y^2}{3} + \frac{3y}{3} - \frac{2}{3}$$

Reduce each fraction, when possible.

$$\frac{{}^{3}\cancel{9}y^2}{\cancel{3}} + \frac{\cancel{3}y}{\cancel{3}} - \frac{2}{3} = 3y^2 + y - \frac{2}{3}$$

Not all the terms were evenly divisible by the denominator.

377. $3y + 1 - \frac{2}{3y}$

Rewrite the fraction as three terms, each with $3y$ in the denominator.

$$\frac{9y^2}{3y} + \frac{3y}{3y} - \frac{2}{3y}$$

Reduce each fraction, when possible.

$$\frac{{}^{3}\cancel{9}y^{\cancel{2}^1}}{\cancel{3}\cancel{y}} + \frac{\cancel{3}\cancel{y}}{\cancel{3}\cancel{y}} - \frac{2}{3y} = 3y + 1 - \frac{2}{3y}$$

Not all the terms were evenly divisible by the denominator.

378. $2x^2 + x + \frac{1}{25}$

Rewrite the fraction as three terms, each with 25 in the denominator.

$$\frac{50x^2}{25} + \frac{25x}{25} + \frac{1}{25}$$

Reduce each fraction, when possible.

$$\frac{{}^{2}\cancel{50}x^2}{\cancel{25}}+\frac{\cancel{25}x}{\cancel{25}}+\frac{1}{25}=2x^2+x+\frac{1}{25}$$

Not all the terms were evenly divisible by the denominator.

379. $2x+1+\frac{1}{25x}$

Rewrite the fraction as three terms, each with $25x$ in the denominator.

$$\frac{50x^2}{25x}+\frac{25x}{25x}+\frac{1}{25x}$$

Reduce each fraction, when possible.

$$\frac{{}^{2}\cancel{50}x^{\cancel{2}1}}{\cancel{25}\cancel{x}}+\frac{\cancel{25}\cancel{x}}{\cancel{25}\cancel{x}}+\frac{1}{25x}=2x+1+\frac{1}{25x}$$

Not all the terms were evenly divisible by the denominator.

380. $2x^3-4x^2+6x-4+\frac{1}{x}$

Rewrite the fraction as five terms, each with x in the denominator.

$$\frac{2x^4}{x}-\frac{4x^3}{x}+\frac{6x^2}{x}-\frac{4x}{x}+\frac{1}{x}$$

Reduce each fraction, when possible.

$$\frac{2x^{\cancel{4}3}}{\cancel{x}}-\frac{4x^{\cancel{3}2}}{\cancel{x}}+\frac{6x^{\cancel{2}1}}{\cancel{x}}-\frac{4\cancel{x}}{\cancel{x}}+\frac{1}{x}$$

$$=2x^3-4x^2+6x-4+\frac{1}{x}$$

Not all the terms were evenly divisible by the denominator.

381. $x^3-2x^2+3x-2+\frac{1}{2x}$

Rewrite the fraction as five terms, each with $2x$ in the denominator.

$$\frac{2x^4}{2x}-\frac{4x^3}{2x}+\frac{6x^2}{2x}-\frac{4x}{2x}+\frac{1}{2x}$$

Reduce each fraction, when possible.

$$\frac{\cancel{2}x^{\cancel{4}3}}{\cancel{2}\cancel{x}}-\frac{{}^{2}\cancel{4}x^{\cancel{3}2}}{\cancel{2}\cancel{x}}+\frac{{}^{3}\cancel{6}x^{\cancel{2}1}}{\cancel{2}\cancel{x}}-\frac{{}^{2}\cancel{4}\cancel{x}}{\cancel{2}\cancel{x}}+\frac{1}{2x}$$

$$=x^3-2x^2+3x-2+\frac{1}{2x}$$

Not all the terms were evenly divisible by the denominator.

382. $1-2xy+3x^2y^2-\frac{5}{xy^2}$

Rewrite the fraction as four terms, each with $5xy^2$ in the denominator.

$$\frac{5xy^2}{5xy^2}-\frac{10x^2y^3}{5xy^2}+\frac{15x^3y^4}{5xy^2}-\frac{25}{5xy^2}$$

Reduce each fraction, when possible.

$$\frac{\cancel{5}\cancel{x}\cancel{y^2}}{\cancel{5}\cancel{x}\cancel{y^2}}-\frac{{}^2\cancel{10}x^{\cancel{2}^1}y^{\cancel{3}^1}}{\cancel{5}\cancel{x}\cancel{y^2}}+\frac{{}^3\cancel{15}x^{\cancel{3}^2}y^{\cancel{4}^2}}{\cancel{5}\cancel{x}\cancel{y^2}}-\frac{{}^5\cancel{25}}{\cancel{5}xy^2}$$

$$=1-2xy+3x^2y^2-\frac{5}{xy^2}$$

Not all the terms were evenly divisible by the denominator.

383. $ax+b+\frac{c}{x}$

Rewrite the fraction as three terms, each with x in the denominator.

$$\frac{ax^2}{x}+\frac{bx}{x}+\frac{c}{x}$$

Reduce each fraction, when possible.

$$\frac{ax^{\cancel{2}^1}}{\cancel{x}}+\frac{b\cancel{x}}{\cancel{x}}+\frac{c}{x}=ax+b+\frac{c}{x}$$

Not all the terms were evenly divisible by the denominator.

384. $x^2+\frac{bx}{a}+\frac{c}{a}$

Rewrite the fraction as three terms, each with a in the denominator.

$$\frac{ax^2}{a}+\frac{bx}{a}+\frac{c}{a}$$

Reduce each fraction, when possible.

$$\frac{\cancel{a}x^2}{\cancel{a}}+\frac{bx}{a}+\frac{c}{a}=x^2+\frac{bx}{a}+\frac{c}{a}$$

Only one term is evenly divisible by the denominator.

385. $x+\frac{b}{a}+\frac{c}{ax}$

Rewrite the fraction as three terms, each with ax in the denominator.

$$\frac{ax^2}{ax}+\frac{bx}{ax}+\frac{c}{ax}$$

Reduce each fraction, when possible.

$$\frac{\cancel{a}x^{\cancel{2}^1}}{\cancel{a}\cancel{x}}+\frac{b\cancel{x}}{a\cancel{x}}+\frac{c}{ax}=x+\frac{b}{a}+\frac{c}{ax}$$

Only one term is evenly divisible by the denominator.

386. $x-2-\frac{4}{x-1}$

Write the denominator as the divisor in long division.

$$x-1\overline{)x^2-3x-2}$$

Write an x above the x^2 term in the dividend. This is what you must multiply the x in the divisor by to get x^2.

$$x-1\overline{)\,x^2-3x-2}\quad\text{quotient: } x$$

Now multiply both terms of the divisor by x and subtract.

$$\begin{array}{r} x \\ x-1\overline{)\,x^2-3x-2} \\ \underline{-(x^2-x)} \\ -2x-2 \end{array}$$

Now multiply both terms of the divisor by –2 and subtract.

$$\begin{array}{r} x\ -2 \\ x-1\overline{)\,x^2-3x-2} \\ \underline{-(x^2-x)} \\ -2x-2 \\ \underline{-(-2x+2)} \\ -4 \end{array}$$

The quotient is $x - 2$ with a remainder of –4, which is written: $x-2+\frac{-4}{x-1}$ or $x-2-\frac{4}{x-1}$.

387. $2x+8+\frac{17}{x-2}$

Write the denominator as the divisor in long division.

$$x-2\overline{)\,2x^2+4x+1}$$

Write $2x$ above the $2x^2$ term in the dividend. This is what you must multiply the x in the divisor by to get $2x^2$.

$$\begin{array}{r} 2x \\ x-2\overline{)\,2x^2+4x+1} \end{array}$$

Now multiply both terms of the divisor by $2x$ and subtract.

$$\begin{array}{r} 2x \\ x-2\overline{)\,2x^2+4x+1} \\ \underline{-(2x^2-4x)} \\ 8x+1 \end{array}$$

Now multiply both terms of the divisor by 8 and subtract.

$$\begin{array}{r} 2x\ +\ 8 \\ x-2\overline{)\,2x^2+4x+1} \\ \underline{-(2x^2-4x)} \\ 8x+1 \\ \underline{-(8x-16)} \\ +17 \end{array}$$

The quotient is $2x + 8$ with a remainder of 17, which is written: $2x+8+\frac{17}{x-2}$.

388. $5x - 12 + \frac{29}{x+3}$

Write the denominator as the divisor in long division.

$$x+3\overline{)5x^2+3x-7}$$

Write $5x$ above the $5x^2$ term in the dividend. This is what you must multiply the x in the divisor by to get $5x^2$.

$$\begin{array}{r} 5x \\ x+3\overline{)5x^2+3x-7} \end{array}$$

Now multiply both terms of the divisor by $5x$ and subtract.

$$\begin{array}{r} 5x \\ x+3\overline{)5x^2+3x-7} \\ -(5x^2+15x) \\ \hline -12x-7 \end{array}$$

Now multiply both terms of the divisor by –12 and subtract.

$$\begin{array}{r} 5x - 12 \\ x+3\overline{)5x^2+3x-7} \\ -(5x^2+15x) \\ \hline -12x-7 \\ -(-12x-36) \\ \hline 29 \end{array}$$

The quotient is $5x - 12$ with a remainder of 29, which is written: $5x-12+\frac{29}{x+3}$.

389. $2x - 6 + \frac{7}{x+1}$

Write the denominator as the divisor in long division.

$$x+1\overline{)2x^2-4x+1}$$

Write $2x$ above the $2x^2$ term in the dividend. This is what you must multiply the x in the divisor by to get $2x^2$.

$$\begin{array}{r} 2x \\ x+1\overline{)2x^2-4x+1} \end{array}$$

Now multiply both terms of the divisor by $5x$ and subtract.

$$\begin{array}{r} 2x \\ x+1\overline{)2x^2-4x+1} \\ -(2x^2+2x) \\ \hline -6x+1 \end{array}$$

Now multiply both terms of the divisor by –6 and subtract.

$$\begin{array}{r} 2x\;-\;6 \\ x+1\overline{)2x^2-4x+1} \\ \underline{-(2x^2+2x)} \\ -6x+1 \\ \underline{-(-6x-6)} \\ 7 \end{array}$$

The quotient is $2x-6$ with a remainder of 7, which is written: $2x-6+\frac{7}{x+1}$.

390. x^2+3x+2

Write the denominator as the divisor in long division.

$$x-1\overline{)x^3+2x^2-x-2}$$

Write x^2 above the x^3 term in the dividend. This is what you must multiply the x in the divisor by to get x^3.

$$\begin{array}{r} x^2\qquad\qquad \\ x-1\overline{)x^3+2x^2-x-2} \end{array}$$

Now multiply both terms of the divisor by x^2 and subtract.

$$\begin{array}{r} x^2\qquad\qquad \\ x-1\overline{)x^3+2x^2-x-2} \\ \underline{-(x^3-x^2)}\qquad \\ 3x^2-x-2 \end{array}$$

Now multiply both terms of the divisor by $3x$ and subtract.

$$\begin{array}{r} x^2\;+\;3x\qquad \\ x-1\overline{)x^3+2x^2-x-2} \\ \underline{-(x^3-x^2)}\qquad \\ 3x^2-x-2 \\ \underline{-(3x^2-3x)} \\ 2x-2 \end{array}$$

Finally, multiply both terms of the divisor by 2 and subtract.

$$\begin{array}{r} x^2\;+\;3x\;+\;2 \\ x-1\overline{)x^3+2x^2-x-2} \\ \underline{-(x^3-x^2)}\qquad \\ 3x^2-x-2 \\ \underline{-(3x^2-3x)} \\ 2x-2 \\ \underline{-(2x-2)} \\ 0 \end{array}$$

The divisor divides evenly — there's no remainder. So the quotient is x^2+3x+2.

391. $2x^2 - 5x - 3$

Write the denominator as the divisor in long division.

$$x+3\overline{)2x^3+x^2-18x-9}$$

Write $2x^2$ above the $2x^3$ term in the dividend. This is what you must multiply the x in the divisor by to get $2x^3$.

$$\begin{array}{r} 2x^2 \qquad\qquad\qquad \\ x+3\overline{)2x^3+x^2-18x-9} \end{array}$$

Now multiply both terms of the divisor by $2x^2$ and subtract.

$$\begin{array}{r} 2x^2 \qquad\qquad\qquad \\ x+3\overline{)2x^3+x^2-18x-9} \\ \underline{-(2x^3+6x^2)} \qquad\qquad \\ -5x^2-18x-9 \end{array}$$

Now multiply both terms of the divisor by $-5x$ and subtract.

$$\begin{array}{r} 2x^2-5x \qquad\quad \\ x+3\overline{)2x^3+x^2-18x-9} \\ \underline{-(2x^3+6x^2)} \qquad\qquad \\ -5x^2-18x-9 \\ \underline{-(-5x^2-15x)} \quad \\ -3x-9 \end{array}$$

Finally, multiply both terms of the divisor by -3 and subtract.

$$\begin{array}{r} 2x^2-5x\ -\ 3 \\ x+3\overline{)2x^3+x^2-18x-9} \\ \underline{-(2x^3+6x^2)} \qquad\qquad \\ -5x^2-18x-9 \\ \underline{-(-5x^2-15x)} \quad \\ -3x-9 \\ \underline{-(-3x-9)} \\ 0 \end{array}$$

The quotient is $2x^2 - 5x - 3$ with no remainder.

392. $3x^2 + 13x + 50 + \dfrac{201}{x-4}$

Write the denominator as the divisor in long division.

$$x-4\overline{)3x^3+x^2-2x+1}$$

Write $3x^2$ above the $3x^3$ term in the dividend. This is what you must multiply the x in the divisor by to get $3x^3$.

$$\begin{array}{r} 3x^2 \qquad\qquad\quad \\ x-4\overline{)3x^3+x^2-2x+1} \end{array}$$

Now multiply both terms of the divisor by $3x^2$ and subtract.

$$\begin{array}{r} 3x^2 \\ x-4\overline{)3x^3+x^2-2x+1} \\ -(3x^3-12x^2) \\ \hline 13x^2-2x+1 \end{array}$$

Now multiply both terms of the divisor by $13x$ and subtract.

$$\begin{array}{r} 3x^2+13x \\ x-4\overline{)3x^3+x^2-2x+1} \\ -(3x^3-12x^2) \\ \hline 13x^2-2x+1 \\ -(13x^2-52x) \\ \hline 50x+1 \end{array}$$

Finally, multiply both terms of the divisor by 50 and subtract.

$$\begin{array}{r} 3x^2+13x+50 \\ x-4\overline{)3x^3+x^2-2x+1} \\ -(3x^3-12x^2) \\ \hline 13x^2-2x+1 \\ -(13x^2-52x) \\ \hline 50x+1 \\ -(50x-200) \\ \hline 201 \end{array}$$

The quotient is $3x^2 + 13x + 50$ with a remainder of 201, which is written: $3x^2+13x+50+\frac{201}{x-4}$.

393. $x^3-4x^2+4x-3+\frac{1}{x+1}$

Write the denominator as the divisor in long division. In the dividend, one power in the listing of decreasing powers is missing: the second power or x^2. Use a 0 as a placeholder and insert that term in the dividend.

$$x+1\overline{)x^4-3x^3+0x^2+x-2}$$

Write x^3 above the x^4 term in the dividend. This is what you must multiply the x in the divisor by to get x^4.

$$\begin{array}{r} x^3 \\ x+1\overline{)x^4-3x^3+0x^2+x-2} \end{array}$$

Now multiply both terms of the divisor by x^3 and subtract.

$$\begin{array}{r} x^3 \\ x+1\overline{)x^4-3x^3+0x^2+x-2} \\ -(x^4+x^3) \\ \hline -4x^3+0x^2+x-2 \end{array}$$

Answers 301–400

Now multiply both terms of the divisor by $-4x^2$ and subtract.

$$\begin{array}{r} x^3-4x^2 \\ x+1\overline{)x^4-3x^3+0x^2+x-2} \\ \underline{-(x^4+x^3)} \\ -4x^3+0x^2+x-2 \\ \underline{-(-4x^3-4x^2)} \\ 4x^2+x-2 \end{array}$$

Now multiply both terms of the divisor by $4x$ and subtract.

$$\begin{array}{r} x^3-4x^2+4x \\ x+1\overline{)x^4-3x^3+0x^2+x-2} \\ \underline{-(x^4+x^3)} \\ -4x^3+0x^2+x-2 \\ \underline{-(-4x^3-4x^2)} \\ 4x^2+x-2 \\ \underline{-(4x^2+4x)} \\ -3x-2 \end{array}$$

Finally, multiply both terms of the divisor by -3 and subtract.

$$\begin{array}{r} x^3-4x^2+4x-3 \\ x+1\overline{)x^4-3x^3+0x^2+x-2} \\ \underline{-(x^4+x^3)} \\ -4x^3+0x^2+x-2 \\ \underline{-(-4x^3-4x^2)} \\ 4x^2+x-2 \\ \underline{-(4x^2+4x)} \\ -3x-2 \\ \underline{-(-3x-3)} \\ 1 \end{array}$$

The quotient is x^3-4x^2+4x-3 with a remainder of 1, which is written: $x^3-4x^2+4x-3+\frac{1}{x+1}$.

394. $2x^3+6x^2+21x+62+\frac{187}{x-3}$

Write the denominator as the divisor in long division. In the dividend, one power in the listing of decreasing powers is missing: the third power or x^3. Use a 0 as a placeholder and insert that term in the dividend.

$$x-3\overline{)2x^4+0x^3+3x^2-x+1}$$

Write $2x^3$ above the $2x^4$ term in the dividend. This is what you must multiply the x in the divisor by to get $2x^4$.

$$\begin{array}{r} 2x^3 \\ x-3\overline{)2x^4+0x^3+3x^2-x+1} \end{array}$$

Now multiply both terms of the divisor by $2x^3$ and subtract.

$$\begin{array}{l} 2x^3 \\ x-3\overline{)2x^4+0x^3+3x^2-x+1} \\ \underline{-(2x^4-6x^3)} \\ 6x^3+3x^2-x+1 \end{array}$$

Now multiply both terms of the divisor by $6x^2$ and subtract.

$$\begin{array}{l} 2x^3+6x^2 \\ x-3\overline{)2x^4+0x^3+3x^2-x+1} \\ \underline{-(2x^4-6x^3)} \\ 6x^3+3x^2-x+1 \\ \underline{-(6x^3-18x^2)} \\ 21x^2-x+1 \end{array}$$

Now multiply both terms of the divisor by $21x$ and subtract.

$$\begin{array}{l} 2x^3+6x^2+21x \\ x-3\overline{)2x^4+0x^3+3x^2-x+1} \\ \underline{-(2x^4-6x^3)} \\ 6x^3+3x^2-x+1 \\ \underline{-(6x^3-18x^2)} \\ 21x^2-x+1 \\ \underline{-(21x^2-63x)} \\ 62x+1 \end{array}$$

Finally, multiply both terms of the divisor by 62 and subtract.

$$\begin{array}{l} 2x^3+6x^2+21x+62 \\ x-3\overline{)2x^4+0x^3+3x^2-x+1} \\ \underline{-(2x^4-6x^3)} \\ 6x^3+3x^2-x+1 \\ \underline{-(6x^3-18x^2)} \\ 21x^2-x+1 \\ \underline{-(21x^2-63x)} \\ 62x+1 \\ \underline{-(62x-186)} \\ 187 \end{array}$$

The quotient is $2x^3 + 6x^2 + 21x + 62$ with a remainder of 187, which is written: $2x^3+6x^2+21x+62+\frac{187}{x-3}$.

395. $2x^2+2x+2+\frac{3}{x-1}$

Write the denominator as the divisor in long division. In the dividend, the x^2 and x terms are missing, so use 0 as a placeholder for each.

$$x-1\overline{)2x^3+0x^2+0x+1}$$

Write $2x^2$ above the $2x^3$ term in the dividend. This is what you must multiply the x in the divisor by to get $2x^3$.

$$\begin{array}{r} 2x^2 \qquad\qquad\qquad \\ x-1\overline{)2x^3+0x^2+0x+1} \end{array}$$

Now multiply both terms of the divisor by $2x^2$ and subtract.

$$\begin{array}{r} 2x^2 \qquad\qquad\qquad \\ x-1\overline{)2x^3+0x^2+0x+1} \\ \underline{-(2x^3-2x^2)} \qquad\quad \\ 2x^2+0x+1 \end{array}$$

Now multiply both terms of the divisor by $2x$ and subtract.

$$\begin{array}{r} 2x^2+2x \qquad\quad \\ x-1\overline{)2x^3+0x^2+0x+1} \\ \underline{-(2x^3-2x^2)} \qquad\quad \\ 2x^2+0x+1 \\ \underline{-(2x^2-2x)} \quad \\ 2x+1 \end{array}$$

Finally, multiply both terms of the divisor by 2 and subtract.

$$\begin{array}{r} 2x^2+2x\ +2 \\ x-1\overline{)2x^3+0x^2+0x+1} \\ \underline{-(2x^3-2x^2)} \qquad\quad \\ 2x^2+0x+1 \\ \underline{-(2x^2-2x)} \quad \\ 2x+1 \\ \underline{-(2x-2)} \\ 3 \end{array}$$

The quotient is $2x^2 + 2x + 2$ with a remainder of 3, which is written: $2x^2+2x+2+\frac{3}{x-1}$.

396. $x-2-\frac{4}{x-1}$

To use synthetic division, write the coefficients of the terms of the dividend to the right of an open box containing the opposite of the constant number in the binomial divisor.

$$\underline{1}| \;\; 1 \quad -3 \quad -2$$

Draw a line under the coefficients, leaving enough room for products.

$$\begin{array}{r|rrr} 1 & 1 & -3 & -2 \\ & & & \\ \hline & & & \end{array}$$

Begin by dropping the first coefficient down below the line, multiply that coefficient by the 1 in the box, put the product under the next coefficient, and add.

$$\begin{array}{r|rrr} 1 & 1 & -3 & -2 \\ & & 1 & \\ \hline & 1 & -2 & \end{array}$$

Repeat the multiply/add process with the –2.

$$\begin{array}{r|rrr} 1 & 1 & -3 & -2 \\ & & 1 & -2 \\ \hline & 1 & -2 & -4 \end{array}$$

The quotient is created from the numbers under the line. The division gives you the answer of $1x-2$ with a remainder of –4, which is written $x-2-\frac{4}{x-1}$. Compare this result with the long division performed in problem 386.

397. $2x+8+\frac{17}{x-2}$

To use synthetic division, write the coefficients of the terms of the dividend to the right of an open box containing the opposite of the constant number in the binomial divisor.

$$\begin{array}{r|rrr} 2 & 2 & 4 & 1 \end{array}$$

Draw a line under the coefficients, leaving enough room for products.

$$\begin{array}{r|rrr} 2 & 2 & 4 & 1 \\ & & & \\ \hline & & & \end{array}$$

Begin by dropping the first coefficient down below the line, multiply that coefficient by the 2 in the box, put the product under the next coefficient, and add.

$$\begin{array}{r|rrr} 2 & 2 & 4 & 1 \\ & & 4 & \\ \hline & 2 & 8 & \end{array}$$

Repeat the multiply/add process with the 8.

$$\begin{array}{r|rrr} 2 & 2 & 4 & 1 \\ & & 4 & 16 \\ \hline & 2 & 8 & 17 \end{array}$$

The quotient is created from the numbers under the line. The division gives you the answer of $2x+8$ with a remainder of 17, which is written $2x+8+\frac{17}{x-2}$. Compare this result with the long division performed in problem 387.

398. $5x-12+\frac{29}{x+3}$

To use synthetic division, write the coefficients of the terms of the dividend to the right of an open box containing the opposite of the constant number in the binomial divisor.

$$\begin{array}{r|rrr} -3 & 5 & 3 & -7 \end{array}$$

Draw a line under the coefficients, leaving enough room for products.

$$\begin{array}{r|rrr} -3 & 5 & 3 & -7 \\ & & & \\ \hline & & & \end{array}$$

Begin by dropping the first coefficient down below the line, multiply that coefficient by the –3 in the box, put the product under the next coefficient, and add.

$$\begin{array}{r|rrr} -3 & 5 & 3 & -7 \\ & & -15 & \\ \hline & 5 & -12 & \end{array}$$

Repeat the multiply/add process with the –12.

$$\begin{array}{r|rrr} -3 & 5 & 3 & -7 \\ & & -15 & 36 \\ \hline & 5 & -12 & 29 \end{array}$$

The quotient is created from the numbers under the line. The division gives you the answer of $5x - 12$ with a remainder of 29, which is written $5x-12+\frac{29}{x+3}$. Compare this result with the long division performed in problem 388.

399. $2x-6+\frac{7}{x+1}$

To use synthetic division, write the coefficients of the terms of the dividend to the right of an open box containing the opposite of the constant number in the binomial divisor.

$$\begin{array}{r|rrr} -1 & 2 & -4 & 1 \end{array}$$

Draw a line under the coefficients, leaving enough room for products.

$$\begin{array}{r|rrr} -1 & 2 & -4 & 1 \\ & & & \\ \hline & & & \end{array}$$

Begin by dropping the first coefficient down below the line, multiply that coefficient by the –1 in the box, put the product under the next coefficient, and add.

$$\begin{array}{r|rrr} -1 & 2 & -4 & 1 \\ & & -2 & \\ \hline & 2 & -6 & \end{array}$$

Repeat the multiply/add process with the –6.

$$\begin{array}{r|rrr} -1 & 2 & -4 & 1 \\ & & -2 & 6 \\ \hline & 2 & -6 & 7 \end{array}$$

The quotient is created from the numbers under the line. The division gives you the answer of $2x - 6$ with a remainder of 7, which is written $2x-6+\frac{7}{x+1}$. Compare this result with the long division performed in problem 389.

400. $x^2 + 3x + 2$

To use synthetic division, write the coefficients of the terms of the dividend to the right of an open box containing the opposite of the constant number in the binomial divisor.

$$\begin{array}{r|rrrr} 1 & 1 & 2 & -1 & -2 \end{array}$$

Draw a line under the coefficients, leaving enough room for products.

$$\begin{array}{r|rrrr} 1 & 1 & 2 & -1 & -2 \\ & & & & \\ \hline \end{array}$$

Begin by dropping the first coefficient down below the line, multiply that coefficient by the 1 in the box, put the product under the next coefficient, and add.

$$\begin{array}{r|rrrr} 1 & 1 & 2 & -1 & -2 \\ & & 1 & & \\ \hline & 1 & 3 & & \end{array}$$

Repeat the multiply/add process.

$$\begin{array}{r|rrrr} 1 & 1 & 2 & -1 & -2 \\ & & 1 & 3 & 2 \\ \hline & 1 & 3 & 2 & 0 \end{array}$$

The quotient is created from the numbers under the line. The division gives you the answer of $x^2 + 3x + 2$ with no remainder. Compare this result with the long division performed in problem 390.

401. $2x^2 - 5x - 3$

To use synthetic division, write the coefficients of the terms of the dividend to the right of an open box containing the opposite of the constant number in the binomial divisor.

$$\begin{array}{r|rrrr} -3 & 2 & 1 & -18 & -9 \end{array}$$

Draw a line under the coefficients, leaving enough room for products. Then begin the process by dropping the first coefficient down below the line, multiply that coefficient by the –3 in the box, put the product under the next coefficient, and add.

$$\begin{array}{r|rrrr} -3 & 2 & 1 & -18 & -9 \\ & & -6 & & \\ \hline & 2 & -5 & & \end{array}$$

Repeat the multiply/add process.

$$\begin{array}{r|rrrr} -3 & 2 & 1 & -18 & -9 \\ & & -6 & 15 & 9 \\ \hline & 2 & -5 & -3 & 0 \end{array}$$

The quotient is created from the numbers under the line. The division gives you the answer of $2x^2 - 5x - 3$ with no remainder. Compare this result with the long division performed in problem 391.

402. $3x^2 + 13x + 50 + \frac{201}{x-4}$

To use synthetic division, write the coefficients of the terms of the dividend to the right of an open box containing the opposite of the constant number in the binomial divisor.

$$\begin{array}{r|rrrr} 4 & 3 & 1 & -2 & 1 \end{array}$$

Draw a line under the coefficients, leaving enough room for products. Then begin the process by dropping the first coefficient down below the line, multiply that coefficient by the 4 in the box, put the product under the next coefficient, and add.

$$\begin{array}{r|rrrr} 4 & 3 & 1 & -2 & 1 \\ & & 12 & & \\ \hline & 3 & 13 & & \end{array}$$

Repeat the multiply/add process.

$$\begin{array}{r|rrrr} 4 & 3 & 1 & -2 & 1 \\ & & 12 & 52 & 200 \\ \hline & 3 & 13 & 50 & 201 \end{array}$$

The quotient is created from the numbers under the line. The division gives you the answer of $3x^2 + 13x + 50$ with a remainder of 201. This is written $3x^2+13x+50+\frac{201}{x-4}$. Compare this result with the long division performed in problem 392.

403. $x^3-4x^2+4x-3+\frac{1}{x+1}$

To use synthetic division, write the coefficients of the terms of the dividend to the right of an open box containing the opposite of the constant number in the binomial divisor. Be sure to use 0 as a placeholder for the missing second degree term.

$$\begin{array}{r|rrrrr} -1 & 1 & -3 & 0 & 1 & -2 \end{array}$$

Draw a line under the coefficients, leaving enough room for products. Then begin the process by dropping the first coefficient down below the line, multiply that coefficient by the –1 in the box, put the product under the next coefficient, and add.

$$\begin{array}{r|rrrrr} -1 & 1 & -3 & 0 & 1 & -2 \\ & & -1 & & & \\ \hline & 1 & -4 & & & \end{array}$$

Repeat the multiply/add process.

$$\begin{array}{r|rrrrr} -1 & 1 & -3 & 0 & 1 & -2 \\ & & -1 & 4 & -4 & 3 \\ \hline & 1 & -4 & 4 & -3 & 1 \end{array}$$

The quotient is created from the numbers under the line. The division gives you the answer of $x^3 - 4x^2 + 4x - 3$ with a remainder of 1. This is written $x^3-4x^2+4x-3+\frac{1}{x+1}$. Compare this result with the long division performed in problem 393.

404. $2x^3+6x^2+21x+62+\frac{187}{x-3}$

To use synthetic division, write the coefficients of the terms of the dividend to the right of an open box containing the opposite of the constant number in the binomial divisor. Be sure to use 0 as a placeholder for the missing third degree term.

$$\begin{array}{r|rrrrr} 3 & 2 & 0 & 3 & -1 & 1 \end{array}$$

Draw a line under the coefficients, leaving enough room for products. Then begin the process by dropping the first coefficient down below the line, multiply that coefficient by the 3 in the box, put the product under the next coefficient, and add.

$$\begin{array}{r|rrrrr} 3 & 2 & 0 & 3 & -1 & 1 \\ & & 6 & & & \\ \hline & 2 & 6 & & & \end{array}$$

Repeat the multiply/add process.

$$\begin{array}{r|rrrrr} 3 & 2 & 0 & 3 & -1 & 1 \\ & & 6 & 18 & 63 & 186 \\ \hline & 2 & 6 & 21 & 62 & 187 \end{array}$$

The quotient is created from the numbers under the line. The division gives you the answer of $2x^3 + 6x^2 + 21x + 62$ with a remainder of 187. This is written $2x^3+6x^2+21x+62+\frac{187}{x-3}$. Compare this result with the long division performed in problem 394.

405. $2x^2+2x+2+\frac{3}{x-1}$

To use synthetic division, write the coefficients of the terms of the dividend to the right of an open box containing the opposite of the constant number in the binomial divisor. Be sure to use placeholders for the missing third and second degree terms.

$$\begin{array}{r|rrrr} 1 & 2 & 0 & 0 & 1 \end{array}$$

Draw a line under the coefficients, leaving enough room for products. Then begin the process by dropping the first coefficient down below the line, multiply that coefficient by the 1 in the box, put the product under the next coefficient, and add.

$$\begin{array}{r|rrrr} 1 & 2 & 0 & 0 & 1 \\ & & 2 & & \\ \hline & 2 & 2 & & \end{array}$$

Repeat the multiply/add process.

$$\begin{array}{r|rrrr} 1 & 2 & 0 & 0 & 1 \\ & & 2 & 2 & 2 \\ \hline & 2 & 2 & 2 & 3 \end{array}$$

The quotient is created from the numbers under the line. The division gives you the answer of $2x^2 + 2x + 2$ with a remainder of 3. This is written $2x^2+2x+2+\frac{3}{x-1}$. Compare this result with the long division performed in problem 395.

406. $x^2 + 1$

Write the denominator as the divisor and the numerator as the dividend of a long-division problem. Use 0 as a placeholder for any missing terms.

$$x^2-1\overline{)x^4+0x^3+0x^2+0x-1}$$

Work from the left-most remaining term in the dividend, and find multipliers of the x^2 part of the divisor to create that left-most term. Repeat multiply/subtract until it either divides evenly or until the remainder has a lower degree than the divisor.

$$\begin{array}{r} x^2 \quad + \quad 1 \\ x^2-1\overline{)x^4+0x^3+0x^2+0x-1} \\ -\left(x^4 \quad - \quad x^2\right) \\ \hline x^2 \quad - \quad 1 \\ -\left(x^2 \quad - \quad 1\right) \\ \hline 0 \end{array}$$

The divisor divides evenly, so there's no remainder. The quotient is $x^2 + 1$.

407. $y^2 + 9$

Write the denominator as the divisor and the numerator as the dividend of a long-division problem. Use 0 as a placeholder for any missing terms.

$$y^2-9\overline{)y^4+0y^3+0y^2+0y-81}$$

Work from the left-most remaining term in the dividend, and find multipliers of the y^2 part of the divisor to create that left-most term. Repeat multiply/subtract until it either divides evenly or until the remainder has a lower degree than the divisor.

$$\begin{array}{r} y^2 \quad + \quad 9 \\ y^2-9\overline{)y^4+0y^3+0y^2+0y-81} \\ -\left(y^4 \quad - \quad 9y^2\right) \\ \hline 9y^2 \quad - \quad 81 \\ -\left(9y^2 \quad - \quad 81\right) \\ \hline 0 \end{array}$$

The divisor divides evenly, so there's no remainder. The quotient is $y^2 + 9$.

408. $x - 1$

Write the denominator as the divisor and the numerator as the dividend of a long-division problem.

$$x^2-2x+1\overline{)x^3-3x^2+3x-1}$$

Work from the left-most remaining term in the dividend, and find multipliers of the x^2 part of the divisor to create that left-most term. Repeat multiply/subtract until it either divides evenly or until the remainder has a lower degree than the divisor.

$$\begin{array}{r} x \quad - \quad 1 \\ x^2-2x+1\overline{)x^3-3x^2+3x-1} \\ -\left(x^3-2x^2+x\right) \\ \hline -x^2+2x-1 \\ -\left(-x^2+2x-1\right) \\ \hline 0 \end{array}$$

The divisor divides evenly, so there's no remainder. The quotient is $x - 1$.

409. $x^3 + 3x^2 + 3x + 1$

Write the denominator as the divisor and the numerator as the dividend of a long-division problem.

$$x+1\overline{)x^4+4x^3+6x^2+4x+1}$$

Work from the left-most remaining term in the dividend, and find multipliers of the x^2 part of the divisor to create that left-most term. Repeat multiply/subtract until it either divides evenly or until the remainder has a lower degree than the divisor.

$$\begin{array}{r} x^3+3x^2+3x+1 \\ x+1\overline{)x^4+4x^3+6x^2+4x+1} \\ \underline{-\left(x^4+x^3\right)} \\ 3x^3+6x^2+4x+1 \\ \underline{-\left(3x^3+3x^2\right)} \\ 3x^2+4x+1 \\ \underline{-\left(3x^2+3x\right)} \\ x+1 \\ \underline{-(x+1)} \\ 0 \end{array}$$

The divisor divides evenly, so there's no remainder. The quotient is $x^3 + 3x^2 + 3x + 1$.

410. $x^2 - 5x + 13$

Write the denominator as the divisor and the numerator as the dividend of a long-division problem.

$$x^2+2x-1\overline{)x^4-3x^3+2x^2+3x-1}$$

Work from the left-most remaining term in the dividend, and find multipliers of the x^2 part of the divisor to create that left-most term. Repeat multiply/subtract until it either divides evenly or until the remainder has a lower degree than the divisor.

$$\begin{array}{r} x^2-5x+13 \\ x^2+2x-1\overline{)x^4-3x^3+2x^2+3x-1} \\ \underline{-\left(x^4+2x^3-x^2\right)} \\ -5x^3+3x^2+3x-1 \\ \underline{-\left(-5x^3-10x^2+5x\right)} \\ 13x^2-2x-1 \\ \underline{-\left(13x^2+26x-13\right)} \\ -28x+12 \end{array}$$

The divisor divides evenly, so there's no remainder. The quotient is $x^2 - 5x + 13$.

411. $x^2+2x+\frac{1}{x^2+2}$

Write the denominator as the divisor and the numerator as the dividend of a long-division problem.

$$x^2+2\overline{)x^4+2x^3+2x^2+4x+1}$$

Work from the left-most remaining term in the dividend, and find multipliers of the x^2 part of the divisor to create that left-most term. Repeat multiply/subtract until it either divides evenly or until the remainder has a lower degree than the divisor.

$$\begin{array}{r} x^2+2x \\ x^2+2\overline{)x^4+2x^3+2x^2+4x+1} \\ \underline{-(x^4 + 2x^2)} \\ 2x^3 + 4x+1 \\ \underline{-(2x^3 + 4x)} \\ 1 \end{array}$$

The quotient is $x^2 + 2x$ with a remainder of 1. This is written $x^2+2x+\frac{1}{x^2+2}$.

412. $x^2+x+2+\frac{18x+9}{x^2-4}$

Write the denominator as the divisor and the numerator as the dividend of a long-division problem.

$$x^2-4\overline{)x^4+x^3-2x^2+14x+1}$$

Work from the left-most remaining term in the dividend, and find multipliers of the x^2 part of the divisor to create that left-most term. Repeat multiply/subtract until it either divides evenly or until the remainder has a lower degree than the divisor.

$$\begin{array}{r} x^2+x+2 \\ x^2-4\overline{)x^4+x^3-2x^2+14x+1} \\ \underline{-(x^4 - 4x^2)} \\ x^3+2x^2+14x+1 \\ \underline{-(x^3 - 4x)} \\ 2x^2+18x+1 \\ \underline{-(2x^2 - 8)} \\ 18x+9 \end{array}$$

The quotient is $x^2 + x + 2$ with a remainder of $18x + 9$. This is written $x^2+x+2+\frac{18x+9}{x^2-4}$.

413. $x^2 + 4$

Write the denominator as the divisor and the numerator as the dividend of a long-division problem.

$$x^2+3x+1\overline{)x^4+3x^3+5x^2+12x+4}$$

Work from the left-most remaining term in the dividend, and find multipliers of the x^2 part of the divisor to create that left-most term. Repeat multiply/subtract until it either divides evenly or until the remainder has a lower degree than the divisor.

$$\begin{array}{r} x^2+4 \\ x^2+3x+1\overline{\smash{)}x^4+3x^3+5x^2+12x+4} \\ -\left(x^4+3x^3+x^2\right) \\ \hline 4x^2+12x+4 \\ -\left(4x^2+12x+4\right) \\ \hline 0 \end{array}$$

The quotient is $x^2 + 4$. It divides evenly, so there's no remainder.

414. $x^2+3x+1-\dfrac{3x-2}{x^4+1}$

Write the denominator as the divisor and the numerator as the dividend of a long-division problem. Replace the missing terms with 0's as placeholders.

$$x^4+1\overline{\smash{)}x^6+3x^5+x^4+0x^3+x^2+0x+3}$$

Work from the left-most remaining term in the dividend, and find multipliers of the x^4 part of the divisor to create that left-most term. Repeat multiply/subtract until it either divides evenly or until the remainder has a lower degree than the divisor.

$$\begin{array}{r} x^2+3x+1 \\ x^4+1\overline{\smash{)}x^6+3x^5+x^4+0x^3+x^2+0x+3} \\ -\left(x^6 \qquad + \qquad x^2\right) \\ \hline 3x^5+x^4 \qquad\qquad 3 \\ -\left(3x^5 \qquad + \qquad 3x\right) \\ \hline x^4 \qquad -3x+3 \\ -\left(x^4 \qquad + \qquad 1\right) \\ \hline -3x+2 \end{array}$$

The quotient is $x^2 + 3x + 1$ with a remainder of $-3x + 2$. This is written $x^2+3x+1+\dfrac{-3x+2}{x^4+1}$ or $x^2+3x+1-\dfrac{3x-2}{x^4+1}$.

415. $x+\dfrac{x-1}{x^5-1}$

Write the denominator as the divisor and the numerator as the dividend of a long-division problem. Replace the missing terms with 0's as placeholders.

$$x^5-1\overline{\smash{)}x^6+0x^5+0x^4+0x^3+0x^2+0x-1}$$

Work from the left-most remaining term in the dividend, and find multipliers of the x^5 part of the divisor to create that left-most term. Repeat multiply/subtract until it either divides evenly or until the remainder has a lower degree than the divisor.

$$x^5-1\overline{)x^6+0x^5+0x^4+0x^3+0x^2+0x-1}^{\;x}$$
$$\underline{-(x^6 \qquad\qquad\qquad\qquad - x)}$$
$$x-1$$

The quotient is x with a remainder of $x-1$. This is written $x+\frac{x-1}{x^5-1}$.

416. Divisible by 2, 4, 8, and 11

Numbers divisible by 2 end in 0, 2, 4, 6, or 8.

Numbers divisible by 4 have the last two digits forming a number divisible by 4.

Numbers divisible by 8 have the last three digits forming a number divisible by 8. (In this case, it's 088.)

Numbers divisible by 11 have sums of alternating digits that have a difference of 0 or 11 or 22… (In this case 8 – 8 = 0.)

417. Divisible by 2 and 5

Numbers divisible by 2 end in 0, 2, 4, 6, or 8.

Numbers divisible by 5 end in 0 or 5.

418. Divisible by 2, 3, 4, 6, and 9

Numbers divisible by 2 end in 0, 2, 4, 6, or 8.

Numbers divisible by 3 have digits that add up to a multiple of 3. (In this case, 3 + 4 + 9 + 2 = 18.)

Numbers divisible by 4 have the last two digits forming a number divisible by 4.

Numbers divisible by 6 are also divisible by *both* 2 and 3.

Numbers divisible by 9 have digits that add up to a multiple of 9. (In this case, 3 + 4 + 9 + 2 = 18.)

419. Divisible by 3, 9, and 11

Numbers divisible by 3 have digits that add up to a multiple of 3. (In this case, 4 + 2 + 5 + 7 = 18.)

Numbers divisible by 9 have digits that add up to a multiple of 9. (In this case, 4 + 2 + 5 + 7 = 18.)

Numbers divisible by 11 have sums of alternating digits that have a difference of 0 or 11 or 22… (In this case 4 + 5 = 9 and 2 + 7 = 9; the difference of the sums is 0.)

420. Divisible by 2, 4, and 5

Numbers divisible by 2 end in 0, 2, 4, 6, or 8.

Numbers divisible by 4 have the last two digits forming a number divisible by 4.

Numbers divisible by 5 end in 0 or 5.

421. **Divisible by 3, 7, and 11**

Numbers divisible by 3 have digits that add up to a multiple of 3. (In this case, $3+0+0+3=6$.)

Numbers divisible by 7 use the rule "double last digit and subtract from remaining — repeating"; final difference is divisible by 7. (In this case $300-2(3)=294$, and $29-2(4)=21$.)

Numbers divisible by 11 have sums of alternating digits that have a difference of 0 or 11 or 22… (In this case $3+0=3$ and $3+0=3$; the difference of the sums is 0.)

422. $2^2 \cdot 7$

Write the number as the product of two other numbers.

$$4 \cdot 7$$

Now write either of these numbers as the product of two numbers, if they aren't prime.

$$=(2 \cdot 2) \cdot 7$$

All three numbers are prime. Write the factorization using exponents.

$$2 \cdot 2 \cdot 7 = 2^2 \cdot 7$$

(There are other multiplications possible; they all end up with the same answer.)

423. $3^2 \cdot 5$

Write the number as the product of two other numbers.

$$5 \cdot 9$$

Now write either of these numbers as the product of two numbers, if they aren't prime.

$$5 \cdot (3 \cdot 3)$$

All three numbers are prime. Write the factorization using exponents.

$$5 \cdot 3 \cdot 3 = 3^2 \cdot 5$$

(There are other multiplications possible; they all end up with the same answer.)

424. $2 \cdot 3 \cdot 5^2$

Write the number as the product of two other numbers.

$$10 \cdot 15$$

Now write both of these numbers as the product of two numbers, since they aren't prime.

$$=(2 \cdot 5) \cdot (3 \cdot 5)$$

All four numbers are prime. Write the factorization using exponents.

$$2 \cdot 5 \cdot 3 \cdot 5 = 2 \cdot 3 \cdot 5^2$$

(There are other multiplications possible; they all end up with the same answer.)

425. $2^2 \cdot 3^3$

Write the number as the product of two other numbers.

$$9 \cdot 12$$

Now write both of these numbers as the product of two numbers, since they aren't prime.

$$= (3 \cdot 3) \cdot (4 \cdot 3)$$

Continue writing products until all the factors are prime.

$$= 3 \cdot 3 \cdot (2 \cdot 2) \cdot 3$$

All five numbers are prime. Write the factorization using exponents.

$$3 \cdot 3 \cdot 2 \cdot 2 \cdot 3 = 2^2 \cdot 3^3$$

(There are other multiplications possible; they all end up with the same answer.)

426. 2^9

Write the number as the product of two other numbers.

$$8 \cdot 64$$

Now write both of these numbers as the product of two numbers, since they aren't prime.

$$2 \cdot 4 \cdot 8 \cdot 8$$

Continue writing products until all the factors are prime.

$$= 2 \cdot (2 \cdot 2) \cdot (2 \cdot 4) \cdot (2 \cdot 4)$$
$$= 2 \cdot 2 \cdot 2 \cdot 2 \cdot (2 \cdot 2) \cdot 2 \cdot (2 \cdot 2)$$

All nine numbers are prime. Write the factorization using exponents.

$$2 \cdot 2 \cdot 2 \cdot 2 \cdot 2 \cdot 2 \cdot 2 \cdot 2 \cdot 2 = 2^9$$

(There are other multiplications possible; they all end up with the same answer.)

427. $2^2 \cdot 5^3$

Write the number as the product of two other numbers.

$$5 \cdot 100$$

Now write 100 as the product of two numbers.

$$= 5 \cdot 10 \cdot 10$$

Continue by writing the 10's as products of prime factors.

$$= 5 \cdot (2 \cdot 5) \cdot (2 \cdot 5)$$

All five numbers are prime. Write the factorization using exponents.

$$= 2^2 \cdot 5^3$$

(There are other multiplications possible; they all end up with the same answer.)

Answers 401–500

428. $2^4 \cdot 11^2$

Write the number as the product of two other numbers.

$$4 \cdot 484$$

Now write both of these numbers as the product of two numbers, since they aren't prime.

$$= (2 \cdot 2) \cdot (4 \cdot 121)$$

Continue writing products until all the factors are prime.

$$= 2 \cdot 2 \cdot (2 \cdot 2) \cdot (11 \cdot 11)$$

All six numbers are prime. Write the factorization using exponents.

$$= 2^4 \cdot 11^2$$

(There are other multiplications possible; they all end up with the same answer.)

429. $2^2 \cdot 3^3 \cdot 5^2$

Write the number as the product of two other numbers.

$$27 \cdot 100$$

Now write both of these numbers as the product of two numbers.

$$= (3 \cdot 9) \cdot (10 \cdot 10)$$

Continue writing products until all the factors are prime.

$$= 3 \cdot (3 \cdot 3) \cdot (2 \cdot 5) \cdot (2 \cdot 5)$$

All seven numbers are prime. Write the factorization using exponents.

$$= 2^2 \cdot 3^3 \cdot 5^2$$

(There are other multiplications possible; they all end up with the same answer.)

430. $6(4x^4 - 5y^8)$

The GCF is 6. Dividing each term by that GCF and putting the quotients as separate terms in the parentheses,

$$24x^4 - 30y^8 = 6(4x^4 - 5y^8).$$

431. $4(11z^5 + 15a - 2)$

The GCF is 4. Dividing each term by that GCF and putting the quotients as separate terms in the parentheses,

$$44z^5 + 60a - 8 = 4(11z^5 + 15a - 2).$$

432. $60(5abc + 7xyz)$

The GCF is 60. Dividing each term by that GCF and putting the quotients as separate terms in the parentheses,

$$300abc + 420xyz = 60(5abc + 7xyz).$$

433. $11(11x^4 - 15z)$

The GCF is 11. Dividing each term by that GCF and putting the quotients as separate terms in the parentheses,

$$121x^4 - 165z = 11(11x^4 - 15z).$$

434. $24x^2y^2(y-2x)$

The GCF is $24x^2y^2$. Divide each term by that GCF and put the quotients as separate terms in parentheses. Be careful to attach the correct signs. $24x^2y^3 - 48x^3y^2 = 24x^2y^2(y-2x)$. Note that the terms in the parentheses have no common factors.

435. $4ab(9a^2 - 6ab - 10b^2)$

The GCF is $4ab$. Divide each term by that GCF and put the quotients as separate terms in parentheses. Be careful to attach the correct signs. $36a^3b - 24a^2b^2 - 40ab^3 = 4ab(9a^2 - 6ab - 10b^2)$

Note that the terms in the parentheses have no common factors shared by all three terms.

436. $3z^{-4}(3+5z-9z^3)$

The GCF is $3z^{-4}$. For the powers of the variable z, you select the power that is smallest, which is –4 in this case. Divide each term by that GCF and put the quotients as separate terms in parentheses. Be careful to attach the correct signs. $9z^{-4} + 15z^{-3} - 27z^{-1} = 3z^{-4}(3+5z-9z^3)$

Note that the terms in the parentheses have no common factors shared by all three terms.

437. $5y^{1/4}(4y^{1/2} - 5)$

The GCF is $5y^{1/4}$. For the powers of the variable y, you select the power that is smallest, which is $\frac{1}{4}$ in this case. Divide each term by that GCF and put the quotients as separate terms in parentheses. Be careful to attach the correct signs. $20y^{3/4} - 25y^{1/4} = 5y^{1/4}(4y^{1/2} - 5)$

Note that the terms in the parentheses have no common factors shared by the terms.

438. $16a^{1/2}b^{3/4}c^{4/5}(1-3abc)$

The GCF is $16a^{1/2}b^{3/4}c^{4/5}$. Choose the smaller power of each variable. Divide each term by that GCF and put the quotients as separate terms in parentheses. Be careful to attach the correct signs.

$$16a^{1/2}b^{3/4}c^{4/5} - 48a^{3/2}b^{7/4}c^{9/5} = 16a^{1/2}b^{3/4}c^{4/5}(1-3abc)$$

Note that the terms in the parentheses have no common factors shared by the terms.

439. $2x^2(5x-1)[4+3x]$

The GCF is $2x^2(5x-1)$. The binomial is a factor of each term. Divide each term by that GCF and put the quotients as separate terms in parentheses. Be careful to attach the correct signs. $8x^2(5x-1)+6x^3(5x-1)=2x^2(5x-1)[4+3x]$

Note that the terms in the parentheses have no common factors shared by the terms.

440. $4x^{-5}y^2(9x^2y^2+5)$

The GCF is $4x^{-5}y^2$. Divide each term by that GCF and put the quotients as separate terms in parentheses. Be careful to attach the correct signs.

$$36x^{-3}y^4+20x^{-5}y^2=4x^{-5}y^2(9x^2y^2+5)$$

Note that the terms in the parentheses have no common factors shared by the terms.

441. $125x^{-5}y^{-4}(x^2+4y^2)$

The GCF is $125x^{-5}y^{-4}$. Choose the smaller power for each variable. Divide each term by that GCF and put the quotients as separate terms in parentheses. Be careful to attach the correct signs. $125x^{-3}y^{-4}+500x^{-5}y^{-2}=125x^{-5}y^{-4}(x^2+4y^2)$

Note that the terms in the parentheses have no common factors shared by the terms.

442. $x(3x-1)[5x-1]$

The GCF is $x(3x-1)$. Divide each term by that GCF and put the quotients as separate terms in the brackets.

$$x(3x-1)^2+2x^2(3x-1)=x(3x-1)[(3x-1)+2x]$$

Now simplify the terms in the brackets.

$$=x(3x-1)[5x-1]$$

443. $-2x^3(x-4)^3(x+8)$

The GCF is $2x^3(x-4)^3$. Divide each term by that GCF and put the quotients as separate terms in the brackets.

$$4x^3(x-4)^4-6x^4(x-4)^3=2x^3(x-4)^3[2(x-4)-3x]$$

Now simplify the terms in the brackets.

$$=2x^3(x-4)^3[2x-8-3x]$$
$$=2x^3(x-4)^3[-x-8]$$
$$=-2x^3(x-4)^3(x+8)$$

Factoring out the negative sign in the brackets wasn't absolutely necessary, but it gives a more consistent form to the binomials.

444. $\dfrac{-4x(1+3x)}{(x-1)^3}$

The GCF of the numerator is $4x(x-1)$. Factor the numerator first.

$$=\frac{4x(x-1)[(x-1)-4x]}{(x-1)^4}$$

Now reduce the fraction by dividing both numerator and denominator by the common factor, $(x-1)$.

$$\frac{4x\cancel{(x-1)}[(x-1)-4x]}{(x-1)^{\cancel{4}3}}=\frac{4x[(x-1)-4x]}{(x-1)^3}$$

Simplify the terms in the brackets.

$$\frac{4x[x-1-4x]}{(x-1)^3}=\frac{4x[-1-3x]}{(x-1)^3}$$

$$=\frac{-4x[1+3x]}{(x-1)^3}$$

445. $x^2(x-3)^2[3+6x-2x^2]$

The GCF of the numerator is $4x^2(x-3)^{-4}$. Remember that you want the smaller negative exponent when factoring out common factors. Factor the numerator first.

$$=\frac{4x^2(x-3)^{-4}[3-2x(x-3)]}{4(x-3)^{-6}}$$

Now reduce the fraction by dividing both numerator and denominator by the common factor, $4(x-3)^{-6}$.

$$=\frac{\cancel{4}x^2(x-3)^{\cancel{-4}2}[3-2x(x-3)]}{\cancel{4}\cancel{(x-3)^{-6}}}$$

$$=\frac{x^2(x-3)^2[3-2x(x-3)]}{1}$$

Simplify the terms in the brackets.

$$=x^2(x-3)^2[3-2x^2+6x]$$

$$=x^2(x-3)^2[3+6x-2x^2]$$

446. $120a^8b$

First write the 7! in an expanded form.

$$=\frac{7\cdot 6\cdot 5\cdot 4!a^5b^{-1}}{4!a^{-3}b^{-2}}$$

Now factor the GCF $4!a^{-3}b^{-2}$ from the numerator.

$$=\frac{\left(4!a^{-3}b^{-2}\right)7\cdot 6\cdot 5\cdot a^{8}b^{1}}{4!a^{-3}b^{-2}}$$

Reduce the fraction and simplify.

$$=\frac{\left(\cancel{4!a^{-3}b^{-2}}\right)7\cdot 6\cdot 5\cdot a^{8}b^{1}}{\cancel{4!a^{-3}b^{-2}}}=210a^{8}b$$

447. $\mathbf{100ab}$

First write 100! in an expanded form.

$$=\frac{100\cdot 99!a^{3/2}b^{1/2}}{99!a^{1/2}b^{-1/2}}$$

Now factor the GCF $99!a^{1/2}b^{-1/2}$ from the numerator.

$$=\frac{\left(99!a^{1/2}b^{-1/2}\right)100a^{1}b^{1}}{99!a^{1/2}b^{-1/2}}$$

Reduce the fraction.

$$=\frac{\left(\cancel{99!a^{1/2}b^{-1/2}}\right)100a^{1}b^{1}}{\cancel{99!a^{1/2}b^{-1/2}}}=100ab$$

448. $\mathbf{\dfrac{2ab-5}{4b^{3}}}$

First factor the numerator.

$$=\frac{7a(2ab-5)}{28ab^{3}}$$

The common factor in the numerator and denominator is $7a$.

$$=\frac{\cancel{7a}(2ab-5)}{{}^{4}\cancel{28a}b^{3}}=\frac{2ab-5}{4b^{3}}$$

449. $\mathbf{\dfrac{3y^{2}(2x-3y^{2})}{10x^{2}}}$

First factor the numerator.

$$=\frac{18xy^{4}\left(2x-3y^{2}\right)}{60x^{3}y^{2}}$$

The common factor of the numerator and denominator is $6xy^{2}$.

$$=\frac{\left(\cancel{6xy^{2}}\right)\left(3y^{2}\right)\left(2x-3y^{2}\right)}{\left(\cancel{6xy^{2}}\right)10x^{2}}$$

$$=\frac{3y^{2}\left(2x-3y^{2}\right)}{10x^{2}}$$

450. $\dfrac{2w\left[2-w^2-w\right]}{(w+1)\left[5w^3-3w-3\right]}$

First factor both the numerator and the denominator.

$$=\frac{8w^3(w+1)^2\left[2-w(w+1)\right]}{4w^2(w+1)^3\left[5w^3-3(w+1)\right]}$$

The GCF of the numerator and denominator is $4w^2(w+1)^2$. Factor that out of the numerator and denominator and simplify.

$$=\frac{\left(4w^2(w+1)^2\right)(2w)\left[2-w(w+1)\right]}{\left(4w^2(w+1)^2\right)(w+1)\left[5w^3-3(w+1)\right]}$$

$$=\frac{(2w)\left[2-w^2-w\right]}{(w+1)\left[5w^3-3w-3\right]}$$

451. $\dfrac{x}{x-1}$

Factor the numerator by dividing by the GCF.

$$=\frac{x(x+1)}{(x+1)(x-1)}$$

The GCF of the numerator and denominator is $(x+1)$.

$$=\frac{x(x+1)}{(x+1)(x-1)}=\frac{x}{x-1}$$

452. $\dfrac{x+1}{x+6}$

Factor the numerator by dividing by the GCF, $(x+1)$.

$$=\frac{(x+1)(x-6)}{(x+6)(x-6)}$$

The GCF of the numerator and denominator is $(x-6)$.

$$=\frac{(x+1)(x-6)}{(x+6)(x-6)}=\frac{x+1}{x+6}$$

453. $\dfrac{3x+1}{x+1}$

Factor the numerator, then factor out the GCF of the denominator.

$$=\frac{(2x-7)(3x+1)}{(x+1)(2x-7)}$$

The GCF of the numerator and denominator is $(2x-7)$.

$$=\frac{(2x-7)(3x+1)}{(x+1)(2x-7)}=\frac{3x+1}{x+1}$$

454. $\frac{x-4}{x+4}$

In both the numerator and denominator, factor out the GCFs and then factor the remaining terms.

$$=\frac{(x+4)\left[x^2-4x\right]}{x\left(x^2+8x+16\right)}=\frac{x(x+4)(x-4)}{x(x+4)^2}$$

The GCF of the numerator and denominator is $x(x + 4)$.

$$=\frac{\cancel{x}\cancel{(x+4)}(x-4)}{\cancel{x}(x+4)^{\cancel{2}1}}=\frac{x-4}{x+4}$$

455. $\frac{(x+3)\left(x^2+2x+4\right)}{x^4}$

Factor the denominator.

$$=\frac{(x-3)(x+3)(x-2)\left(x^2+2x+4\right)}{x^4(x-2)[x-3]}$$

The GCF of the numerator and denominator is $(x - 2)(x - 3)$.

$$=\frac{\cancel{(x-3)}(x+3)\cancel{(x-2)}\left(x^2+2x+4\right)}{x^4\cancel{(x-2)}\cancel{(x-3)}}$$

$$=\frac{(x+3)\left(x^2+2x+4\right)}{x^4}$$

456. $(x+6)(x-6)$

The difference of perfect squares factors into the sum and difference of the roots of the terms. The square root of x^2 is x, and the square root of 36 is 6.

$$x^2-36=(x+6)(x-6)$$

457. $(3y+10)(3y-10)$

The difference of perfect squares factors into the sum and difference of the roots of the terms. The square root of $9y^2$ is $3y$, and the square root of 100 is 10.

$$9y^2-100=(3y+10)(3y-10)$$

458. $(9a+y)(9a-y)$

The difference of perfect squares factors into the sum and difference of the roots of the terms. The square root of $81a^2$ is $9a$, and the square root of y^2 is y.

$$81a^2-y^2=(9a+y)(9a-y)$$

459. $(2x+7z)(2x-7z)$

The difference of perfect squares factors into the sum and difference of the roots of the terms. The square root of $4x^2$ is $2x$, and the square root of $49z^2$ is $7z$.

$$4x^2-49z^2=(2x+7z)(2x-7z)$$

460. $(8xy+5zw^2)(8xy-5zw^2)$

The difference of perfect squares factors into the sum and difference of the roots of the terms. The square root of $64x^2y^2$ is $8xy$, and the square root of $25z^2w^4$ is $5zw^2$.

$$64x^2y^2-25z^2w^4=\left(8xy+5zw^2\right)\left(8xy-5zw^2\right)$$

461. $(6a^2b^3+11)(6a^2b^3-11)$

The difference of perfect squares factors into the sum and difference of the roots of the terms. The square root of $36a^4b^6$ is $6a^2b^3$, and the square root of 121 is 11.

$$36a^4b^6-121=\left(6a^2b^3+11\right)\left(6a^2b^3-11\right)$$

462. $(11x^{1/4}+12y^{1/8})(11x^{1/4}-12y^{1/8})$

The difference of perfect squares factors into the sum and difference of the roots of the terms. The square root of $121x^{1/2}$ is $11x^{1/4}$; remember, when you multiply factors with the same base, you add exponents, so you need half of $\frac{1}{2}$ for the power on x. The square root of $144y^{1/4}$ is $12y^{1/8}$; again, you need half of the starting exponent.

$$121x^{1/2}-144y^{1/4}=\left(11x^{1/4}+12y^{1/8}\right)\left(11x^{1/4}-12y^{1/8}\right)$$

463. $(5x^{-1}+3y^{-2})(5x^{-1}-3y^{-2})$

The difference of perfect squares factors into the sum and difference of the roots of the terms. The square root of $25x^{-2}$ is $5x^{-1}$; remember, when you multiply factors with the same base, you add exponents, so you need half of -2 for the power on x. The square root of $9y^4$ is $3y^{-2}$; again, you need half of the starting exponent.

$$25x^{-2}-9y^{-4}=\left(5x^{-1}+3y^{-2}\right)\left(5x^{-1}-3y^{-2}\right)$$

464. $(4+xy^{-1/8})(4-xy^{-1/8})$

The difference of perfect squares factors into the sum and difference of the roots of the terms. The square root of 16 is 4. The square root of $x^2y^{-1/4}$ is $xy^{-1/8}$; remember, when you multiply factors with the same base, you add exponents, so you need half of $-\frac{1}{4}$ for the power on y.

$$16-x^2y^{-1/4}=\left(4+xy^{-1/8}\right)\left(4-xy^{-1/8}\right)$$

465. $(z^{-2/9}+7w^{1/4})(z^{-2/9}-7w^{1/4})$

The difference of perfect squares factors into the sum and difference of the roots of the terms. The square root of $z^{-4/9}$ is $z^{-2/9}$; remember, when you multiply factors with the same base, you add exponents, so you need half of $-\frac{4}{9}$ for the power on z. The square root of $49w^{1/2}$ is $7w^{1/4}$; again, you need half of the starting exponent.

$$z^{-4/9}-49w^{1/2}=\left(z^{-2/9}+7w^{1/4}\right)\left(z^{-2/9}-7w^{1/4}\right)$$

466. $(x+2)(x^2-2x+4)$

The sum of perfect cubes factors into a binomial containing the two cube roots of the original terms and a trinomial containing squares of the roots and a multiple of the roots. The cube root of x^3 is x, and the cube root of +8 is +2.

Write the binomial containing the cube roots: $(x + 2)$.

Now square the first term and put it in the first position of the trinomial; square the second term and put it in the last position of the trinomial. Put the opposite of the product of the two roots in the middle.

$$\left(x^2-2x+4\right)$$

The complete factorization is the product of the binomial and trinomial.

$$x^3+8=(x+2)\left(x^2-2x+4\right)$$

467. $(x+7)(x^2-7x+49)$

The sum of perfect cubes factors into a binomial containing the two cube roots of the original terms and a trinomial containing squares of the roots and a multiple of the roots. The cube root of x^3 is x, and the cube root of +343 is +7.

Write the binomial containing the cube roots: $(x + 7)$.

Now square the first term and put it in the first position of the trinomial; square the second term and put it in the last position of the trinomial. Put the opposite of the product of the two roots in the middle.

$$\left(x^2-7x+49\right)$$

The complete factorization is the product of the binomial and trinomial.

$$x^3+343=(x+7)\left(x^2-7x+49\right)$$

468. $(a-6z)(a^2+6az+36z^2)$

The difference between perfect cubes factors into a binomial containing the two cube roots of the original terms and a trinomial containing squares of the roots and a multiple of the roots. The cube root of a^3 is a, and the cube root of $-216z^3$ is $-6z$.

Write the binomial containing the cube roots: $(a - 6z)$.

Now square the first term and put it in the first position of the trinomial; square the second term and put it in the last position of the trinomial. Put the opposite of the product of the two roots in the middle.

$$\left(a^2+6az+36z^2\right)$$

The complete factorization is the product of the binomial and trinomial.

$$a^3-216z^3=(a-6z)\left(a^2+6az+36z^2\right)$$

469. $(1-y)(1+y+y^2)$

The difference between perfect cubes factors into a binomial containing the two cube roots of the original terms and a trinomial containing squares of the roots and a multiple of the roots. The cube root of 1 is 1, and the cube root of $-y^3$ is $-y$.

Write the binomial containing the cube roots: $(1-y)$.

Now square the first term and put it in the first position of the trinomial; square the second term and put it in the last position of the trinomial. Put the opposite of the product of the two roots in the middle.

$$\left(1+y+y^2\right)$$

The complete factorization is the product of the binomial and trinomial.

$$1-y^3=(1-y)\left(1+y+y^2\right)$$

470. $(5z+7)(25z^2-35z+49)$

The sum of perfect cubes factors into a binomial containing the two cube roots of the original terms and a trinomial containing squares of the roots and a multiple of the roots. The cube root of $125z^3$ is $5z$, and the cube root of +343 is +7.

Write the binomial containing the cube roots: $(5z+7)$.

Now square the first term and put it in the first position of the trinomial; square the second term and put it in the last position of the trinomial. Put the opposite of the product of the two roots in the middle.

$$\left(25z^2-35z+49\right)$$

The complete factorization is the product of the binomial and trinomial.

$$125z^3+343=(5z+7)\left(25z^2-35z+49\right)$$

471. $(2a+3b)(4a^2-6ab+9b^2)$

The sum of perfect cubes factors into a binomial containing the two cube roots of the original terms and a trinomial containing squares of the roots and a multiple of the roots. The cube root of $8a^3$ is $2a$, and the cube root of $27b^3$ is $3b$.

Write the binomial containing the cube roots: $(2a+3b)$.

Now square the first term and put it in the first position of the trinomial; square the second term and put it in the last position of the trinomial. Put the opposite of the product of the two roots in the middle.

$$(4a^2-6ab+9b^2)$$

The complete factorization is the product of the binomial and trinomial.

$$8a^3+27b^3=(2a+3b)(4a^2-6ab+9b^2)$$

472. $(9x-10y^2)(81x^2+90xy^2+100y^4)$

The difference between perfect cubes factors into a binomial containing the two cube roots of the original terms and a trinomial containing squares of the roots and a multiple of the roots. The cube root of $729x^3$ is $9x$, and the cube root of $-1000y^6$ is $-10y^2$.

Write the binomial containing the cube roots: $(9x-10y^2)$.

Now square the first term and put it in the first position of the trinomial; square the second term and put it in the last position of the trinomial. Put the opposite of the product of the two roots in the middle.

$$(81x^2+90xy^2+100y^4)$$

The complete factorization is the product of the binomial and trinomial.

$$729x^3-1000y^6=(9x-10y^2)(81x^2+90xy^2+100y^4)$$

473. $(8x^3-5y^9)(64x^6+40x^3y^9+25y^{18})$

The difference between perfect cubes factors into a binomial containing the two cube roots of the original terms and a trinomial containing squares of the roots and a multiple of the roots. The cube root of $512x^9$ is $8x^3$, and the cube root of $-125y^{27}$ is $-5y^9$.

Write the binomial containing the cube roots: $(8x^3-5y^9)$.

Now square the first term and put it in the first position of the trinomial; square the second term and put it in the last position of the trinomial. Put the opposite of the product of the two roots in the middle.

$$(64x^6+40x^3y^9+25y^{18})$$

The complete factorization is the product of the binomial and trinomial.

$$512x^9-125y^{27}=(8x^3-5y^9)(64x^6+40x^3y^9+25y^{18})$$

474. $(3x^{1/9}-1)(9x^{2/9}+3x^{1/9}+1)$

The difference between perfect cubes factors into a binomial containing the two cube roots of the original terms and a trinomial containing squares of the roots and a multiple of the roots. The cube root of $27x^{1/3}$ is $3x^{1/9}$. Remember, you add exponents when multiplying like bases, and $(x^{1/9})^3=x^{1/9}\cdot x^{1/9}\cdot x^{1/9}=x^{3/9}=x^{1/3}$. The cube root of -1 is -1.

Write the binomial containing the cube roots: $(3x^{1/9}-1)$.

Now square the first term and put it in the first position of the trinomial; square the second term and put it in the last position of the trinomial. Put the opposite of the product of the two roots in the middle.

$$\left(9x^{2/9}+3x^{1/9}+1\right)$$

The complete factorization is the product of the binomial and trinomial.

$$27x^{1/3}-1=\left(3x^{1/9}-1\right)\left(9x^{2/9}+3x^{1/9}+1\right)$$

475. $\left(2y^{-2}+7z^{-1/3}\right)\left(4y^{-4}-14y^{-2}z^{-1/3}+49z^{-2/3}\right)$

The sum of perfect cubes factors into a binomial containing the two cube roots of the original terms and a trinomial containing squares of the roots and a multiple of the roots. The cube root of $8y^{-6}$ is $2y^{-2}$. Remember, you add exponents when multiplying like bases, and $\left(y^{-2}\right)^3=y^{-2}\cdot y^{-2}\cdot y^{-2}=y^{-6}$. The cube root of $343z^{-1}$ is $7z^{-1/3}$. Again, you add exponents when multiplying like bases, and $\left(z^{-1/3}\right)^3=z^{-1/3}\cdot z^{-1/3}\cdot z^{-1/3}=z^{-3/3}=z^{-1}$.

Write the binomial containing the cube roots: $\left(2y^{-2}+7z^{-1/3}\right)$.

Now square the first term and put it in the first position of the trinomial; square the second term and put it in the last position of the trinomial. Put the opposite of the product of the two roots in the middle.

$$\left(4y^{-4}-14y^{-2}z^{-1/3}+49z^{-2/3}\right)$$

The complete factorization is the product of the binomial and trinomial.

$$8y^{-6}+343z^{-1}=\left(2y^{-2}+7z^{-1/3}\right)\left(4y^{-4}-14y^{-2}z^{-1/3}+49z^{-2/3}\right)$$

476. $3x^2y^3(x+5)(x-5)$

First factor out the GCF, $3x^2y^3$.

$$3x^4y^3-75x^2y^3=3x^2y^3\left(x^2-25\right)$$

Next, factor the binomial in the parentheses as the difference of two squares.

$$=3x^2y^3(x+5)(x-5)$$

477. $6x^2y^2(x+4y)(x-4y)$

First factor out the GCF, $6x^2y^2$.

$$6x^4y^2-96x^2y^4=6x^2y^2\left(x^2-16y^2\right)$$

Next, factor the binomial in the parentheses as the difference of two squares.

$$=6x^2y^2(x+4y)(x-4y)$$

478. $36(z+10w)(z-10w)$

First factor out the GCF, 36.

$$36z^2-3600w^2=36\left(z^2-100w^2\right)$$

Next, factor the binomial in the parentheses as the difference of two squares.

$$= 36(z + 10w)(z - 10w)$$

479. $\mathbf{100x(x+3)(x-3)}$

First factor out the GCF, $100x$.

$$100x^3 - 900x = 100x(x^2 - 9)$$

Next, factor the binomial in the parentheses as the difference of two squares.

$$= 100x(x+3)(x-3)$$

480. $\mathbf{4y(2y+1)(4y^2-2y+1)}$

First factor out the GCF, $4y$.

$$32y^4 + 4y = 4y(8y^3 + 1)$$

Next, factor the binomial in the parentheses as the sum of two cubes.

$$= 4y(2y+1)(4y^2 - 2y + 1)$$

481. $\mathbf{4xy^2(x+2)(x^2-2x+4)}$

First factor out the GCF, $4xy^2$.

$$4x^4y^2 + 32xy^2 = 4xy^2(x^3 + 8)$$

Next, factor the binomial in the parentheses as the sum of two cubes.

$$= 4xy^2(x+2)(x^2 - 2x + 4)$$

482. $\mathbf{(25x^2+1)(5x+1)(5x-1)}$

First factor the binomial as the difference of two squares.

$$625x^4 - 1 = (25x^2 + 1)(25x^2 - 1)$$

Now factor the second binomial, which is also the difference of two squares.

$$= (25x^2 + 1)(5x+1)(5x-1)$$

483. $\mathbf{(4x^2+9y^4)(2x+3y^2)(2x-3y^2)}$

First factor the binomial as the difference of two squares.

$$16x^4 - 81y^8 = (4x^2 + 9y^4)(4x^2 - 9y^4)$$

Now factor the second binomial, which is also the difference of two squares.

$$= (4x^2 + 9y^4)(2x + 3y^2)(2x - 3y^2)$$

484. $x^{-7}(x-1)(x^2+x+1)$

First factor out the GCF, x^{-7}.

$$x^{-4}-x^{-7}=x^{-7}(x^3-1)$$

Now factor the binomial in the parentheses as the difference of two cubes.

$$=x^{-7}(x-1)(x^2+x+1)$$

485. $y^{-12}(y^2+1)(y+1)(y-1)$

First factor out the GCF, y^{-12}.

$$y^{-8}-y^{-12}=y^{-12}(y^4-1)$$

Now factor the binomial in the parentheses as the difference of two squares.

$$=y^{-12}(y^2+1)(y^2-1)$$

Finally, factor the right-most binomial as the difference of two squares.

$$=y^{-12}(y^2+1)(y+1)(y^2-1)$$

486. $216(ab+c^2)(a^2b^2-abc^2+c^4)$

First factor out the GCF, 216.

$$216a^3b^3+216c^6=216(a^3b^3+c^6)$$

Now factor the binomial in the parentheses as the sum of two cubes.

$$=216(ab+c^2)(a^2b^2-abc^2+c^4)$$

487. $125(ab^2+2c^3)(ab^2-2c^3)$

First factor out the GCF, 125.

$$125a^2b^4-500c^6=125(a^2b^4-4c^6)$$

Now factor the binomial in the parentheses as the difference of two squares.

$$=125(ab^2+2c^3)(ab^2-2c^3)$$

488. $b^6(a-2)(a^2+2a+4)$

First factor out the GCF, b^6.

$$a^3b^6-8b^6=b^6(a^3-8)$$

Now factor the binomial as the difference of cubes.

$$=b^6(a-2)(a^2+2a+4)$$

489. $3x^2(3y+1)(9y^2-3y+1)$

First factor out the GCF, $3x^2$.

$$81x^2y^3+3x^2=3x^2(27y^3+1)$$

Now factor the binomial as the sum of cubes.

$$=3x^2(3y+1)(9y^2-3y+1)$$

490. $4xy(2x-1)(4x^2+2x+1)$

First factor out the GCF, $4xy$.

$$32x^4y-4xy=4xy(8x^3-1)$$

Now factor the binomial as the difference of cubes.

$$=4xy(2x-1)(4x^2+2x+1)$$

491. $9b^2(a^2b+2z)(a^4b^2-2a^2bz+4z^2)$

First factor out the GCF, $9b^2$.

$$9a^6b^5+72z^3b^2=9b^2(a^6b^3+8z^3)$$

Now factor the binomial as the sum of cubes. Remember that $(a^2)^3=a^6$.

$$=9b^2(a^2b+2z)(a^4b^2-2a^2bz+4z^2)$$

492. $2x^2(x^2+9)(x+3)(x-3)$

First factor out the GCF, $2x^2$.

$$2x^6-162x^2=2x^2(x^4-81)$$

Next, factor the binomial as the difference of squares.

$$=2x^2(x^2+9)(x^2-9)$$

The last binomial is still the difference of squares and needs to be factored.

$$=2x^2(x^2+9)(x+3)(x-3)$$

493. $(x+1)(x^2-x+1)(x-1)(x^2+x+1)$

Factor the binomial as the difference of squares.

$$x^6-1=(x^3+1)(x^3-1)$$

The first binomial is the sum of cubes, and the second binomial is the difference of cubes.

$$=(x+1)(x^2-x+1)(x-1)(x^2+x+1)$$

The original problem could have been factored as the difference of cubes, but the resulting trinomial, (x^4+x^2+1), is very difficult to factor.

494. $(y^2+2)(y^4-2y^2+4)(y^2-2)(y^4+2y^2+4)$

First factor the binomial as the difference of squares.

$$y^{12}-64=(y^6+8)(y^6-8)$$

The first binomial is the sum of cubes, and the second binomial is the difference of cubes.

$$=(y^2+2)(y^4-2y^2+4)(y^2-2)(y^4+2y^2+4)$$

495. $10x^2(a^2+10b^4)(a^2-10b^4)$

First factor out the GCF, $10x^2$.

$$10a^4x^2-1000b^8x^2=10x^2(a^4-100b^8)$$

Now factor the binomial as the difference of squares.

$$=10x^2(a^2+10b^4)(a^2-10b^4)$$

496. $3x^2y^2(4x^2-2xy+7y^2)$

The GCF is $3x^2y^2$. Divide each term by that factor and place the quotients within parentheses.

497. $7a^2bc(10b^2+9abc-3a^2c^2)$

The GCF is $7a^2bc$. Divide each term by that factor and place the quotients within parentheses.

498. $48(x-4)(1-x)$

The GCF is $3(x-4)$. Divide each term by that factor and place the quotients within brackets.

$$3(x-4)\left[(x-4)^2+2x(x-4)-3x^2\right]$$

Simplify what's in the brackets by squaring the binomial and distributing the $2x$.

$$=3(x-4)\left[x^2-8x+16+2x^2-8x-3x^2\right]$$
$$=3(x-4)[-16x+16]$$

Now factor 16 out of the second binomial and multiply it times the lead 3.

$$=48(x-4)(1-x)$$

499. $12x^2y(5x^3-4x^4y+3y^2)$

The GCF is $12x^2y$. Divide each term by that factor and place the quotients within parentheses.

500. $(x-10)(x+2)$

The last terms in the binomials must have a product of –20 and a sum of –8. Use –10 and 2.

$$x^2-8x-20=(x-10)(x+2)$$

501. $(x+9)(x+1)$

The last terms in the binomials must have a product of 9 and a sum of 10. Use 9 and 1. All the signs are positive.

$$x^2+10x+9=(x+9)(x+1)$$

502. $(y-8)(y+2)$

The last terms in the binomials must have a product of –16 and a sum of –6. Use –8 and 2.

$$y^2-6y-16=(y-8)(y+2)$$

503. $(z-6)(z+8)$

The last terms in the binomials must have a product of –48 and a sum of 2. Use 8 and –6.

$$z^2+2z-48=(z-6)(z+8)$$

504. $(2x-3)(x+2)$

The last terms in the binomials must have a product of –6 and the coefficients of the first terms must have a product of 2. The sum of the outer and inner products must be 1. An outer product of 4 and inner product of –3 gives you the sum of 1.

$$2x^2+x-6=(2x-3)(x+2)$$

505. $(3x-4)(x+3)$

The last terms in the binomials must have a product of –12 and the coefficients of the first terms must have a product of 3. The sum of the outer and inner products must be 5. An outer product of 9 and inner product of –4 gives you the sum of 5.

$$3x^2+5x-12=(3x-4)(x+3)$$

506. $(3z+4)^2$

The last terms in the binomials must have a product of 16, and the coefficients of the first terms must have a product of 9. The sum of the outer and inner products must be 24. Notice that the first and last terms in the trinomial are perfect squares. Watch for a binomial squared! All the signs are positive.

$$9z^2+24z+16=(3z+4)(3z+4)=(3z+4)^2$$

507. $(4x-5)^2$

The last terms in the binomials must have a product of 25, and the coefficients of the first terms must have a product of 16. The sum of the outer and inner products must be –40. Notice that the first and last terms in the trinomial are perfect squares. Watch for a binomial squared! Since the middle term is negative, there's a minus sign in the binomials.

$$16x^2-40x+25=(4x-5)(4x-5)=(4x-5)^2$$

508. $(w-64)(w+1)$

The last terms in the binomials must have a product of –64 and a sum of –63. Use –64 and 1.

$$w^2-63w-64=(w-64)(w+1)$$

509. $(4x-5)(x+5)$

The last terms in the binomials must have a product of –25 and the coefficients of the first terms must have a product of 4. The sum of the outer and inner products must be 15. An outer product of 20 and inner product of –5 gives you the sum of 15.

$$4x^2+15x-25=(4x-5)(x+5)$$

510. $(5x-6)(8x+9)$

The last terms in the binomials must have a product of –54 and the coefficients of the first terms must have a product of 40. The sum of the outer and inner products must be –3. An outer product of 45 and inner product of –48 gives you the sum of –3.

$$40x^2-3x-54=(5x-6)(8x+9)$$

511. $(8x+5)(2x-3)$

The last terms in the binomials must have a product of –15 and the coefficients of the first terms must have a product of 16. The sum of the outer and inner products must be –14. An outer product of –24 and inner product of 10 gives you the sum of –14.

$$16x^2-14x-15=(8x+5)(2x-3)$$

512. $(x^5-4)(x^5-1)$

It's often helpful to picture the trinomial as a corresponding quadratic, such as y^2-5y+4. The last terms in the binomials must have a product of 4 and a sum of –5. Use –4 and –1. Start by factoring the quadratic.

$$=(y-4)(y-1)$$

And then replace the y's with the powers of x. Remember that the powers of x in the factored form will be half the highest power in the original trinomial.

513. $\left(y^3-7\right)\left(y^3+3\right)$

It's often helpful to picture the trinomial as a corresponding quadratic, such as $x^2-4x-21$. The last terms in the binomials must have a product of –21 and a sum of –4. Use –7 and 3. Start by factoring the quadratic.

$$=(x-7)(x+3)$$

And then replace the *x*'s with the powers of *y*. Remember that the powers of *y* in the factored form will be half the highest power in the original trinomial.

514. $\left(y^8+5\right)\left(y^8-5\right)$

It's often helpful to picture the binomial as a corresponding quadratic, such as z^2-25. Notice that you have the difference of perfect squares, which factors into the sum and difference of the roots. Start by writing the factorization of the quadratic.

$$=(z+5)(z-5)$$

And then replace the *z*'s with the powers of *y*. Be careful to use half of the exponent 16 and not take the square root of 16. Just remember that the powers of *y* in the factored form will be half the highest power in the original binomial.

515. $\left(5a^2+7b^5\right)\left(5a^2-7b^5\right)$

It's often helpful to picture the binomial as a corresponding quadratic, such as $25z^2-49w^2$. Notice that you have the difference of perfect squares, which factors into the sum and difference of the roots. Start by writing the factorization of the quadratic.

$$=(5z+7w)(5z-7w)$$

And then replace the *z*'s and *w*'s with the powers of *a* and *b*. Remember that the powers of *a* and *b* in the factored form will be half the highest power in the original binomial.

516. $\left(x^{-4}-6\right)\left(x^{-4}+3\right)$

It's often helpful to picture the trinomial as a corresponding quadratic, such as $y^2-3y-18$. The last terms in the binomials must have a product of –18 and a sum of –3. Use –6 and 3. Start by factoring the quadratic.

$$=(y-6)(y+3)$$

And then replace the *y*'s with the powers of *x*. Remember that the powers of *x* in the factored form will be half the highest power in the original trinomial.

517. $\left(x^{-3}+4\right)\left(x^{-3}+1\right)$

It's often helpful to picture the trinomial as a corresponding quadratic, such as y^2+5y+4. The last terms in the binomials must have a product of 4 and a sum of 5. Use 4 and 1. Start by factoring the quadratic.

$$=(y+4)(y+1)$$

And then replace the *y*'s with the powers of *x*. Remember that the powers of *x* in the factored form will be half the highest power in the original trinomial.

518. $(5x^{1/6}-1)(x^{1/6}-2)$

It's often helpful to picture the trinomial as a corresponding quadratic, such as $5y^2-11y+2$. The last terms in the binomials must have a product of 2 and the coefficients of the first terms must have a product of 5. The sum of the inner and outer terms must be –11. Put 5 and –2 as the outer terms. Start by factoring the quadratic.

$$=(5y-1)(y-2)$$

And then replace the y's with the powers of x. Remember that the powers of x in the factored form will be half the highest power in the original trinomial, and half of $\frac{1}{3}$ is $\frac{1}{6}$.

519. $(3x^{1/5}+4)(2x^{1/5}-3)$

It's often helpful to picture the trinomial as a corresponding quadratic, such as $6y^2-y-12$. The last terms in the binomials must have a product of –12 and the coefficients of the first terms must have a product of 6. The difference between the inner and outer terms must be –1. Use 3 and –3 for the outer terms, and use 4 and 2 for the inner terms. Start by factoring the quadratic.

$$=(3y+4)(2y-3)$$

And then replace the y's with the powers of x. Remember that the powers of x in the factored form will be half the highest power in the original trinomial.

520. $5(z+3)^2$

First factor out the GCF, 5.

$$=5(z^2+6z+9)$$

The trinomial in the parentheses is a perfect square.

$$=5(z+3)(z+3)=5(z+3)^2$$

521. $2x(3x+1)^2$

First factor out the GCF, $2x$.

$$=2x(9x^2+6x+1)$$

The trinomial in the parentheses is a perfect square.

$$=2x(3x+1)(3x+1)=2x(3x+1)^2$$

522. $4y(y-3)(y+1)$

First factor out the GCF, $4y$.

$$=4y(y^2-2y-3)$$

The trinomial in the parentheses factors into the product of two binomials.

$$=4y(y-3)(y+1)$$

523. $x^4(3x-1)(2x+1)$

First factor out the GCF, x^4.

$$= x^4\left(6x^2+x-1\right)$$

The trinomial in the parentheses factors into the product of two binomials.

$$= x^4(3x-1)(2x+1)$$

524. $x^3(x+4)^2$

First factor out the GCF, x^3.

$$= x^3\left(x^2+8x+16\right)$$

The trinomial in the parentheses is a perfect square.

$$= x^3(x+4)(x+4) = x^3(x+4)^2$$

525. $6y(4-y)^2$

First factor out the GCF, $6y$.

$$= 6y\left(16-8y+y^2\right)$$

The trinomial in the parentheses is a perfect square.

$$= 6y(4-y)(4-y) = 6y(4-y)^2$$

526. $(w+2)(w-2)(w+3)(w-3)$

First factor the *quadratic-like* expression into the product of two binomials.

$$= \left(w^2-4\right)\left(w^2-9\right)$$

Next factor the binomials by writing them as the sum and difference of square roots.

$$= (w+2)(w-2)(w+3)(w-3)$$

527. $(x-1)(x^2+x+1)(x-2)(x^2+2x+4)$

First factor the *quadratic-like* trinomial by writing it as the product of two binomials.

$$= \left(x^3-8\right)\left(x^3-1\right)$$

Next factor the binomials as the difference of perfect cubes.

$$= (x-1)\left(x^2+x+1\right)(x-2)\left(x^2+2x+4\right)$$

528. $\mathbf{5(x+3)^3(x+5)(x-2)}$

First factor out the GCF, $5(x+3)^3$.

$$=5(x+3)^3(x^2+3x-10)$$

Now factor the trinomial.

$$=5(x+3)^3(x+5)(x-2)$$

529. $\mathbf{2(x+1)(x-2)(2x+1)}$

First factor out the GCF, $2(x+1)$.

$$=2(x+1)(2x^2-3x-2)$$

Now factor the trinomial.

$$=2(x+1)(x-2)(2x+1)$$

530. $\mathbf{(x+9)(x-9)(a+11)(a+2)}$

First factor out the GCF, (x^2-81).

$$=(x^2-81)(a^2+13a+22)$$

Now factor the binomial, which is the difference of squares, and the trinomial.

$$=(x+9)(x-9)(a+11)(a+2)$$

531. $\mathbf{(x-1)(x^2+x+1)(2y-1)(2y+3)}$

First factor out the GCF, (x^3-1).

$$=(x^3-1)(4y^2+4y-3)$$

Now factor the binomial as the difference of squares and the trinomial as the product of two binomials.

$$=(x-1)(x^2+x+1)(2y-1)(2y+3)$$

532. $\mathbf{10(6y^{1/8}+5)(5y^{1/8}-3)}$

First factor out the GCF, 10.

$$=10(30y^{1/4}+7y^{1/8}-15)$$

Now factor the trinomial by writing it as the product of two binomials.

$$=10(6y^{1/8}+5)(5y^{1/8}-3)$$

Answers 501–600

533. $6(y^{1/2}-2)(y^{1/2}+1)$

First factor out the GCF, 6.

$$=6(y-y^{1/2}-2)$$

Now factor the trinomial by writing it as the product of two binomials. Remember that the power of the variables in the binomials is half the highest power of the original trinomial.

$$=6(y^{1/2}-2)(y^{1/2}+1)$$

534. $x^{-3}(1-5x)(3-4x)$

First factor out the GCF, x^{-3}. Remember that you factor out the *smallest* power of the variable.

$$=x^{-3}(3-19x^1+20x^2)$$

Now factor the trinomial in the parentheses.

$$=x^{-3}(1-5x)(3-4x)$$

535. $x^{-4}(4x-1)(3x+2)$

First factor out the GCF, x^{-4}. Remember that you factor out the *smallest* power of the variable.

$$=x^{-4}(12x^2+5x^1-2)$$

Now factor the trinomial in the parentheses.

$$=x^{-4}(4x-1)(3x+2)$$

536. $(c-3)(b+2)$

Factor b out of the first two terms, and factor 2 out of the second two terms.

$$=b(c-3)+2(c-3)$$

Now the two terms have the binomial GCF of $(c-3)$. Factor it out.

$$=(c-3)(b+2)$$

537. $(x-ab)(x+yz)$

Factor x out of the first two terms, and factor yz out of the second two terms.

$$=x(x-ab)+yz(x-ab)$$

Now the two terms have the binomial GCF of $(x-ab)$. Factor it out.

$$=(x-ab)(x+yz)$$

538. $(2x-3)(x^2+1)$

Factor x^2 out of the first two terms. The second two terms are *relatively prime*, so their GCF is 1.

$$=x^2(2x-3)+1(2x-3)$$

Now the two terms have the binomial GCF of $(2x-3)$. Factor it out.

$$=(2x-3)(x^2+1)$$

539. $(z^2+4)(2x-3)$

Factor $2x$ out of the first two terms, and factor -3 out of the second two terms.

$$=2x(z^2+4)-3(z^2+4)$$

Now the two terms have the binomial GCF of (z^2+4). Factor it out.

$$=(z^2+4)(2x-3)$$

540. $(n^{1/2}+2)(n-4)$

Factor n out of the first two terms, and factor -4 out of the second two terms.

$$=n(n^{1/2}+2)-4(n^{1/2}+2)$$

Now the two terms have the binomial GCF of $(n^{1/2}+2)$. Factor it out.

$$=(n^{1/2}+2)(n-4)$$

541. $(y^{1/2}-3)(y^2+2)$

Factor y^2 out of the first two terms, and factor 2 out of the second two terms.

$$=y^2(y^{1/2}-3)+2(y^{1/2}-3)$$

Now the two terms have the binomial GCF of $(y^{1/2}-3)$. Factor it out.

$$=(y^{1/2}-3)(y^2+2)$$

542. $(x-3)(4+y-z)$

The first two terms have a common factor of 4, the second two terms have a common factor of y, and the last two terms have a common factor of $-z$. Factor out the groupings.

$$=4(x-3)+y(x-3)-z(x-3).$$

Notice that $-z$ had to be factored out of the last grouping so that the signs of the terms in the parentheses would be the same as in the other two binomials.

Now the three terms have the binomial GCF of $(x-3)$. Factor that out.

$$=(x-3)(4+y-z)$$

543. $(k+4)(x+y+z)$

The first two terms have a common factor of x, the second two terms have a common factor of y, and the last two terms have a common factor of z. Factor out the groupings.

$$= x(k+4) + y(k+4) + z(k+4)$$

Now the three terms have the binomial GCF of $(k+4)$. Factor that out.

$$= (k+4)(x+y+z)$$

544. $(y^2+3)(x+3)(x-4)$

The first two terms have a GCF of x^2, the second two terms have a GCF of $-x$, and the last two terms have a common factor of -12. Notice that $-x$ and -12 had to be factored out of the second and last grouping so that the signs of the terms in the parentheses would be the same as in the first binomial.

$$= x^2(y^2+3) - x(y^2+3) - 12(y^2+3)$$

Now the three terms have the binomial GCF of (y^2+3). Factor that out.

$$= (y^2+3)(x^2-x-12)$$

The trinomial can be factored and written as the product of two binomials.

$$= (y^2+3)(x+3)(x-4)$$

545. $(x^2-2)(x+1)(2x+1)$

The first two terms have a GCF of $2x^2$, the second two terms have a GCF of $3x$, and the last two terms are relatively prime, so their common factor is 1.

$$= 2x^2(x^2-2) + 3x(x^2-2) + 1(x^2-2)$$

Now the three terms have the binomial GCF of (x^2-2). Factor that out.

$$= (x^2-2)(2x^2+3x+1)$$

The trinomial can be factored and written as the product of two binomials.

$$= (x^2-2)(x+1)(2x+1)$$

546. $(n+3)(m-5)(m+5)$

The first two terms have a GCF of m^2, and the last two terms have a common factor of -25.

$$= m^2(n+3) - 25(n+3)$$

Now the two terms have the binomial GCF of $(n+3)$. Factor that out.

$$= (n+3)(m^2-25)$$

The second binomial is the difference of squares.

$$= (n+3)(m-5)(m+5)$$

547. $(x+4)(2x-5)(2x+5)$

The first two terms have a GCF of $4x^2$, and the last two terms have a common factor of –25.

$$= 4x^2(x+4) - 25(x+4)$$

Now the two terms have the binomial GCF of $(x + 4)$. Factor that out.

$$= (x+4)(4x^2 - 25)$$

The second binomial is the difference of squares.

$$= (x+4)(2x-5)(2x+5)$$

548. $4x(x+7)(x-7)$

First factor out the GCF, $4x$.

$$= 4x(x^2 - 49)$$

The binomial is the difference of squares.

$$= 4x(x+7)(x-7)$$

549. $6x^2(x-2)(x^2+2x+4)$

First factor out the GCF, $6x^2$.

$$= 6x^2(x^3 - 8)$$

The binomial is the difference of cubes.

$$= 6x^2(x-2)(x^2+2x+4)$$

550. $(y+2)(y-2)(y-3)(y^2+3y+9)$

The first two terms have a common factor of y^3, and the last two terms have a common factor of –27.

$$= y^3(y^2-4) - 27(y^2-4)$$

Now the two terms have a common factor of $(y^2 - 4)$. Factor that out.

$$= (y^2-4)(y^3-27)$$

The first binomial is the difference of squares, and the second binomial is the difference of cubes.

$$= (y+2)(y-2)(y-3)(y^2+3y+9)$$

551. $x(x+2)(x-2)(x+3)(x-3)$

First factor out the GCF, x.

$$= x(x^4 - 13x^2 + 36)$$

Now factor the *quadratic-like* trinomial by writing it as the product of two binomials.

$$= x(x^2-4)(x^2-9)$$

Each of the binomials is the difference of perfect squares.

$$= x(x+2)(x-2)(x+3)(x-3)$$

552. $(4x+5)(4x-5)(x^3+3)$

Factor the *quadratic-like* trinomial by writing it as the product of two binomials.

$$= (16x^2-25)(x^2+3)$$

The first binomial is the difference of perfect squares.

$$= (4x+5)(4x-5)(x^3+3)$$

553. $4x^2(x-1)(x+1)(x^2+1)$

First factor out the GCF, $4x^2$.

$$= 4x^2(x^4-1)$$

The binomial is the difference of perfect squares.

$$= 4x^2(x^2-1)(x^2+1)$$

The left binomial is still the difference of squares.

$$= 4x^2(x-1)(x+1)(x^2+1)$$

554. $(z+3)(z^2-3z+9)(z-3)(z^2+3z+9)$

The binomial is the difference of perfect squares.

$$= (z^3+27)(z^3-27)$$

The first binomial is the sum of cubes, and the second is the difference of cubes.

$$= (z+3)(z^2-3z+9)(z-3)(z^2+3z+9)$$

555. $(y^4+1)(y^2+1)(y+1)(y-1)$

The binomial is the difference of squares.

$$= (y^4+1)(y^4-1)$$

The second binomial is still the difference of squares.

$$= (y^4+1)(y^2+1)(y^2-1)$$

And the last binomial is still the difference of squares.

$$= (y^4+1)(y^2+1)(y+1)(y-1)$$

556. $(b+1)(b-1)(4b+1)(16b^2-4b+1)$

The first two terms have a GCF of $64b^3$; the last two terms are *relatively prime*, so their GCF is 1.

$$=64b^3(b^2-1)+1(b^2-1)$$

Now the GCF of the two terms is the binomial (b^2-1). Factor it out.

$$=(b^2-1)(64b^3+1)$$

The first binomial is the difference of perfect squares, and the second is the sum of cubes.

$$=(b+1)(b-1)(4b+1)(16b^2-4b+1)$$

557. $(z+3)(z-3)(3z-2)(9z^2+6z+4)$

The first two terms have a GCF of $27z^3$, and the last two terms have a GCF of –8.

$$=27z^3(z^2-9)-8(z^2-9)$$

Now factor out the GCF, (z^2-9).

$$=(z^2-9)(27z^3-8)$$

The first binomial is the difference of squares, and the second binomial is the difference of cubes.

$$=(z+3)(z-3)(3z-2)(9z^2+6z+4)$$

558. $(z^2+1)(z+1)(z-1)(z^2+4)(z+2)(z-2)$

Factor the *quadratic-like* trinomial.

$$=(z^4-1)(z^4-16)$$

Both of the binomials are the difference of perfect squares.

$$=(z^2+1)(z^2-1)(z^2+4)(z^2-4)$$

And, finally, the second and last binomials are the difference of squares.

$$=(z^2+1)(z+1)(z-1)(z^2+4)(z+2)(z-2)$$

559. $x^3(x-1)^2$

First factor out the GCF, x^3.

$$=x^3(x^2-2x+1)$$

The trinomial is a perfect square.

$$=x^3(x-1)^2$$

560. $(x-2)^2(x+2)^2$

Factor the *quadratic-like* trinomial by writing it as the product of two binomials.

$$=(x^2-4)(x^2-4)$$

The binomials can now be factored as the difference of two squares.

$$=(x-2)(x+2)(x-2)(x+2)$$
$$=(x-2)^2(x+2)^2$$

561. $y^{-6}(y-3)(y^2+3y+9)$

First factor out the GCF, y^{-6}.

$$=y^{-6}(y^3-27)$$

The binomial is the difference of cubes.

$$=y^{-6}(y-3)(y^2+3y+9)$$

562. $(x-1)(x+1)(3x+4)^2[x^2+3x+3]$

First factor the GCF, $(x^2-1)(3x+4)^2$, from the two terms.

$$=(x^2-1)(3x+4)^2[(x^2-1)+(3x+4)]$$

Next, simplify the expression in the brackets.

$$=(x^2-1)(3x+4)^2[x^2+3x+3]$$

Finally, factor the first binomial as the difference of two squares.

$$=(x-1)(x+1)(3x+4)^2[x^2+3x+3]$$

563. $[(y+2)(y^2-2y+4)]^3[(y-3)(y+3)][y^3-y^2+17]$

First, factor out the GCF, $(y^3+8)^3(y^2-9)$.

$$=(y^3+8)^3(y^2-9)[(y^3+8)-(y^2-9)]$$

Now simplify the terms in the brackets.

$$=(y^3+8)^3(y^2-9)[y^3-y^2+17]$$

Finally, factor the first binomial using the sum of cubes and the second binomial using the difference of squares.

$$=[(y+2)(y^2-2y+4)]^3[(y-3)(y+3)][y^3-y^2+17]$$

564. $(z+1)^{1/2}(z-1)(z^2+z+1)[z^3-z-2]$

First, factor out the GCF, $(z+1)^{1/2}(z^3-1)$.

$$=(z+1)^{1/2}(z^3-1)[(z^3-1)-(z+1)]$$

Simplify the terms in the brackets.

$$=(z+1)^{1/2}(z^3-1)[z^3-z-2]$$

Now factor the second binomial using the difference of cubes.

$$=(z+1)^{1/2}(z-1)(z^2+z+1)[z^3-z-2]$$

565. $x[(2x-5)(2x+5)(x+5)(x^2-5x+25)]$

First factor x from each term.

$$=x(4x^5-25x^3+500x^2-3125)$$

Next, use grouping and factor x^3 from the first two terms in the parentheses and 125 from the last two terms.

$$=x[x^3(4x^2-25)+125(4x^2-25)]$$

Now factor the common factor $(4x^2-25)$ from the two terms in the brackets.

$$=x[(4x^2-25)(x^3+125)]$$

Finally, factor the difference of squares and sum of cubes in the brackets.

$$=x[(2x-5)(2x+5)(x+5)(x^2-5x+25)]$$

566. 8

Add –7 to each side of the equation.

$$\begin{array}{rcrcr} x & + & 7 & = & 15 \\ & & -7 & & -7 \\ \hline x & & & = & 8 \end{array}$$

567. 13

Add 3 to each side of the equation.

$$\begin{array}{rcrcr} y & - & 3 & = & 10 \\ & & 3 & & 3 \\ \hline y & & & = & 13 \end{array}$$

568. –12

Add –14 to each side of the equation.

$$\begin{array}{rcrcr} z & + & 14 & = & 2 \\ & & -14 & & -14 \\ \hline z & & & = & -12 \end{array}$$

569. –1

Add 3 to each side of the equation.

$$\begin{array}{rcrcr} x & - & 3 & = & -4 \\ & & 3 & & 3 \\ \hline x & & & = & -1 \end{array}$$

570. –2

First add $-3x$ to each side of the equation.

$$\begin{array}{rcrcrcr} 5x & + & 3 & = & 3x & - & 1 \\ -3x & & & & -3x & & \\ \hline 2x & + & 3 & = & & - & 1 \end{array}$$

Next, add –3 to each side of the equation.

$$\begin{array}{rcrcr} 2x & + & 3 & = & -1 \\ & & -3 & & -3 \\ \hline 2x & & & = & -4 \end{array}$$

Now divide each side of the equation by 2.

$$\frac{2x}{2} = \frac{-4}{2}$$

$$x = -2$$

571. 8

First add $-3x$ to each side of the equation.

$$\begin{array}{rcrcrcr} 3x & + & 9 & = & 5x & - & 7 \\ -3x & & & & -3x & & \\ \hline & & 9 & = & 2x & - & 7 \end{array}$$

Next, add 7 to each side of the equation.

$$\begin{array}{rcrcr} 9 & = & 2x & - & 7 \\ 7 & = & & & 7 \\ \hline 16 & = & 2x & & \end{array}$$

Now divide each side of the equation by 2.

$$\frac{16}{2} = \frac{2x}{2}$$

$$8 = x$$

572. 1

First combine the like terms on each side of the equation.

$$9y + 3 = 7y + 5$$

Now add $-7y$ to each side of the equation.

$$\begin{array}{rcrcrcr} 9y & + & 3 & = & 7y & + & 5 \\ -7y & & & & -7y & & \\ \hline 2y & + & 3 & = & & + & 5 \end{array}$$

Next add –3 to each side of the equation.

$$\begin{array}{rcl} 2y + 3 & = & 5 \\ -3 & = & -3 \\ \hline 2y & = & 2 \end{array}$$

And now divide each side by 2.

$$\frac{2y}{2} = \frac{2}{2}$$
$$y = 1$$

573. 1

First combine the like terms on the left side of the equation.

$$2z + 6 = 12 - 4z$$

Now add $4z$ to each side of the equation.

$$\begin{array}{rcl} 2z + 6 & = & 12 - 4z \\ 4z & & \quad\;\; 4z \\ \hline 6z + 6 & = & 12 \end{array}$$

Now add –6 to each side of the equation.

$$\begin{array}{rcl} 6z + 6 & = & 12 \\ -6 & = & -6 \\ \hline 6z & = & 6 \end{array}$$

Finally, divide each side by 6.

$$\frac{6z}{6} = \frac{6}{6}$$
$$z = 1$$

574. –3

Divide each side of the equation by –8.

$$\frac{-8x}{-8} = \frac{24}{-8}$$
$$x = -3$$

575. $\frac{1}{2}$

Divide each side of the equation by –4.

$$\frac{-4x}{-4} = \frac{-2}{-4}$$
$$x = \frac{1}{2}$$

576. –4

First combine the terms on the left.

$$-5y = 20$$

Now divide each side of the equation by –5.

$$\frac{-5y}{-5} = \frac{20}{-5}$$
$$y = -4$$

577. –4

Divide each side of the equation by 3.

$$\frac{-12}{3} = \frac{3z}{3}$$
$$-4 = z$$

578. 5

Add 4 to each side of the equation.

$$\begin{array}{rcl} 3x - 4 & = & 11 \\ 4 & & 4 \\ \hline 3x & = & 15 \end{array}$$

Now divide each side of the equation by 3.

$$\frac{3x}{3} = \frac{15}{3}$$
$$x = 5$$

579. –1

Add –7 to each side of the equation.

$$\begin{array}{rcl} 5x + 7 & = & 2 \\ -7 & & -7 \\ \hline 5x & = & -5 \end{array}$$

Now divide each side of the equation by 5.

$$\frac{5x}{5} = \frac{-5}{5}$$
$$x = -1$$

580. –4

Add –7*x* to each side of the equation.

$$\begin{array}{rcl} 4x - 3 & = & 7x + 9 \\ -7x & & -7x \\ \hline -3x - 3 & = & 9 \end{array}$$

Next add 3 to each side of the equation.

$$\begin{array}{rcl} -3x - 3 & = & 9 \\ 3 & & 3 \\ \hline -3x & = & 12 \end{array}$$

And now divide each side by –3.

$$\frac{-3x}{-3} = \frac{12}{-3}$$
$$x = -4$$

Another choice would be to add $-4x$ to each side of the original equation in order to keep a positive sign on the coefficient. The answer comes out the same, of course.

581. **–15**

Add $-3x$ to each side of the equation.

$$\begin{array}{rcrcl} 4x + 7 & = & 3x & - & 8 \\ -3x & & -3x & & \\ \hline x + 7 & = & & - & 8 \end{array}$$

Now add –7 to each side of the equation.

$$\begin{array}{rcl} x + 7 & = & -8 \\ -7 & & -7 \\ \hline x & = & -15 \end{array}$$

582. $-\frac{9}{2}$

First combine the like terms on the left side of the equation.

$$5y + 14 = 5 + 3y$$

Now add $-3y$ to each side.

$$\begin{array}{rcl} 5y + 14 & = & 5 + 3y \\ -3y & & -3y \\ \hline 2y + 14 & = & 5 \end{array}$$

Add –14 to each side.

$$\begin{array}{rcl} 2y + 14 & = & 5 \\ -14 & & -14 \\ \hline 2y & = & -9 \end{array}$$

And now divide each side by 2.

$$\frac{2y}{2} = \frac{-9}{2}$$
$$y = -\frac{9}{2}$$

583. **–1**

First combine the like terms on the left side of the equation.

$$4x + 4 = 3 + 3x$$

Now add $-3x$ to each side.

$$\begin{array}{rcl} 4x + 4 & = & 3 + 3x \\ -3x & & -3x \\ \hline x + 4 & = & 3 \end{array}$$

Answers 501–600

Add –4 to each side.

$$\begin{array}{rcl} x + 4 & = & 3 \\ -4 & & -4 \\ \hline x & = & -1 \end{array}$$

584. 27

Multiply each side of the equation by 3.

$$\begin{aligned} \cancel{3}\cdot\frac{z}{\cancel{3}} &= 9\cdot 3 \\ z &= 27 \end{aligned}$$

585. –72

Multiply each side of the equation by 6.

$$\begin{aligned} \cancel{6}\cdot\frac{w}{\cancel{6}} &= -12\cdot 6 \\ w &= -72 \end{aligned}$$

586. 32

Multiply each side of the equation by –4.

$$\begin{aligned} \cancel{-4}\cdot\frac{w}{\cancel{-4}} &= -8(-4) \\ w &= 32 \end{aligned}$$

587. –50

Multiply each side of the equation by –5.

$$\begin{aligned} \cancel{-5}\cdot\frac{x}{\cancel{-5}} &= 10(-5) \\ x &= -50 \end{aligned}$$

588. –10

First add –6 to each side of the equation.

$$\begin{array}{rcl} 6 - \frac{y}{2} & = & 11 \\ -6 & & -6 \\ \hline -\frac{y}{2} & = & 5 \end{array}$$

Now multiply each side of the equation by –2.

$$\begin{aligned} \cancel{-2}\left(-\frac{y}{\cancel{2}}\right) &= 5(-2) \\ x &= -10 \end{aligned}$$

589. –6

First add –7 to each side of the equation.

$$\begin{aligned} \frac{x}{3} + 7 &= 5 \\ -7 \quad &\quad -7 \\ \hline \frac{x}{3} &= -2 \end{aligned}$$

Now multiply each side of the equation by 3.

$$\begin{aligned} \cancel{3} \cdot \frac{x}{\cancel{3}} &= 3(-2) \\ x &= -6 \end{aligned}$$

590. –16

First add 3 to each side of the equation.

$$\begin{aligned} \frac{z}{4} - 3 &= z + 9 \\ 3 \quad &\quad 3 \\ \hline \frac{z}{4} &= z + 12 \end{aligned}$$

Now add $-z$ to each side of the equation. Remember that $1z$ can be written as $\frac{4}{4}z$.

$$\begin{aligned} \frac{z}{4} &= z + 12 \\ -z \quad &\quad -z \\ \hline -\frac{3}{4}z &= 12 \end{aligned}$$

Now multiply each side of the equation by the reciprocal of $-\frac{3}{4}$, which is $-\frac{4}{3}$.

$$\begin{aligned} \left(-\frac{\cancel{4}}{\cancel{3}}\right)\left(-\frac{\cancel{3}}{\cancel{4}}\right)z &= {}^{4}\cancel{12}\left(-\frac{4}{\cancel{3}}\right) \\ z &= -16 \end{aligned}$$

591. 6

First add –4 to each side of the equation.

$$\begin{aligned} 4 + \frac{w}{2} &= 13 - w \\ -4 \quad &\quad -4 \\ \hline \frac{w}{2} &= 9 - w \end{aligned}$$

Now add w to each side of the equation. Remember that $1w$ can be written as $\frac{2}{2}w$.

$$\begin{aligned} \frac{w}{2} &= 9 - w \\ w \quad &\quad w \\ \hline \frac{3}{2}w &= 9 \end{aligned}$$

Now multiply each side of the equation by the reciprocal of $\frac{3}{2}$, which is $\frac{2}{3}$.

$$\left(\frac{2}{\cancel{3}}\right)\left(\frac{\cancel{3}}{\cancel{2}}\right)w = {}^{3}\cancel{9}\left(\frac{2}{\cancel{3}}\right)$$
$$w = 6$$

592. −12

You could add 4 to each side to start, but this time a good choice would be to multiply every term in the equation by 4 to eliminate the fractions.

$$\cancel{4}\left(\frac{5}{\cancel{4}}x\right) - 4\cdot 4 = \cancel{4}\left(\frac{9}{\cancel{4}}x\right) + 4\cdot 8$$
$$5x - 16 = 9x + 32$$

Now add 16 to each side of the equation.

$$\begin{array}{rcl} 5x - 16 & = & 9x + 32 \\ 16 & & 16 \\ \hline 5x & = & 9x + 48 \end{array}$$

Now add $-9x$ to each side.

$$\begin{array}{rcl} 5x & = & 9x + 48 \\ -9x & & -9x \\ \hline -4x & = & 48 \end{array}$$

Now divide each side by −4.

$$\frac{\cancel{-4}x}{\cancel{-4}} = \frac{48}{-4}$$
$$x = -12$$

593. $-\frac{1}{5}$

You could add −1 to each side to start, but this time a good choice would be to multiply every term in the equation by 2 to eliminate the fractions.

$$\cancel{2}\left(\frac{5}{\cancel{2}}x\right) + 2\cdot 1 = 2\cdot 2 + \cancel{2}\left(\frac{15}{\cancel{2}}x\right)$$
$$5x + 2 = 4 + 15x$$

Now add −2 to each side of the equation.

$$\begin{array}{rcl} 5x + 2 & = & 4 + 15x \\ -2 & & -2 \\ \hline 5x & = & 2 + 15x \end{array}$$

Now add $-15x$ to each side.

$$\begin{array}{rcl} 5x & = & 2 + 15x \\ -15x & & -15x \\ \hline -10x & = & 2 \end{array}$$

Now divide each side by −10.

$$\frac{\cancel{-10}x}{\cancel{-10}} = \frac{2}{-10}$$
$$x = -\frac{2}{10} = -\frac{1}{5}$$

594. -7

Multiply each side of the equation by 4. Be sure to distribute the 4 over both terms on the right.

$$\cancel{4}\left(\frac{3x+1}{\cancel{4}}\right)=4(x+2)$$
$$3x+1=4x+8$$

Now add $-3x$ to each side of the equation.

$$\begin{array}{rcrcl} 3x+1 & = & 4x & + & 8 \\ -3x & & -3x & & \\ \hline 1 & = & x & + & 8 \end{array}$$

And now add –8 to each side.

$$\begin{array}{rcrcr} 1 & = & x & + & 8 \\ -8 & & & & -8 \\ \hline -7 & = & x & & \end{array}$$

595. 6

First multiply each side of the equation by 4. Be sure to distribute the 4 over both terms on the left.

$$4(w-2)=\cancel{4}\left(\frac{3w-2}{\cancel{4}}\right)$$
$$4w-8=3w-2$$

Add 8 to each side of the equation.

$$\begin{array}{rcrcr} 4w-8 & = & 3w & - & 2 \\ 8 & & & & 8 \\ \hline 4w & = & 3w & + & 6 \end{array}$$

Now add $-3w$ to each side.

$$\begin{array}{rcrcr} 4w & = & 3w & + & 6 \\ -3w & & -3w & & \\ \hline w & = & & & 6 \end{array}$$

596. 21

First add 1 to each side of the equation.

$$\begin{array}{rcr} \frac{x-3}{6}-1 & = & 2 \\ 1 & & 1 \\ \hline \frac{x-3}{6} & = & 3 \end{array}$$

Multiply each side of the equation by 6.

$$\cancel{6}\left(\frac{x-3}{\cancel{6}}\right)=6\cdot 3$$
$$x-3=18$$

Now add 3 to each side.

$$\begin{array}{rcl} x - 3 &=& 18 \\ 3 && 3 \\ \hline x &=& 21 \end{array}$$

597. -5

First add –5 to each side of the equation.

$$\begin{array}{rcl} \frac{2z+1}{3} + 5 &=& 2 \\ -5 && -5 \\ \hline \frac{2z+1}{3} &=& -3 \end{array}$$

Multiply each side of the equation by 3.

$$\cancel{3}\left(\frac{2z+1}{\cancel{3}}\right) = 3(-3)$$

$$2z+1 = -9$$

Now add –1 to each side.

$$\begin{array}{rcl} 2z + 1 &=& -9 \\ -1 && -1 \\ \hline 2z &=& -10 \end{array}$$

Finally, divide each side by 2.

$$\frac{\cancel{2}z}{\cancel{2}} = \frac{-10}{2}$$

$$z = -5$$

598. 15

Multiply each side (each term) of the equation by 15, the LCM of 3 and 5, to eliminate the fractions. Then simplify and combine like terms.

$$15\left(\frac{4x}{3} - \frac{3x}{5}\right) = 15(11)$$

$$^{5}\cancel{15}\left(\frac{4x}{\cancel{3}}\right) - {}^{3}\cancel{15}\left(\frac{3x}{\cancel{5}}\right) = 165$$

$$5(4x) - 3(3x) = 165$$

$$20x - 9x = 165$$

$$11x = 165$$

Now divide each side of the equation by 11.

$$\frac{\cancel{11}x}{\cancel{11}} = \frac{165}{11}$$

$$x = 15$$

599. –40

First add –3 to each side of the equation.

$$\begin{array}{rcrcl} \frac{x}{5} & + & 6 & = & 3 & + & \frac{x}{8} \\ & & -3 & & -3 & & \\ \hline \frac{x}{5} & + & 3 & = & & & \frac{x}{8} \end{array}$$

Now multiply each side (all three terms) by 40, the LCM of the denominators, to eliminate the fractions. This could have been done before adding the –3 to each side, but removing the extra term early simplifies the multiplications later.

$$\overset{8}{\cancel{40}}\left(\frac{x}{\cancel{5}}\right)+40\cdot 3=\overset{5}{\cancel{40}}\left(\frac{x}{\cancel{8}}\right)$$

$$8x+120=5x$$

Add $-8x$ to each side.

$$\begin{array}{rcrcl} 8x & + & 120 & = & 5x \\ -8x & & & & -8x \\ \hline & & 120 & = & -3x \end{array}$$

And now divide each side by –3.

$$\frac{\overset{40}{\cancel{120}}}{\cancel{-3}}=\frac{\cancel{-3}x}{\cancel{-3}}$$

$$-40=x$$

600. 9

First add 1 to each side of the equation.

$$\begin{array}{rcrcl} \frac{2(y+3)}{3} & - & 1 & = & \frac{5(y-2)}{7} & + & 2 \\ & & 1 & & & & 1 \\ \hline \frac{2(y+3)}{3} & & & = & \frac{5(y-2)}{7} & + & 3 \end{array}$$

Now multiply each term (both sides of the equation) by 21, the LCM of the denominators, to eliminate the fractions. This could have been done before adding the 1 to each side, but removing the extra term early simplifies the multiplications later. After multiplying by 21, simplify and combine like terms.

$$\overset{7}{\cancel{21}}\left(\frac{2(y+3)}{\cancel{3}}\right)=\overset{3}{\cancel{21}}\left(\frac{5(y-2)}{\cancel{7}}\right)+21\cdot 3$$

$$14(y+3)=15(y-2)+63$$

$$14y+42=15y-30+63$$

$$14y+42=15y+33$$

Now add –33 to each side.

$$\begin{array}{rcrcl} 14y & + & 42 & = & 15y & + & 33 \\ & & -33 & & & & -33 \\ \hline 14y & + & 9 & = & 15y & & \end{array}$$

Finally, add $-14y$ to each side.

$$\begin{array}{rcl} 14y + 9 &=& 15y \\ -14y &=& -14y \\ \hline 9 &=& y \end{array}$$

601. 9

First add -3 to each side of the equation.

$$\begin{array}{rcl} \frac{4(x-2)}{7} + 3 &=& \frac{x+3}{6} + 5 \\ -3 && -3 \\ \hline \frac{4(x-2)}{7} &=& \frac{x+3}{6} + 2 \end{array}$$

Now multiply each term (both sides of the equation) by 42, the LCM of the denominators, to eliminate the fractions. This could have been done before adding the -3 to each side, but removing the extra term early simplifies the multiplications later. After multiplying by 42, simplify and add like terms.

$$^{6}\cancel{42}\left(\frac{4(x-2)}{\cancel{7}}\right) = {}^{7}\cancel{42}\left(\frac{x+3}{\cancel{6}}\right) + 42 \cdot 2$$

$$24(x-2) = 7(x+3) + 84$$

$$24x - 48 = 7x + 21 + 84$$

$$24x - 48 = 7x + 105$$

Now add 48 to each side, and then add $-7x$ to each side.

$$\begin{array}{rcl} 24x - 48 &=& 7x + 105 \\ 48 && 48 \\ \hline 24x &=& 7x + 153 \end{array}$$

$$\begin{array}{rcl} 24x &=& 7x + 153 \\ -7x && -7x \\ \hline 17x &=& 153 \end{array}$$

Now divide each side by 17.

$$\frac{\cancel{17}x}{\cancel{17}} = \frac{153}{17}$$

$$x = 9$$

602. 20

Multiply each side of the equation by 20, the LCM of the three denominators, to eliminate the fractions. After multiplying by 20, simplify and add like terms.

$$20\left(\frac{x}{5} + \frac{x}{4} + \frac{x}{2}\right) = 20 \cdot 19$$

$$^{4}\cancel{20}\left(\frac{x}{\cancel{5}}\right) + {}^{5}\cancel{20}\left(\frac{x}{\cancel{4}}\right) + {}^{10}\cancel{20}\left(\frac{x}{\cancel{2}}\right) = 380$$

$$4x + 5x + 10x = 380$$

$$19x = 380$$

Now divide each side by 19.

$$\frac{\cancel{19}x}{\cancel{19}} = \frac{380}{19}$$

$$x = 20$$

603. 60

Multiply each side of the equation by 60, the LCM of the three denominators, to eliminate the fractions. After multiplying by 60, simplify and combine like terms.

$$60\left(\frac{y}{4} + \frac{y}{5} - \frac{y}{6}\right) = 60 \cdot 17$$

$$^{15}\cancel{60}\left(\frac{y}{\cancel{4}}\right) + {}^{12}\cancel{60}\left(\frac{y}{\cancel{5}}\right) - {}^{10}\cancel{60}\left(\frac{y}{\cancel{6}}\right) = 1020$$

$$15y + 12y - 10y = 1020$$

$$17y = 1020$$

Now divide both sides by 17.

$$\frac{\cancel{17}y}{\cancel{17}} = \frac{1020}{17}$$

$$y = 60$$

604. 3

First add the two fractions on the left. They have the same denominator, so you just find the sum of the numerators.

$$\frac{27}{y} = 9$$

Now multiply each side by y, and then divide each side by 9.

$$\cancel{y} \cdot \frac{27}{\cancel{y}} = 9 \cdot y$$

$$27 = 9y$$

$$\frac{27}{9} = \frac{\cancel{9}y}{\cancel{9}}$$

$$3 = y$$

605. 5

First subtract the two fractions on the left. They have the same denominator, so you just find the difference of the numerators.

$$\frac{5}{x} = 1$$

Now multiply each side by x.

$$\cancel{x} \cdot \frac{5}{\cancel{x}} = 1 \cdot x$$

$$5 = x$$

606. ± 5

Find the square root of both sides.

$$\sqrt{x^2} = \pm\sqrt{25}$$
$$x = \pm 5$$

Technically, you should put the ± in front of both radicals, but that would just give you four different equations with the two answers each repeated. Just don't forget to use the ± on one side.

607. ± 11

Find the square root of both sides.

$$\sqrt{x^2} = \pm\sqrt{121}$$
$$x = \pm 11$$

Technically, you should put the ± in front of both radicals, but that would just give you four different equations with the two answers each repeated. Just don't forget to use the ± on one side.

608. ± 3

First divide both sides of the equation by 3.

$$\frac{\cancel{3}y^2}{\cancel{3}} = \frac{27}{3}$$
$$y^2 = 9$$

Now find the square root of both sides.

$$\sqrt{y^2} = \pm\sqrt{9}$$
$$y = \pm 3$$

609. ± 4

First divide both sides of the equation by 5.

$$\frac{\cancel{5}z^2}{\cancel{5}} = \frac{80}{5}$$
$$z^2 = 16$$

Now find the square root of both sides.

$$\sqrt{z^2} = \pm\sqrt{16}$$
$$z = \pm 4$$

610. ± 10

First add 100 to each side of the equation.

$$\begin{array}{rcrcr} n^2 & - & 100 & = & 0 \\ & & 100 & & 100 \\ \hline n^2 & & & = & 100 \end{array}$$

Now find the square root of both sides.

$$\sqrt{n^2} = \pm\sqrt{100}$$
$$n = \pm 10$$

611. ± 1

First add 1 to each side of the equation.

$$\begin{array}{rcrcr} m^2 & - & 1 & = & 0 \\ & & 1 & & 1 \\ \hline m^2 & & & = & 1 \end{array}$$

Now find the square root of both sides.

$$\sqrt{m^2} = \pm\sqrt{1}$$
$$m = \pm 1$$

612. $\pm\frac{3}{2}$

First add 9 to each side of the equation.

$$\begin{array}{rcrcr} 4x^2 & - & 9 & = & 0 \\ & & 9 & & 9 \\ \hline 4x^2 & & & = & 9 \end{array}$$

Next divide both sides of the equation by 4.

$$\frac{\cancel{4}x^2}{\cancel{4}} = \frac{9}{4}$$
$$x^2 = \frac{9}{4}$$

Now find the square root of both sides.

$$\sqrt{x^2} = \pm\sqrt{\frac{9}{4}} = \pm\frac{\sqrt{9}}{\sqrt{4}}$$
$$x = \pm\frac{3}{2}$$

613. $\pm\frac{5}{2}$

First add 150 to each side of the equation.

$$\begin{array}{rcrcr} 24x^2 & - & 150 & = & 0 \\ & & 150 & & 150 \\ \hline 24x^2 & & & = & 150 \end{array}$$

Next divide both sides of the equation by 24 and reduce the fraction.

$$\frac{24x^2}{24} = \frac{150}{24}$$

$$x^2 = \frac{150}{24} = \frac{25}{4}$$

Now find the square root of both sides.

$$\sqrt{x^2} = \pm\sqrt{\frac{25}{4}} = \pm\frac{\sqrt{25}}{\sqrt{4}}$$

$$x = \pm\frac{5}{2}$$

614. **–3 or 5**

Factor the quadratic into the product of two binomials.

$$(x + 3)(x - 5) = 0$$

Using the *multiplication property of zero,* set each binomial equal to 0 and solve for x.

$$x + 3 = 0, x = -3$$

$$x - 5 = 0, x = 5$$

615. **–4 or –11**

Factor the quadratic into the product of two binomials.

$$(y + 4)(y + 11) = 0$$

Using the *multiplication property of zero,* set each binomial equal to 0 and solve for y.

$$y + 4 = 0, y = -4$$

$$y + 11 = 0, y = -11$$

616. $\frac{3}{2}$ **or –2**

Factor the quadratic into the product of two binomials.

$$(2x - 3)(x + 2) = 0$$

Using the *multiplication property of zero,* set each binomial equal to 0 and solve for x.

$$2x - 3 = 0, x = \frac{3}{2}$$

$$x + 2 = 0, x = -2$$

617. $\frac{5}{3}$ **or 1**

Factor the quadratic into the product of two binomials.

$$(3x - 5)(x - 1) = 0$$

Using the *multiplication property of zero,* set each binomial equal to 0 and solve for x.

$$3x - 5 = 0, x = \frac{5}{3}$$

$$x - 1 = 0, x = 1$$

618. **3 or 0**

Factor the quadratic by factoring out the common factor, y.

$y(y - 3) = 0$

Using the *multiplication property of zero,* set each factor equal to 0 and solve for y.

$y = 0$

$y - 3 = 0, y = 3$

619. **7 or 0**

First set the equation equal to 0 by adding $-7z$ to each side.

$z^2 - 7z = 0$

Factor the quadratic by factoring out the common factor, z.

$z(z - 7) = 0$

Using the *multiplication property of zero,* set each factor equal to 0 and solve for z.

$z = 0$

$z - 7 = 0, z = 7$

620. $-\frac{1}{2}$ **or 0**

Factor the quadratic by factoring out the common factor, x.

$x(2x + 1) = 0$

Using the *multiplication property of zero,* set each factor equal to 0 and solve for x. $x = 0$.

$2x + 1 = 0, x = -\frac{1}{2}$

621. **0 or** $\frac{2}{3}$

First set the equation equal to 0 by adding $-2y$ to each side.

$3y^2 - 2y = 0$

Factor the quadratic by factoring out the common factor, y.

$y(3y - 2) = 0$

Using the *multiplication property of zero,* set each factor equal to 0 and solve for y.

$y = 0$

$3y - 2 = 0, y = \frac{2}{3}$

622. $\frac{3}{2}$ **or** $-\frac{3}{4}$

Factor the quadratic into the product of two binomials.

$(2x - 3)(4x + 3) = 0$

Using the *multiplication property of zero,* set each binomial equal to 0 and solve for x.

$2x - 3 = 0, x = \frac{3}{2}$

$4x + 3 = 0, x = -\frac{3}{4}$

623. $-\frac{2}{5}$ **or** $-\frac{5}{2}$

Factor the quadratic into the product of two binomials.

$(5x + 2)(2x + 5) = 0$

Using the *multiplication property of zero,* set each binomial equal to 0 and solve for x.

$5x + 2 = 0, x = -\frac{2}{5}$

$2x + 5 = 0, x = -\frac{5}{2}$

624. $\frac{1}{4}$ **or** $-\frac{1}{2}$

First factor out the GCF, 2.

$2(8x^2 + 2x - 1) = 0$

Now factor the quadratic into the product of two binomials.

$2(4x - 1)(2x + 1) = 0$

Using the *multiplication property of zero,* set each binomial equal to 0 and solve for x.

$4x - 1 = 0, x = \frac{1}{4}$

$2x + 1 = 0, x = -\frac{1}{2}$

You don't set the factor 2 equal to 0, because there is no variable in the factor, and $2 \neq 0$.

625. **–1 or** $\frac{5}{2}$

First factor out the GCF, 3.

$3(2x^2 - 3x - 5) = 0$

Now factor the quadratic into the product of two binomials.

$3(x + 1)(2x - 5) = 0$

Using the *multiplication property of zero,* set each binomial equal to 0 and solve for x.

$x + 1 = 0, x = -1$

$2x - 5 = 0, x = \frac{5}{2}$

You don't set the factor 3 equal to 0, because there is no variable in the factor, and $3 \neq 0$.

626. **–1 or 3**

First multiply each term by 2; this eliminates the fractions.

$$2\left(\frac{1}{2}x^2 - x - \frac{3}{2}\right) = 2 \cdot 0$$

$$x^2 - 2x - 3 = 0$$

Now factor the quadratic into the product of two binomials.

$$(x+1)(x-3)=0$$

Using the *multiplication property of zero,* set each binomial equal to 0 and solve for x.

$$x+1=0,\ x=-1$$

$$x-3=0,\ x=3$$

627. $\frac{1}{2}$ or $\frac{4}{3}$

First multiply each term by 6; this eliminates the fractions.

$$6\left(x^2-\frac{11}{6}x+\frac{2}{3}\right)=6\cdot 0$$

$$6x^2-11x+4=0$$

Now factor the quadratic into the product of two binomials.

$$(2x-1)(3x-4)=0$$

Using the *multiplication property of zero,* set each binomial equal to 0 and solve for x.

$$2x-1=0,\ x=\frac{1}{2}$$

$$3x-4=0,\ x=\frac{4}{3}$$

628. 0 or $\frac{3}{2}$

First set the equation equal to 0 by adding $-\frac{3}{2}y$ to each side.

$$y^2-\frac{3}{2}y=0$$

Factor the quadratic by factoring out the common factor, y: $y\left(y-\frac{3}{2}\right)=0$.

Using the *multiplication property of zero,* set each factor equal to 0. Solve for y.

$$y=0$$

$$y-\frac{3}{2}=0,\ y=\frac{3}{2}$$

629. 0 or 16

First set the equation equal to 0 by adding $-4x$ to each side.

$$\frac{1}{4}x^2-4x=0$$

Next, multiply both sides of the equation by 4.

$$4\left(\frac{1}{4}x^2-4x\right)=4\cdot 0$$

$$x^2-16x=0$$

Factor the quadratic by factoring out the common factor, x.

$$x(x-16)=0$$

Using the *multiplication property of zero,* set each factor equal to 0 and solve for x.

$x = 0$

$x - 16 = 0, x = 16$

630. **1 or –4**

Using the standard form of the quadratic equation, $ax^2 + bx + c = 0$, identify a, b, and c.

$a = 1, b = 3, c = -4$

Substitute these values into the quadratic formula and simplify.

$$x = \frac{-b \pm \sqrt{b^2 - 4ac}}{2a}$$

$$x = \frac{-3 \pm \sqrt{3^2 - 4(1)(-4)}}{2(1)}$$

$$= \frac{-3 \pm \sqrt{9 - (-16)}}{2}$$

$$= \frac{-3 \pm \sqrt{9 + 16}}{2}$$

$$= \frac{-3 \pm \sqrt{25}}{2}$$

$$= \frac{-3 \pm 5}{2}$$

So $x = \frac{-3+5}{2} = \frac{2}{2} = 1$

or $x = \frac{-3-5}{2} = \frac{-8}{2} = -4$

You could also have solved this problem using factoring.

631. **6 or 2**

Using the standard form of the quadratic equation, $ax^2 + bx + c = 0$, identify a, b, and c.

$a = 1, b = -8, c = 12$

Substitute these values into the quadratic formula and simplify.

$$x = \frac{-b \pm \sqrt{b^2 - 4ac}}{2a}$$

$$x = \frac{-(-8) \pm \sqrt{(-8)^2 - 4(1)(12)}}{2(1)}$$

$$= \frac{8 \pm \sqrt{64 - 48}}{2}$$

$$= \frac{8 \pm \sqrt{16}}{2}$$

$$= \frac{8 \pm 4}{2}$$

So $x = \frac{8+4}{2} = \frac{12}{2} = 6$

or $x = \frac{8-4}{2} = \frac{4}{2} = 2$

You could also have solved this problem using factoring.

632. $\frac{3}{2}$ **or** -2

Using the standard form of the quadratic equation, $ax^2 + bx + c = 0$, identify a, b, and c.

$a = 2, b = 1, c = -6$

Substitute these values into the quadratic formula and simplify.

$$x = \frac{-b \pm \sqrt{b^2 - 4ac}}{2a}$$

$$x = \frac{-1 \pm \sqrt{1^2 - 4(2)(-6)}}{2(2)}$$

$$= \frac{-1 \pm \sqrt{1-(-48)}}{4}$$

$$= \frac{-1 \pm \sqrt{1+48}}{4}$$

$$= \frac{-1 \pm \sqrt{49}}{4}$$

$$= \frac{-1 \pm 7}{4}$$

So $x = \frac{-1+7}{4} = \frac{6}{4} = \frac{3}{2}$

or $x = \frac{-1-7}{4} = \frac{-8}{4} = -2$

You could also have solved this problem using factoring.

633. $-\frac{1}{2}$ **or** $-\frac{4}{5}$

Using the standard form of the quadratic equation, $ax^2 + bx + c = 0$, identify a, b, and c.

$a = 10, b = 13, c = 4$

Substitute these values into the quadratic formula and simplify.

$$x = \frac{-b \pm \sqrt{b^2 - 4ac}}{2a}$$

$$x = \frac{-13 \pm \sqrt{13^2 - 4(10)(4)}}{2(10)}$$

$$= \frac{-13 \pm \sqrt{169-(160)}}{20}$$

$$= \frac{-13 \pm \sqrt{9}}{20}$$

$$= \frac{-13 \pm 3}{20}$$

So $x = \frac{-13+3}{20} = \frac{-10}{20} = -\frac{1}{2}$

or $x = \frac{-13-3}{20} = \frac{-16}{20} = -\frac{4}{5}$

You could also have solved this problem using factoring.

634. $\frac{3 \pm \sqrt{13}}{2}$

Using the standard form of the quadratic equation, $ax^2 + bx + c = 0$, identify a, b, and c.

$a = 1, b = -3, c = -1$

Substitute these values into the quadratic formula and simplify.

$$x = \frac{-b \pm \sqrt{b^2 - 4ac}}{2a}$$

$$x = \frac{-(-3) \pm \sqrt{(-3)^2 - 4(1)(1)}}{2(1)}$$

$$= \frac{3 \pm \sqrt{9-(-4)}}{2}$$

$$= \frac{3 \pm \sqrt{9+4}}{2}$$

$$= \frac{3 \pm \sqrt{13}}{2}$$

$$\text{So } x = \frac{3+\sqrt{13}}{2}$$

$$\text{or } x = \frac{3-\sqrt{13}}{2}$$

635. $\frac{-5 \pm \sqrt{17}}{2}$

Using the standard form of the quadratic equation, $ax^2 + bx + c = 0$, identify a, b, and c.

$a = 1, b = 5, c = 2$

Substitute these values into the quadratic formula and simplify.

$$x = \frac{-b \pm \sqrt{b^2 - 4ac}}{2a}$$

$$x = \frac{-5 \pm \sqrt{5^2 - 4(1)(2)}}{2(1)}$$

$$= \frac{-5 \pm \sqrt{25-8}}{2}$$

$$= \frac{-5 \pm \sqrt{17}}{2}$$

$$\text{So } x = \frac{-5+\sqrt{17}}{2}$$

$$\text{or } x = \frac{-5-\sqrt{17}}{2}$$

636. $\frac{1 \pm \sqrt{41}}{4}$

Using the standard form of the quadratic equation, $ax^2 + bx + c = 0$, identify a, b, and c.

$a = 2, b = -1, c = -5$

Substitute these values into the quadratic formula and simplify.

$$x = \frac{-b \pm \sqrt{b^2 - 4ac}}{2a}$$

$$x = \frac{-(-1) \pm \sqrt{(-1)^2 - 4(2)(-5)}}{2(2)}$$

$$= \frac{1 \pm \sqrt{1-(-40)}}{4}$$

$$= \frac{1 \pm \sqrt{1+40}}{4}$$

$$= \frac{1 \pm \sqrt{41}}{4}$$

So $x = \frac{1+\sqrt{41}}{4}$

or $x = \frac{1-\sqrt{41}}{4}$

637. $\frac{2 \pm \sqrt{14}}{2}$

Using the standard form of the quadratic equation, $ax^2 + bx + c = 0$, identify a, b, and c.

$a = 2, b = -4, c = -5$

Substitute these values into the quadratic formula and simplify.

$$x = \frac{-b \pm \sqrt{b^2 - 4ac}}{2a}$$

$$x = \frac{-(-4) \pm \sqrt{(-4)^2 - 4(2)(-5)}}{2(2)}$$

$$= \frac{4 \pm \sqrt{16-(-40)}}{4}$$

$$= \frac{4 \pm \sqrt{16+40}}{4}$$

$$= \frac{4 \pm \sqrt{56}}{4}$$

$$= \frac{4 \pm \sqrt{4 \cdot 14}}{4} = \frac{4 \pm 2\sqrt{14}}{4}$$

$$= \frac{{}^2\not{4} \pm \not{2}\sqrt{14}}{{}_2\not{4}} = \frac{2 \pm \sqrt{14}}{2}$$

You can't reduce the fraction any further, because the radical term doesn't have a factor of 2 in the front as a multiplier.

So $x = \frac{2+\sqrt{14}}{2}$

or $x = \frac{2-\sqrt{14}}{2}$

638. $\frac{-3 \pm \sqrt{6}}{3}$

Using the standard form of the quadratic equation, $ax^2 + bx + c = 0$, identify a, b, and c.

$a = 3, b = 6, c = 1$

Substitute these values into the quadratic formula and simplify.

$$x = \frac{-b \pm \sqrt{b^2 - 4ac}}{2a}$$

$$x = \frac{-6 \pm \sqrt{6^2 - 4(3)(1)}}{2(3)}$$

$$= \frac{-6 \pm \sqrt{36-12}}{6}$$

$$= \frac{-6 \pm \sqrt{24}}{6}$$

$$= \frac{-6 \pm \sqrt{4 \cdot 6}}{6} = \frac{-6 \pm 2\sqrt{6}}{6}$$

$$= \frac{-{}^{3}\not{6} \pm \not{2}\sqrt{6}}{{}^{3}\not{6}} = \frac{-3 \pm \sqrt{6}}{3}$$

You can't reduce the fraction any further, because the radical term doesn't have a factor of 3 in the front as a multiplier.

$$\text{So } x = \frac{-3 + \sqrt{6}}{3}$$

$$\text{or } x = \frac{-3 - \sqrt{6}}{3}$$

639. $\frac{7 \pm 3\sqrt{13}}{2}$

Using the standard form of the quadratic equation, $ax^2 + bx + c = 0$, identify a, b, and c.

$a = 1, b = -7, c = -17$

Substitute these values into the quadratic formula and simplify.

$$x = \frac{-b \pm \sqrt{b^2 - 4ac}}{2a}$$

$$x = \frac{-(-7) \pm \sqrt{(-7)^2 - 4(1)(-17)}}{2(1)}$$

$$= \frac{7 \pm \sqrt{49 - (-68)}}{2}$$

$$= \frac{7 \pm \sqrt{49 + 68}}{2}$$

$$= \frac{7 \pm \sqrt{117}}{2}$$

$$= \frac{7 \pm \sqrt{9 \cdot 13}}{2} = \frac{7 \pm 3\sqrt{13}}{2}$$

$$\text{So } x = \frac{7 + 3\sqrt{13}}{2}$$

$$\text{or } x = \frac{7 - 3\sqrt{13}}{2}$$

640. $\frac{-4 \pm \sqrt{10}}{2}$

Using the standard form of the quadratic equation, $ax^2 + bx + c = 0$, identify a, b, and c.

$a = 2, b = 8, c = 3$

Substitute these values into the quadratic formula and simplify.

$$x = \frac{-b \pm \sqrt{b^2 - 4ac}}{2a}$$

$$x = \frac{-8 \pm \sqrt{8^2 - 4(2)(3)}}{2(2)}$$

$$= \frac{-8 \pm \sqrt{64 - 24}}{4}$$

$$= \frac{-8 \pm \sqrt{40}}{4}$$

$$= \frac{-8 \pm \sqrt{4 \cdot 10}}{4} = \frac{-8 \pm 2\sqrt{10}}{4}$$

$$= \frac{-\cancel{8}^{\,4} \pm \cancel{2}\sqrt{10}}{{}^{2}\cancel{4}} = \frac{-4 \pm \sqrt{10}}{2}$$

You can't reduce the fraction any further, because the radical term doesn't have a factor of 2 in the front as a multiplier.

So $x = \dfrac{-4 + \sqrt{10}}{2}$

or $x = \dfrac{-4 - \sqrt{10}}{2}$

641. $6 \pm 3\sqrt{3}$

Using the standard form of the quadratic equation, $ax^2 + bx + c = 0$, identify a, b, and c.

$a = 1, b = -12, c = 9$

Substitute these values into the quadratic formula and simplify.

$$x = \frac{-b \pm \sqrt{b^2 - 4ac}}{2a}$$

$$x = \frac{-(-12) \pm \sqrt{(-12)^2 - 4(1)(9)}}{2(1)}$$

$$= \frac{12 \pm \sqrt{144 - 36}}{2}$$

$$= \frac{12 \pm \sqrt{108}}{2}$$

$$= \frac{12 \pm \sqrt{36 \cdot 3}}{2} = \frac{12 \pm 6\sqrt{3}}{2}$$

$$= \frac{{}^{6}\cancel{12} \pm {}^{3}\cancel{6}\sqrt{3}}{\cancel{2}} = 6 \pm 3\sqrt{3}$$

So $x = 6 + 3\sqrt{3}$

or $x = 6 - 3\sqrt{3}$

642. 4 or –6

First add 24 to each side of the equation.

$x^2 + 2x = 24$

Now find half of the coefficient of x, 1, and square it – which is still 1. Add that to each side of the equation.

$$x^2 + 2x + 1 = 24 + 1 = 25$$

Now you can factor the trinomial on the left, writing it as the square of a binomial.

$$(x + 1)^2 = 25$$

Take the square root of both sides.

$$\sqrt{(x+1)^2} = \pm\sqrt{25}$$

Technically, you should use a ± in front of the radical on the left, also, but that would just give you four different equations – two of which are repeats. Having the ± on the right gives you the two answers you need.

Simplifying the equation, you get

$$x + 1 = \pm 5$$

Add –1 to each side to get the two answers.

$$x = -1 + 5 = 4$$

$$\text{or } x = -1 - 5 = -6$$

643. $\frac{5}{2}$ **or** -8

First divide each term by 2 so that you have a coefficient of 1 on the x^2 term.

$$x^2 + \frac{11}{2}x - 20 = 0$$

Now add 20 to each side of the equation.

$$x^2 + \frac{11}{2}x = 20$$

Next find half of the coefficient of x, which is $\frac{1}{2}\cdot\frac{11}{2} = \frac{11}{4}$, and square it – which gives you $\frac{121}{16}$. Add that to each side of the equation.

$$x^2 + \frac{11}{2}x + \frac{121}{16} = 20 + \frac{121}{16}$$

$$= \frac{320}{16} + \frac{121}{16}$$

$$x^2 + \frac{11}{2}x + \frac{121}{16} = \frac{441}{16}$$

Now you can factor the trinomial on the left, writing it as the square of a binomial.

$$\left(x + \frac{11}{4}\right)^2 = \frac{441}{16}$$

Notice that the second term in the binomial is the same as the "half of" that was found in an earlier step.

Take the square root of both sides.

$$\sqrt{\left(x + \frac{11}{4}\right)^2} = \pm\sqrt{\frac{441}{16}}$$

Technically, you should use a ± in front of the radical on the left, also, but that would just give you four different equations – two of which are repeats. Having the ± on the right gives you the two answers you need.

Simplifying the equation, you get

$$x+\frac{11}{4}=\pm\frac{21}{4}$$

Add $-\frac{11}{4}$ to each side to get the two answers.

$$x=-\frac{11}{4}+\frac{21}{4}=\frac{10}{4}=\frac{5}{2}$$

$$\text{or } x=-\frac{11}{4}-\frac{21}{4}=\frac{-32}{4}=-8$$

644. $2\pm\sqrt{2}$

First add –2 to each side of the equation.

$$x^2-4x=-2$$

Now find half of the coefficient of x, –2, and square it – which is 4. Add that to each side of the equation.

$$x^2-4x+4=-2+4=2$$

Now you can factor the trinomial on the left, writing it as the square of a binomial.

$$(x-2)^2=2$$

Take the square root of both sides.

$$\sqrt{(x-2)^2}=\pm\sqrt{2}$$

Technically, you should use a ± in front of the radical on the left, also, but that would just give you four different equations – two of which are repeats. Having the ± on the right gives you the two answers you need.

Simplifying the equation, you get

$$x-2=\pm\sqrt{2}$$

Add 2 to each side to get the two answers.

$$x=2+\sqrt{2}$$

$$\text{or } x=2-\sqrt{2}$$

645. $6\pm3\sqrt{5}$

First add 9 to each side of the equation.

$$x^2-12x=9$$

Now find half of the coefficient of x, –6, and square it – which is 36. Add that to each side of the equation.

$$x^2-12x+36=9+36=45$$

Now you can factor the trinomial on the left, writing it as the square of a binomial.

$$(x-6)^2=45$$

Take the square root of both sides.

$$\sqrt{(x-6)^2} = \pm\sqrt{45} = \pm\sqrt{9\cdot 5} = \pm 3\sqrt{5}$$

Technically, you should use a ± in front of the radical on the left, also, but that would just give you four different equations – two of which are repeats. Having the ± on the right gives you the two answers you need.

Simplifying the equation, you get

$$x-6=\pm 3\sqrt{5}$$

Add 6 to each side to get the two answers.

$$x=6+3\sqrt{5}$$

$$\text{or } x=6-3\sqrt{5}$$

646. **$3i$**

Using the rule for a product under a radical, $\sqrt{-9}$ can be written $\sqrt{9(-1)} = \sqrt{9}\sqrt{-1} = 3\sqrt{-1}$. Since $\sqrt{-1} = i$, the radical $\sqrt{-9} = 3i$.

647. **$5i$**

Using the rule for a product under a radical, $\sqrt{-25}$ can be written $\sqrt{25(-1)} = \sqrt{25}\sqrt{-1} = 5\sqrt{-1}$. Since $\sqrt{-1} = i$, the radical $\sqrt{-25} = 5i$.

648. **$4-6i$**

Using the rule for a product under a radical, $\sqrt{-36}$ can be written $\sqrt{36(-1)} = \sqrt{36}\sqrt{-1} = 6\sqrt{-1}$. Since $\sqrt{-1} = i$, the expression $4-\sqrt{-36} = 4-6i$.

649. **$-5+7i$**

Using the rule for a product under a radical, $\sqrt{-49}$ can be written $\sqrt{49(-1)} = \sqrt{49}\sqrt{-1} = 7\sqrt{-1}$. Since $\sqrt{-1} = i$, the expression $-5+\sqrt{-49} = -5+7i$.

650. **$-2+2i\sqrt{3}$**

Using the rule for a product under a radical, $\sqrt{-12}$ can be written $\sqrt{12(-1)} = \sqrt{4\cdot 3(-1)} = \sqrt{4}\sqrt{3}\sqrt{-1} = 2\sqrt{3}\sqrt{-1}$. Since $\sqrt{-1} = i$, the expression $-2+\sqrt{-12} = -2+2i\sqrt{3}$.

Even though the format is technically $a + bi$, with the i at the end, it's usually a better idea to put the i in front of the radical.

651. **$3-3i\sqrt{3}$**

Using the rule for a product under a radical, $\sqrt{-27}$ can be written $\sqrt{27(-1)} = \sqrt{9\cdot 3(-1)} = \sqrt{9}\sqrt{3}\sqrt{-1} = 3\sqrt{3}\sqrt{-1}$. Since $\sqrt{-1} = i$, the expression $3-\sqrt{-27} = 3-3i\sqrt{3}$. Even though the format is technically $a + bi$, with the i at the end, it's usually a better idea to put the i in front of the radical.

652. $-2+10i\sqrt{2}$

Using the rule for a product under a radical, $\sqrt{-200}$ can be written $\sqrt{200(-1)}=\sqrt{100\cdot 2(-1)}=\sqrt{100}\sqrt{2}\sqrt{-1}=10\sqrt{2}\sqrt{-1}$. Since $\sqrt{-1}=i$, the expression $-2+\sqrt{-200}=-2+10i\sqrt{2}$. Even though the format is technically $a+bi$, with the i at the end, it's usually a better idea to put the i in front of the radical.

653. $4-5i\sqrt{3}$

Using the rule for a product under a radical, $\sqrt{-75}$ can be written $\sqrt{75(-1)}=\sqrt{25\cdot 3(-1)}=\sqrt{25}\sqrt{3}\sqrt{-1}=5\sqrt{3}\sqrt{-1}$. Since $\sqrt{-1}=i$, the expression $4-\sqrt{-75}=4-5i\sqrt{3}$. Even though the format is technically $a+bi$, with the i at the end, it's usually a better idea to put the i in front of the radical.

654. $-2\pm 2i$

Using the standard form of the quadratic equation, $ax^2+bx+c=0$, identify a, b, and c.

$a=1$, $b=4$, $c=8$

Substitute these values into the quadratic formula and simplify.

$$x=\frac{-b\pm\sqrt{b^2-4ac}}{2a}$$

$$x=\frac{-4\pm\sqrt{4^2-4(1)(8)}}{2(1)}$$

$$=\frac{-4\pm\sqrt{16-32}}{2}$$

$$=\frac{-4\pm\sqrt{-16}}{2}$$

$$=\frac{-4\pm 4i}{2}=-2\pm 2i$$

655. $\frac{-1\pm 3i\sqrt{11}}{2}$

Using the standard form of the quadratic equation, $ax^2+bx+c=0$, identify a, b, and c.

$a=1$, $b=1$, $c=25$

Substitute these values into the quadratic formula and simplify.

$$x=\frac{-b\pm\sqrt{b^2-4ac}}{2a}$$

$$x=\frac{-1\pm\sqrt{1^2-4(1)(25)}}{2(1)}$$

$$=\frac{-1\pm\sqrt{1-100}}{2}$$

$$=\frac{-1\pm\sqrt{-99}}{2}$$

$$=\frac{-1\pm 3i\sqrt{11}}{2}$$

656. **3 or 1 positive, 1 negative**

Counting the number of possible positive real roots:

The sign changes from positive to negative to positive to negative. That's three changes, so there could be as many as three positive real roots. If there aren't three, then there is one.

Counting the number of possible negative real roots:

Replacing each x with $-x$ in the equation, and simplifying, you have $x^4 + 3x^3 + 2x^2 + 4x - 9 = 0$. There is one sign change, so there is one negative real root.

657. **2 or 0 positive, 3 or 1 negative**

Counting the number of possible positive real roots:

The sign changes from positive to negative to positive. That's two changes, so there could be as many as two positive real roots. If there aren't two, then there are none.

Counting the number of possible negative real roots:

Replacing each x with $-x$ in the equation, and simplifying, you have $-x^5 + x^3 - 4x + 1 = 0$. There are three sign changes, so there are either three negative real roots, or if not three, there is one.

658. **3 or 1 positive, 1 negative**

Counting the number of possible positive real roots:

The sign changes from positive to negative to positive to negative. That's three changes, so there could be as many as three positive real roots; if not three, then there is one.

Counting the number of possible negative real roots: Replacing each x with $-x$ in the equation, and simplifying, you have $5x^4 + 3x^3 - 6x - 2 = 0$. There is one sign change, so there is one negative real root.

659. **4 or 2 or 0 positive, 0 negative**

Counting the number of possible positive real roots:

The sign changes from positive to negative to positive to negative to positive. That's four changes, so there could be as many as four positive real roots; if not four, then there are 2 or none.

Counting the number of possible negative real roots:

Replacing each x with $-x$ in the equation, and simplifying, you have $x^6 + x^4 + x^3 + 6x^2 + x + 9 = 0$. There are no sign changes, so there are no negative real roots.

660. **±1, ±2, ±4, ±8**

The divisors of the constant term are:

±1, ±2, ±4, ±8.

The lead coefficient is 1, so there are no additional possibilities.

661. $\pm1, \pm2, \pm3, \pm6, \pm\frac{1}{5}, \pm\frac{2}{5}, \pm\frac{3}{5}, \pm\frac{6}{5}$

The divisors of the constant term are:

$\pm1, \pm2, \pm3, \pm6.$

The lead coefficient is 5, and its divisors are:

$\pm1, \pm5$

Dividing the factors of 6 by 1 makes no change, but dividing the factors of 6 by 5 adds eight more possible roots.

662. $\pm1, \pm2, \pm4, \pm\frac{1}{2}$

The divisors of the constant term are:

$\pm1, \pm2, \pm4.$

The lead coefficient is 2, and its divisors are:

$\pm1, \pm2$

Dividing the factors of 4 by 1 makes no change, but dividing the factors of 4 by 2 adds two more possible roots. (Divisions that result in a number already listed are not included.)

663. $\pm1, \pm3, \pm\frac{1}{2}, \pm\frac{3}{2}, \pm\frac{1}{3}, \pm\frac{1}{6}$

The divisors of the constant term are:

$\pm1, \pm3.$

The lead coefficient is 6, and its divisors are:

$\pm1, \pm2, \pm3, \pm6$

Dividing the factors of 3 by 1 makes no change, but dividing the factors of 3 by each of the other factors of 6 adds eight more possible roots. (Divisions that result in a number already listed are not included.)

664. **Using synthetic division: –2 is a root**

$$\begin{array}{r|rrrr} 2 & 1 & -3 & 2 & 24 \\ & & 2 & -2 & 0 \\ \hline & 1 & -1 & 0 & 24 \end{array}$$

The remainder is not 0, so 2 is not a root.

$$\begin{array}{r|rrrr} -2 & 1 & -3 & 2 & 24 \\ & & -2 & 10 & -24 \\ \hline & 1 & -5 & 12 & 0 \end{array}$$

The remainder is 0, so –2 is a root.

$$\begin{array}{r|rrrr} 3 & 1 & -3 & 2 & 24 \\ & & 3 & 0 & 6 \\ \hline & 1 & 0 & 2 & 30 \end{array}$$

The remainder is not 0, so 3 is not a root.

$$\begin{array}{r|rrrr} 4 & 1 & -3 & 2 & 24 \\ & & 4 & 4 & 24 \\ \hline & 1 & 1 & 6 & 48 \end{array}$$

The remainder is not 0, so 4 is not a root.

665. Using synthetic division: 3 is a root

$$\begin{array}{r|rrrrr} 1 & 1 & -5 & 3 & 8 & 3 \\ & & 1 & -4 & -1 & 7 \\ \hline & 1 & -4 & -1 & 7 & 10 \end{array}$$

The remainder is not 0, so 1 is not a root.

$$\begin{array}{r|rrrrr} -1 & 1 & -5 & 3 & 8 & 3 \\ & & -1 & 6 & -9 & 1 \\ \hline & 1 & -6 & 9 & -1 & 4 \end{array}$$

The remainder is not 0, so –1 is not a root.

$$\begin{array}{r|rrrrr} 3 & 1 & -5 & 3 & 8 & 3 \\ & & 3 & -6 & -9 & -3 \\ \hline & 1 & -2 & -3 & -1 & 0 \end{array}$$

The remainder is 0, so 3 is a root.

$$\begin{array}{r|rrrrr} -3 & 1 & -5 & 3 & 8 & 3 \\ & & -3 & 24 & -81 & 219 \\ \hline & 1 & -8 & 27 & -73 & 222 \end{array}$$

The remainder is not 0, so –3 is not a root.

666. Using synthetic division: 1 is a root

$$\begin{array}{r|rrrrrr} 1 & 1 & -4 & -3 & 0 & 4 & 2 \\ & & 1 & -3 & -6 & -6 & -2 \\ \hline & 1 & -3 & -6 & -6 & -2 & 0 \end{array}$$

The remainder is 0, so 1 is a root.

$$\begin{array}{r|rrrrrr} -1 & 1 & -4 & -3 & 0 & 4 & 2 \\ & & -1 & 5 & -2 & 2 & -6 \\ \hline & 1 & -5 & 2 & -2 & 6 & -4 \end{array}$$

The remainder is not 0, so –1 is not a root.

$$\begin{array}{r|rrrrrr} 2 & 1 & -4 & -3 & 0 & 4 & 2 \\ & & 2 & -4 & -14 & -28 & -48 \\ \hline & 1 & -2 & -7 & -14 & -24 & -46 \end{array}$$

The remainder is not 0, so 2 is not a root.

$$\begin{array}{r|rrrrrr} -2 & 1 & -4 & -3 & 0 & 4 & 2 \\ & & -2 & 12 & -18 & 36 & -80 \\ \hline & 1 & -6 & 9 & -18 & 40 & -78 \end{array}$$

The remainder is not 0, so –2 is not a root.

667. **Using synthetic division: 1 is a root**

$$\begin{array}{r|rrrrrrr} 1 & 1 & -1 & 0 & 1 & 0 & -2 & 1 \\ & & 1 & 0 & 0 & 1 & 1 & -1 \\ \hline & 1 & 0 & 0 & 1 & 1 & -1 & 0 \end{array}$$

The remainder is 0, so 1 is a root.

$$\begin{array}{r|rrrrrrr} -1 & 1 & -1 & 0 & 1 & 0 & -2 & 1 \\ & & -1 & 2 & -2 & 1 & -1 & 3 \\ \hline & 1 & -2 & 2 & -1 & 1 & -3 & 4 \end{array}$$

The remainder is not 0, so –1 is not a root.

668. ±2, –3

Looking at sign changes, you see that there's 1 possible positive real root and 2 or 0 possible negative real roots. Checking for rational roots, you see that the factors of 12 are ±1, ±2, ±3, ±4, ±6, ±12. Lots of possibilities.

Using synthetic division, and checking to see if 2 is a solution,

$$\begin{array}{r|rrrr} 2 & 1 & 3 & -4 & -12 \\ & & 2 & 10 & 12 \\ \hline & 1 & 5 & 6 & 0 \end{array}$$

The remainder is 0, so 2 must be a root.

Now, using the numbers in the bottom row, write the quotient (the result of the division).

You get $x^2 + 5x + 6 = 0$, which factors into $(x + 2)(x + 3) = 0$. The solutions from that factorization are $x = -2$ or $x = -3$. So the two roots of the original polynomial are ±2 or –3.

Another way to solve this equation is to apply grouping to the original cubic polynomial:

$$\begin{aligned} & x^3 + 3x^2 - 4x - 12 \\ &= x^2(x + 3) - 4(x + 3) \\ &= (x + 3)(x^2 - 4) \\ &= (x + 3)(x - 2)(x + 2) = 0 \end{aligned}$$

Then, setting each factor equal to 0, you get the same three roots.

669. 1, ±5

Looking at sign changes, you see that there are 2 or 0 possible positive real roots and 1 possible negative real root. Checking for rational roots, you see that the factors of 25 are ±1, ±5, ±25.

Using synthetic division, and checking to see if 1 is a solution,

$$\begin{array}{r|rrrr} 1 & 1 & -1 & -25 & 25 \\ & & 1 & 0 & -25 \\ \hline & 1 & 0 & -25 & 0 \end{array}$$

The remainder is 0, so 1 must be a root.

Now, using the numbers in the bottom row, write the quotient (the result of the division). You get $x^2 - 25 = 0$, which factors into $(x + 5)(x - 5) = 0$. The solutions from that factorization are $x = -5$ or $x = 5$. So the three roots of the original polynomial are 1, –5, or 5.

Another way to solve this equation is to apply grouping to the original cubic polynomial:

$$\begin{aligned} &x^3 - x^2 - 25x + 25 \\ &= x^2(x-1) - 25(x-1) \\ &= (x-1)(x^2-25) \\ &= (x-1)(x-5)(x+5) = 0 \end{aligned}$$

Then, setting each factor equal to 0, you get the same three roots.

670. **–2, –3, 1**

Looking at sign changes, you see that there's 1 possible positive real root and 2 or 0 possible negative real roots. Checking for rational roots, you see that the factors of 6 are ±1, ±2, ±3, ±6.

Using synthetic division, and checking to see if –2 is a solution,

$$\begin{array}{r|rrrr} -2 & 1 & 4 & 1 & -6 \\ & & -2 & -4 & 6 \\ \hline & 1 & 2 & -3 & 0 \end{array}$$

The remainder is 0, so –2 must be a root.

Now, using the numbers in the bottom row, write the quotient (the result of the division). You get $x^2 + 2x - 3 = 0$, which factors into $(x + 3)(x - 1) = 0$. The solutions from that factorization are $x = -3$ or $x = 1$. So the three roots of the original polynomial are –2, –3, or 1.

671. **–1, –4, 6**

Looking at sign changes, you see that there's 1 possible positive real root and 2 or 0 possible negative real roots. Checking for rational roots, you see that the factors of 24 are ±1, ±2, ±3, ±4, ±6, ±8, ±12, or ±24. That's quite a few to choose from when checking for roots.

Using synthetic division, and checking to see if –1 is a solution,

$$\begin{array}{r|rrrr} -1 & 1 & -1 & -26 & -24 \\ & & -1 & 2 & 24 \\ \hline & 1 & -2 & -24 & 0 \end{array}$$

The remainder is 0, so –1 must be a root.

Now, using the numbers in the bottom row, write the quotient (the result of the division). You get $x^2 - 2x - 24 = 0$, which factors into $(x + 4)(x - 6) = 0$. The solutions from that factorization are $x = -4$ or $x = 6$. So the three roots of the original polynomial are –1, –4, or 6.

672. **±3**

With one sign change, you'll find only one positive real root and one negative real root. The possible roots are ±1, ±3, ±9, ±27, or ±81. Since the polynomial is the difference of two perfect squares, the most efficient way of solving this is to factor the polynomial.

$$x^4 - 81 = (x^2 + 9)(x^2 - 9) = (x^2 + 9)(x + 3)(x - 3) = 0.$$

The first binomial doesn't factor any more, and, when you set it equal to 0, you don't get any real roots. The second two factors give you $x = -3$ and $x = 3$.

673. ± 2

With one sign change, you'll find only one positive real root and one negative real root. The possible roots are ±1, ±2, ±4, ±8, ±16, ±32, or ±64. Since the polynomial is the difference of two perfect squares, the most efficient way of solving this is to factor the polynomial.

$$x^6 - 64 = (x^3 + 8)(x^3 - 8).$$

The two binomials are both factorable – one as the sum of cubes and the other as the difference of cubes.

$$= (x + 2)(x^2 - 2x + 4)(x - 2)(x^2 + 2x + 4) = 0.$$

The two binomials give you roots of $x = -2$ and $x = 2$. The two trinomials don't factor, and, even using the quadratic formula, you won't find any real roots.

674. 1, –4, –4

Looking at sign changes, you see that there's 1 possible positive real root and 2 or 0 possible negative real roots. Checking for rational roots, you see that the factors of 16 are ±1, ±2, ±4, ±8, or ±16.

Using synthetic division, and checking to see if 1 is a solution,

$$\begin{array}{r|rrrr} 1 & 1 & 7 & 8 & -16 \\ & & 1 & 8 & 16 \\ \hline & 1 & 8 & 16 & 0 \end{array}$$

The remainder is 0, so 1 must be a root.

Now, using the numbers in the bottom row, write the quotient (the result of the division). You get $x^2 + 8x + 16 = 0$, which factors into $(x + 4)(x + 4) = 0$; it's a perfect square trinomial. The solutions from that factorization are $x = -4$ twice. So the three roots of the original polynomial are 1, –4, and –4.

675. 2, 2, 5

Looking at sign changes, you see that there are 3 or 1 possible positive real roots and 0 possible negative real roots. Checking for rational roots, you see that the factors of 20 are ±1, ±2, ±4, ±5, ±10, or ±20.

Using synthetic division, and checking to see if 2 is a solution,

$$\begin{array}{r|rrrr} 2 & 1 & -9 & 24 & -20 \\ & & 2 & -14 & 20 \\ \hline & 1 & -7 & 10 & 0 \end{array}$$

The remainder is 0, so 2 must be a root.

Now, using the numbers in the bottom row, write the quotient (the result of the division). You get $x^2 - 7x + 10 = 0$, which factors into $(x - 2)(x - 5) = 0$. The solutions from that factorization are $x = 2$ or $x = 5$. So the three roots of the original polynomial are 2, 2, and 5.

676. **±1, ±6**

With two sign changes, there are either 2 or 0 positive real roots and 2 or 0 negative real roots. The possible rational roots are ±1, ±2, ±3, ±4, ±6, ±9, ±12, ±18, or ±36. That's quite a large number of roots to choose from if using synthetic division. But, since the polynomial is *quadratic-like*, the most efficient way of solving this is to factor the polynomial.

$$x^4 - 37x^2 + 36 = (x^2 - 1)(x^2 - 36) = 0.$$

Each of the binomials factors as the difference of squares.

$$= (x + 1)(x - 1)(x + 6)(x - 6) = 0$$

So there are two positive and two negative roots: ±1 or ±6.

677. **±3, ±8**

With two sign changes, there are either 2 or 0 positive real roots and 2 or 0 negative real roots. The possible rational roots are ±1, ±2, ±3, ±4, ±6, ±8, ±9, ±12, ±16, ±18, ±24, ±32, ±36, ±48, ±64, ±72, ±96, ±144, ±192, ±288, or ±576. That's huge – especially when making choices using synthetic division. But, since the polynomial is *quadratic-like*, the most efficient way of solving this is to factor the polynomial.

$$x^4 - 73x^2 + 576 = (x^2 - 9)(x^2 - 64) = 0.$$

Each of the binomials factors as the difference of squares.

$$= (x + 3)(x - 3)(x + 8)(x - 8) = 0$$

So there are two positive and two negative roots: ±3 or ±8.

678. **2, 4, –1, –1**

Looking at sign changes, you see that there are 2 or 0 possible positive real roots and 2 or 0 possible negative real roots. Checking for rational roots, you see that the factors of 8 are ±1, ±2, ±4, or ±8.

Using synthetic division, and checking to see if 2 is a solution,

$$\begin{array}{r|rrrrr} 2 & 1 & -4 & -3 & 10 & 8 \\ & & 2 & -4 & -14 & -8 \\ \hline & 1 & -2 & -7 & -4 & 0 \end{array}$$

The remainder is 0, so 2 must be a root.

The quotient is represented in the bottom row and can be written as $x^3 - 2x^2 - 7x - 4 = 0$. Now look for rational roots of this polynomial by checking the factors of 4. Trying the number 4,

$$\begin{array}{r|rrrr} 4 & 1 & -2 & -7 & -4 \\ & & 4 & 8 & 4 \\ \hline & 1 & 2 & 1 & 0 \end{array}$$

The remainder is 0, so 4 must be a root.

The quotient can be written as $x^2 + 2x + 1 = 0$ which factors into $(x + 1)^2 = 0$. This gives you the two identical roots $x = -1$ and $x = -1$. So, putting all the roots together, you have $x = 2$, $x = 4$, $x = -1$, or $x = -1$.

You get the same final answers if you make choices for roots in different orders. Just keep using the quotients from the choices that work to make the work simpler.

679. **0, –1, –2, –2, 6**

First factor out an x.

$$x(x^4 - x^3 - 22x^2 - 44x - 24) = 0$$

Now, checking signs, there's only 1 positive real root and 3 or 1 negative real roots. Looking at the constant term, 24, you have possible rational roots of ±1, ±2, ±3, ±4, ±6, ±8, ±12, or ±24. Using synthetic division and checking to see if –1 is a root:

$$\begin{array}{r|rrrrr} -1 & 1 & -1 & -22 & -44 & -24 \\ & & -1 & 2 & 20 & 24 \\ \hline & 1 & -2 & -20 & -24 & 0 \end{array}$$

The remainder is 0, so –1 must be a root.

The quotient is $x^3 - 2x^2 - 20x - 24$, so the choices for possible rational roots stays the same. Now trying –2,

$$\begin{array}{r|rrrr} -2 & 1 & -2 & -20 & -24 \\ & & -2 & 8 & 24 \\ \hline & 1 & -4 & -12 & 0 \end{array}$$

The remainder is 0, so –2 must be a root.

The quotient factors: $x^2 - 4x - 12 = (x - 6)(x + 2) = 0$, providing roots of 6 and –2. So the five roots are: 0, –1, –2, 6, and –2. Notice that the 0 must be included and that the –2 is a double root.

680. $\pm 1, \frac{9}{4}$

Looking at the sign changes, you see that there are 2 or 0 positive real roots and 1 negative real root. The constant is 9, so possible rational roots are ±1, ±3, and ±9. Twelve other possibilities for rational roots come from dividing the divisors of 9 by the divisors of 4: $\pm\frac{1}{2}, \pm\frac{3}{2}, \pm\frac{9}{2}, \pm\frac{1}{4}, \pm\frac{3}{4}, \pm\frac{9}{4}$

Using synthetic division and checking to see if 1 is a root:

$$\begin{array}{r|rrrr} 1 & 4 & -9 & -4 & 9 \\ & & 4 & -5 & -9 \\ \hline & 4 & -5 & -9 & 0 \end{array}$$

The remainder is 0, so 1 must be a root.

Factoring the polynomial representing the quotient, $4x^2 - 5x - 9 = (4x - 9)(x + 1) = 0$.

The roots are $\frac{9}{4}$ and –1.

So the three roots are ±1 or $\frac{9}{4}$.

An easier way of solving this problem is to recognize that the original polynomial can be factored using grouping.

$$4x^3 - 9x^2 - 4x + 9$$
$$= x^2(4x - 9) - 1(4x - 9)$$
$$= (4x - 9)(x^2 - 1)$$
$$= (4x - 9)(x - 1)(x + 1) = 0$$

You get the same roots when you set the individual factors equal to 0.

681. $-3, \pm\frac{3}{2}$

Looking at the sign changes, you see that there is 1 positive real root and 2 or 0 negative real roots. The constant is 27, so possible rational roots are ±1, ±3, ±9, and ±27. Even more possibilities for rational roots come from dividing the divisors of 27 by the divisors of 4: $\pm\frac{1}{2}, \pm\frac{3}{2}, \pm\frac{9}{2}, \pm\frac{27}{2}, \pm\frac{1}{4}, \pm\frac{3}{4}, \pm\frac{9}{4}, \pm\frac{27}{4}$

Using synthetic division and checking to see if –3 is a root:

$$\begin{array}{r|rrrr} -3 & 4 & 12 & -9 & -27 \\ & & -12 & 0 & 27 \\ \hline & 4 & 0 & -9 & 0 \end{array}$$

The remainder is 0, so –3 must be a root.

Factoring the polynomial representing the quotient, $4x^2 - 9 = (2x - 3)(2x + 3) = 0$. The roots are $\pm\frac{3}{2}$. So the roots are –3 or $\pm\frac{3}{2}$.

An easier way of solving this problem is to recognize that the original polynomial can be factored using grouping.

$$4x^3 + 12x^2 - 9x - 27$$
$$= 4x^2(x + 3) - 9(x + 3)$$
$$= (x + 3)(4x^2 - 9)$$
$$= (x + 3)(2x - 3)(2x + 3) = 0$$

You get the same roots when you set the individual factors equal to 0.

682. $0, \pm 1, \frac{1}{2}, -3$

First factor out an x.

$$x(2x^4 + 5x^3 - 5x^2 - 5x + 3) = 0$$

Now, checking signs, there are 2 or 0 possible positive real roots and 2 or 0 negative real roots. Looking at the constant term, 3, you have possible rational roots of ±1 or ±3. The other possibilities come from dividing the first possibilities by 2, the coefficient of the lead term, giving you the additional possible roots of $\pm\frac{1}{2}, \pm\frac{3}{2}$.

Using synthetic division and checking to see if 1 is a root:

$$\begin{array}{r|rrrrr} 1 & 2 & 5 & -5 & -5 & 3 \\ & & 2 & 7 & 2 & -3 \\ \hline & 2 & 7 & 2 & -3 & 0 \end{array}$$

The remainder is 0, so 1 must be a root.

Using the quotient and trying –1 this time:

–1	2	7	2	–3
		–2	–5	3
	2	5	–3	0

The remainder is 0, so –1 must be a root.

Now, factoring the polynomial formed from the quotient: $2x^2 + 5x - 3 = (2x - 1)(x + 3) = 0$

Solving for x gives you the roots of $\frac{1}{2}$ or –3. So the roots, including the 0 from factoring out x at the beginning, are: 0, ±1, $\frac{1}{2}$, or –3.

683. $\pm 5, -\frac{1}{3}, 2$

Looking at the sign changes, you see that there are 2 or 0 positive real roots and 2 or 0 negative real roots. The constant is 50, so possible rational roots are ±1, ±2, ±5, ±10, ±25, and ±50. Even more possibilities for rational roots come from dividing the divisors of 50 by 3: $\pm\frac{1}{3}, \pm\frac{2}{3}, \pm\frac{5}{3}, \pm\frac{10}{3}, \pm\frac{25}{3}, \pm\frac{50}{3}$.

Using synthetic division and checking to see if 5 is a root:

5	3	–5	–77	125	50
		15	50	–135	–50
	3	10	–27	–10	0

The remainder is 0, so 5 must be a root.

Using the quotient and trying –5 this time:

–5	3	10	–27	–10
		–15	25	10
	3	–5	–2	0

The remainder is 0, so –5 must be a root.

The quadratic representing the quotient can be factored: $3x^2 - 5x - 2 = (3x + 1)(x - 2) = 0$. This gives the last two roots: $-\frac{1}{3}$ or 2. So the roots are ±5, $-\frac{1}{3}$, or 2.

684. $2, -3, \frac{5}{4}, \frac{7}{2}$

Looking at the sign changes, you see that there are 3 or 1 positive real roots and 1 negative real root. The constant is 210, so possible rational roots are ±1, ±2, ±3, ±5, ±6, ±7, ±10, ±14, ±15, ±21, ±30, ±35, ±42, ±70, ±105, and ±210. Whew. And that doesn't even include the possibilities for fractional roots. Rather than list all the possibilities formed by dividing the current list by 2, 4, or 8, let's hope that we can find enough integer roots first. Using synthetic division and checking to see if 2 is a root:

2	8	–30	–51	263	–210
		16	–28	–158	210
	8	–14	–79	105	0

Wonderful. Now, using the quotient and trying –3 this time:

$$\begin{array}{r|rrrr} -3 & 8 & -14 & -79 & 105 \\ & & -24 & 114 & -105 \\ \hline & 8 & -38 & 35 & 0 \end{array}$$

The quadratic represented by the quotient, $8x^2 - 38x + 35$ can be factored into $(4x-5)(2x-7)$. If you had tried $\frac{5}{4}$ or $\frac{7}{2}$ in the synthetic division, they would have come out as roots. But the factoring is so much easier than trying a lot of fractions. The four roots (three positive and one negative) are: $2, -3, \frac{5}{4}, \frac{7}{2}$.

685. $1, 1, 1, -2, \frac{1}{5}$

Looking at the sign changes, you see that there are 4, 2, or 0 positive real roots and 1 negative real root. The constant is 2, so possible rational roots are ±1, ±2.

A few more possibilities for rational roots come from dividing the divisors of 2 by 5: $\pm\frac{1}{5}, \pm\frac{2}{5}$. Using synthetic division and checking to see if 1 is a root:

$$\begin{array}{r|rrrrrr} 1 & 5 & -6 & -14 & 28 & -15 & 2 \\ & & 5 & -1 & -15 & 13 & -2 \\ \hline & 5 & -1 & -15 & 13 & -2 & 0 \end{array}$$

That was so easy to compute that it warrants trying the number 1 again with the quotient.

$$\begin{array}{r|rrrrr} 1 & 5 & -1 & -15 & 13 & -2 \\ & & 5 & 4 & -11 & 2 \\ \hline & 5 & 4 & -11 & 2 & 0 \end{array}$$

Okay – one more time with the number 1 and the new quotient.

$$\begin{array}{r|rrrr} 1 & 5 & 4 & -11 & 2 \\ & & 5 & 9 & -2 \\ \hline & 5 & 9 & -2 & 0 \end{array}$$

The quotient is finally down to three terms and can be written as $5x^2 + 9x - 2$, which factors into $(5x-1)(x+2)$, giving you roots of $\frac{1}{5}$ and –2. So the five roots are 1, 1, 1, $\frac{1}{5}$, and –2.

686. $x = 14$

Square both sides of the equation to remove the radical.

$$\left(\sqrt{x+2}\right)^2 = 4^2$$
$$x+2=16$$

Add –2 to each side of the equation to solve for x.

$$\begin{array}{rcl} x + 2 &=& 16 \\ -2 & & -2 \\ \hline x &=& 14 \end{array}$$

Check to be sure the solution isn't extraneous by substituting the answer back into the original equation.

$$\sqrt{14+2} \stackrel{?}{=} 4$$

$$\sqrt{16} \stackrel{?}{=} 4$$

$$4 = 4$$

The solution is $x = 14$.

687. $y = -7$

Square both sides of the equation to remove the radical.

$$\left(\sqrt{2-y}\right)^2 = 3^2$$

$$2 - y = 9$$

Add –2 to each side of the equation.

$$\begin{array}{rcrcr} 2 & - & y & = & 9 \\ -2 & & & & -2 \\ \hline & - & y & = & 7 \end{array}$$

Now divide each side by –1 to solve for y.

$$\frac{-y}{-1} = \frac{7}{-1}$$

$$y = -7$$

Check to be sure the solution isn't extraneous by substituting the answer back into the original equation.

$$\sqrt{2-(-7)} \stackrel{?}{=} 3$$

$$\sqrt{2+7} \stackrel{?}{=} 3$$

$$\sqrt{9} \stackrel{?}{=} 3$$

$$3 = 3$$

The solution is $y = -7$.

688. $z = \pm 3$

Square both sides of the equation to remove the radical.

$$\left(\sqrt{z^2+7}\right)^2 = 4^2$$

$$z^2 + 7 = 16$$

Add –7 to each side of the equation.

$$\begin{array}{rcrcr} z^2 & + & 7 & = & 16 \\ & & -7 & & -7 \\ \hline z^2 & & & = & 9 \end{array}$$

Now take the square root of both sides.

$$\sqrt{z^2} = \pm\sqrt{9}$$
$$z = \pm 3$$

Technically, you should have the ± on both sides – in front of both radicals. But, since that just gives you repeats of the same answers, you only need put the ± on one side.

Check to be sure one or both of the solutions aren't extraneous by substituting the answers back into the original equation.

First check $z = 3$.

$$\sqrt{3^2+7} \stackrel{?}{=} 4$$
$$\sqrt{9+7} \stackrel{?}{=} 4$$
$$\sqrt{16} \stackrel{?}{=} 4$$
$$4 = 4$$

That solution is okay. Now check $z = -3$.

$$\sqrt{(-3)^2+7} \stackrel{?}{=} 4$$
$$\sqrt{9+7} \stackrel{?}{=} 4$$
$$\sqrt{16} \stackrel{?}{=} 4$$
$$4 = 4$$

Using the order of operations correctly shows you that the –3 also works.

The solutions are $z = \pm 3$.

689. $x = \pm 6$

Square both sides of the equation to remove the radical.

$$\left(\sqrt{x^2-11}\right)^2 = 5^2$$
$$x^2 - 11 = 25$$

Add 11 to each side of the equation.

$$\begin{array}{rcrcr} x^2 & - & 11 & = & 25 \\ & & +11 & & +11 \\ \hline x^2 & & & = & 36 \end{array}$$

Now take the square root of both sides.

$$\sqrt{x^2} = \pm\sqrt{36}$$
$$x = \pm 6$$

Technically, you should have the ± on both sides – in front of both radicals. But, since that just gives you repeats of the same answers, you only need put the ± on one side.

Check to be sure one or both of the solutions aren't extraneous by substituting the answers back into the original equation.

First check $x = 6$.

$$\sqrt{6^2 - 11} \stackrel{?}{=} 5$$
$$\sqrt{36 - 11} \stackrel{?}{=} 5$$
$$\sqrt{25} \stackrel{?}{=} 5$$
$$5 = 5$$

That solution is okay. Now check $x = -6$.

$$\sqrt{(-6)^2 - 11} \stackrel{?}{=} 5$$
$$\sqrt{36 - 11} \stackrel{?}{=} 5$$
$$\sqrt{25} \stackrel{?}{=} 5$$
$$5 = 5$$

Using the order of operations correctly shows you that the –6 also works.

The solutions are $x = \pm 6$.

690. $x = 6$ only

First add $-x$ to both sides of the equation to isolate the radical on the left.

$$\begin{array}{rcl} \sqrt{x+3} + x &=& 9 \\ -x & & -x \\ \hline \sqrt{x+3} &=& 9 - x \end{array}$$

Square both sides of the equation to remove the radical. Be careful when squaring the binomial on the right.

$$\left(\sqrt{x+3}\right)^2 = (9 - x)^2$$
$$x + 3 = 81 - 18x + x^2$$

Add $-x$ and -3 to each side of the equation.

$$\begin{array}{rcl} x + 3 &=& 81 - 18x + x^2 \\ -x \quad -3 & & -3 \quad -x \\ \hline 0 &=& 78 - 19x + x^2 \end{array}$$

Now rewrite the quadratic equation in the standard form and factor it.

$$x^2 - 19x + 78 = (x - 6)(x - 13) = 0$$

Setting the binomials equal to 0, you get the two solutions $x = 6$ or $x = 13$.

Check to be sure one or both of the solutions aren't extraneous by substituting the answers back into the original equation.

First check $x = 6$.

$$\sqrt{6+3} + 6 \stackrel{?}{=} 9$$
$$\sqrt{9} + 6 \stackrel{?}{=} 9$$
$$3 + 6 \stackrel{?}{=} 9$$
$$9 = 9$$

That solution is okay. Now check $x = 13$.

$$\sqrt{13+3}+13\stackrel{?}{=}9$$

$$\sqrt{16}+13\stackrel{?}{=}9$$

$$4+13\stackrel{?}{=}9$$

$$17\neq 9$$

The solution $x = 13$ is extraneous, so the only solution is $x = 6$.

691. $x = 8$ **only**

First add $-x$ to both sides of the equation to isolate the radical on the left.

$$\begin{array}{rcccl} \sqrt{2x+9} & + & x & = & 13 \\ & & -x & & -x \\ \hline \sqrt{2x+9} & & & = & 13-x \end{array}$$

Square both sides of the equation to remove the radical. Be careful when squaring the binomial on the right.

$$\left(\sqrt{2x+9}\right)^2=(13-x)^2$$

$$2x+9=169-26x+x^2$$

Add $-2x$ and -9 to each side of the equation.

$$\begin{array}{ccccccccc} 2x & + & 9 & = & 169 & - & 26x & + & x^2 \\ -2x & & -9 & & -9 & & -2x & & \\ \hline & & 0 & = & 160 & - & 28x & + & x^2 \end{array}$$

Now rewrite the quadratic equation in the standard form and factor it.

$$x^2-28x+160=(x-20)(x-8)=0$$

Setting the binomials equal to 0, you get the two solutions $x = 20$ and $x = 8$.

Check to be sure one or both of the solutions aren't extraneous by substituting the answers back into the original equation.

First check $x = 20$.

$$\sqrt{2(20)+9}+20\stackrel{?}{=}13$$

$$\sqrt{40+9}+20\stackrel{?}{=}13$$

$$\sqrt{49}+20\stackrel{?}{=}13$$

$$7+20\stackrel{?}{=}13$$

$$27\neq 13$$

That solution is extraneous. Now check $x = 8$.

$$\sqrt{2(8)+9}+8\stackrel{?}{=}13$$

$$\sqrt{16+9}+8\stackrel{?}{=}13$$

$$\sqrt{25}+8\stackrel{?}{=}13$$

$$5+8\stackrel{?}{=}13$$

$$13=13$$

The solution $x = 8$ works, so it's the only solution.

692. $x = 9$ **only**

First add –7 to both sides of the equation to isolate the radical on the left.

$$\begin{array}{rcrcl} \sqrt{x-5} & + & 7 & = & x \\ & & -7 & & -7 \\ \hline \sqrt{x-5} & & & = & x-7 \end{array}$$

Square both sides of the equation to remove the radical. Be careful when squaring the binomial on the right.

$$\left(\sqrt{x-5}\right)^2=(x-7)^2$$

$$x-5=x^2-14x+49$$

Add $-x$ and 5 to each side of the equation.

$$\begin{array}{rcrcrcrcr} x & - & 5 & = & x^2 & - & 14x & + & 49 \\ -x & & +5 & & & & -x & & +5 \\ \hline & & 0 & = & x^2 & - & 15x & + & 54 \end{array}$$

Now rewrite the quadratic equation in the standard form and factor it.

$$x^2 - 15x + 54 = (x-6)(x-9) = 0$$

Setting the binomials equal to 0, you get the two solutions $x = 6$ or $x = 9$.

Check to be sure one or both of the solutions aren't extraneous by substituting the answers back into the original equation.

First check $x = 6$.

$$\sqrt{6-5}+7\stackrel{?}{=}6$$

$$\sqrt{1}+7\stackrel{?}{=}6$$

$$8\neq 6$$

That solution is extraneous. Now check $x = 9$.

$$\sqrt{9-5}+7\stackrel{?}{=}9$$

$$\sqrt{4}+7\stackrel{?}{=}9$$

$$2+7\stackrel{?}{=}9$$

$$9=9$$

The solution $x = 9$ works, so it's the only solution.

693. **$x = 5$ only**

First add –2 to both sides of the equation to isolate the radical on the left.

$$\begin{array}{rcl} \sqrt{2x-5} + 2 & = & x \\ -2 & & -2 \\ \hline \sqrt{2x-1} & = & x - 2 \end{array}$$

Square both sides of the equation to remove the radical. Be careful when squaring the binomial on the right.

$$\left(\sqrt{2x-1}\right)^2 = (x-2)^2$$

$$2x-1 = x^2-4x+4$$

Add $-2x$ and $+1$ to each side of the equation.

$$\begin{array}{rcl} 2x - 1 & = & x^2 - 4x + 4 \\ -2x + 1 & & -2x + 1 \\ \hline 0 & = & x^2 - 6x + 5 \end{array}$$

Now rewrite the quadratic equation in the standard form and factor it.

$$x^2 - 6x + 5 = (x-5)(x-1) = 0$$

Setting the binomials equal to 0, you get the two solutions $x = 5$ or $x = 1$.

Check to be sure one or both of the solutions aren't extraneous by substituting the answers back into the original equation.

First check $x = 5$.

$$\sqrt{2(5)-1}+2 \stackrel{?}{=} 5$$

$$\sqrt{10-1}+2 \stackrel{?}{=} 5$$

$$\sqrt{9}+2 \stackrel{?}{=} 5$$

$$3+2 \stackrel{?}{=} 5$$

$$5 = 5$$

That solution works. Now check $x = 1$.

$$\sqrt{2(1)-1}+2 \stackrel{?}{=} 1$$

$$\sqrt{2-1}+2 \stackrel{?}{=} 1$$

$$\sqrt{1}+2 \stackrel{?}{=} 1$$

$$1+2 \stackrel{?}{=} 1$$

$$3 \neq 1$$

The solution is extraneous, so $x = 5$ is the only solution.

694. $x = 0$ **only**

First add x to both sides of the equation to isolate the radical on the left.

$$\begin{array}{rcl} \sqrt{x+9} - x &=& 3 \\ +x && +x \\ \hline \sqrt{x+9} &=& 3 + x \end{array}$$

Square both sides of the equation to remove the radical. Be careful when squaring the binomial on the right.

$$\left(\sqrt{x+9}\right)^2 = (3+x)^2$$

$$x+9 = 9+6x+x^2$$

Add $-x$ and -9 to each side of the equation.

$$\begin{array}{rcl} x + 9 &=& 9 + 6x + x^2 \\ -x \quad -9 && -9 \quad -x \\ \hline 0 &=& 5x + x^2 \end{array}$$

Now factor the quadratic on the right.

$$5x + x^2 = x(5 + x) = 0$$

Setting the binomials equal to 0, you get the two solutions $x = 0$ or $x = -5$.

Check to be sure one or both of the solutions aren't extraneous by substituting the answers back into the original equation.

First check $x = 0$.

$$\sqrt{0+9} - 0 \stackrel{?}{=} 3$$

$$\sqrt{9} \stackrel{?}{=} 3$$

$$3 = 3$$

That solution works. Now check $x = -5$.

$$\sqrt{-5+9} - (-5) \stackrel{?}{=} 3$$

$$\sqrt{4} + 5 \stackrel{?}{=} 3$$

$$2 + 5 \stackrel{?}{=} 3$$

$$7 \neq 3$$

The solution is extraneous, so $x = 0$ is the only solution.

695. $x = -3$ **and** $x = -5$

First add x to both sides of the equation to isolate the radical on the left.

$$\begin{array}{rcl} \sqrt{2x+10} - x &=& 5 \\ +x && +x \\ \hline \sqrt{2x+10} &=& 5 + x \end{array}$$

Square both sides of the equation to remove the radical. Be careful when squaring the binomial on the right.

$$\left(\sqrt{2x+10}\right)^2 = (5+x)^2$$
$$2x+10 = 25+10x+x^2$$

Add $-2x$ and -10 to each side of the equation.

$$\begin{array}{rcrcrcrcr} 2x & + & 10 & = & 25 & + & 10x & + & x^2 \\ -2x & & -10 & & -10 & & -2x & & \\ \hline & & 0 & = & 15 & + & 8x & + & x^2 \end{array}$$

Now rewrite the quadratic equation in the standard form and factor it.

$x^2 + 8x + 15 = (x + 3)(x + 5) = 0$

Setting the binomials equal to 0, you get the two solutions $x = -3$ or $x = -5$.

Check to be sure one or both of the solutions aren't extraneous by substituting the answers back into the original equation.

First check $x = -3$.

$$\sqrt{2(-3)+10} - (-3) \stackrel{?}{=} 5$$
$$\sqrt{-6+10} + 3 \stackrel{?}{=} 5$$
$$\sqrt{4} + 3 \stackrel{?}{=} 5$$
$$2+3 \stackrel{?}{=} 5$$
$$5 = 5$$

That solution works. Now check $x = -5$.

$$\sqrt{2(-5)+10} - (-5) \stackrel{?}{=} 5$$
$$\sqrt{-10+10} + 5 \stackrel{?}{=} 5$$
$$\sqrt{0} + 5 \stackrel{?}{=} 5$$
$$5 = 5$$

Both solutions work.

696. **$x = 3$ and $x = 4$**

Square both sides of the equation to remove the radical. Be careful when squaring the binomial on the right.

$$\left(\sqrt{x-3}\right)^2 = (x-3)^2$$
$$x-3 = x^2 - 6x + 9$$

Add $-x$ and 3 to each side of the equation.

$$\begin{array}{rcrcrcrcr} x & - & 3 & = & x^2 & - & 6x & + & 9 \\ -x & & +3 & & & & -x & & +3 \\ \hline & & 0 & = & x^2 & - & 7x & + & 12 \end{array}$$

Now factor the quadratic equation.

$x^2 - 7x + 12 = (x - 3)(x - 4) = 0$

Setting the binomials equal to 0, you get the two solutions $x = 3$ or $x = 4$.

Check to be sure one or both of the solutions aren't extraneous by substituting the answers back into the original equation.

First check $x = 3$.

$$\sqrt{3-3} \stackrel{?}{=} 3-3$$

$$\sqrt{0} \stackrel{?}{=} 0$$

$$0 = 0$$

That solution works. Now check $x = 4$.

$$\sqrt{4-3} \stackrel{?}{=} 4-3$$

$$\sqrt{1} \stackrel{?}{=} 1$$

$$1 = 1$$

Both solutions work.

697. $x = 7$ **only**

Square both sides of the equation to remove the radical. Be careful when squaring the binomial on the right.

$$\left(\sqrt{x-7}\right)^2 = (7-x)^2$$

$$x-7 = 49-14x+x^2$$

Add $-x$ and 7 to each side of the equation.

$$\begin{array}{rcrcrcrcr} x & - & 7 & = & 49 & - & 14x & + & x^2 \\ -x & & +7 & & +7 & & -x & & \\ \hline & & 0 & = & 56 & - & 15x & + & x^2 \end{array}$$

Now rewrite the quadratic equation in the standard form and factor it.

$$x^2 - 15x + 56 = (x-7)(x-8) = 0$$

Setting the binomials equal to 0, you get the two solutions $x = 7$ or $x = 8$.

Check to be sure one or both of the solutions aren't extraneous by substituting the answers back into the original equation.

First check $x = 7$.

$$\sqrt{7-7} \stackrel{?}{=} 7-7$$

$$\sqrt{0} \stackrel{?}{=} 0$$

$$0 = 0$$

That solution works. Now check $x = 8$.

$$\sqrt{8-7} \stackrel{?}{=} 7-8$$

$$\sqrt{1} \stackrel{?}{=} -1$$

$$1 \neq -1$$

The solution $x = 8$ is extraneous, so the only solution is $x = 7$.

698. $x = 3$ **and** $x = -1$

Square both sides of the equation. Be careful, on the left, to treat the product as the square of a binomial (don't forget the middle term).

$$\left(\sqrt{x+1}+1\right)^2 = \left(\sqrt{2x+3}\right)^2$$

$$\left(\sqrt{x+1}\right)^2 + 2\left(\sqrt{x+1}\right)(1) + 1^2 = \left(\sqrt{2x+3}\right)^2$$

$$x+1+2\sqrt{x+1}+1 = 2x+3$$

Simplify the equation by combining like terms on the left.

$$x+2+2\sqrt{x+1} = 2x+3$$

Now isolate the radical on the left by adding $-x$ and -2 to each side of the equation.

$$\begin{array}{ccccccccc} x & + & 2 & + & 2\sqrt{x+1} & = & 2x & + & 3 \\ -x & & -2 & & & & -x & & -2 \\ \hline & & & & 2\sqrt{x+1} & = & x & + & 1 \end{array}$$

Next, square both sides of the equation.

$$\left(2\sqrt{x+1}\right)^2 = (x+1)^2$$

$$4(x+1) = (x+1)^2$$

$$4x+4 = x^2+2x+1$$

Now add $-4x$ and -4 to each side of the equation.

$$\begin{array}{ccccccccc} 4x & + & 4 & = & x^2 & + & 2x & + & 1 \\ -4x & & -4 & & & & -4x & & -4 \\ \hline & & 0 & = & x^2 & - & 2x & - & 3 \end{array}$$

Factor the quadratic.

$$x^2 - 2x - 3 = (x-3)(x+1) = 0$$

Setting the binomials equal to 0, the solutions are $x = 3$ and $x = -1$.

Check to be sure one or both of the solutions aren't extraneous by substituting the answers back into the original equation.

First check $x = 3$.

$$\sqrt{3+1}+1 \stackrel{?}{=} \sqrt{2(3)+3}$$

$$\sqrt{4}+1 \stackrel{?}{=} \sqrt{6+3}$$

$$2+1 \stackrel{?}{=} \sqrt{9}$$

$$3 = 3$$

It works. Now try $x = -1$.

$$\sqrt{-1+1}+1 \stackrel{?}{=} \sqrt{2(-1)+3}$$

$$\sqrt{0}+1 \stackrel{?}{=} \sqrt{-2+3}$$

$$0+1 \stackrel{?}{=} \sqrt{1}$$

$$1 = 1$$

They both work.

Answers 601–700

699. $x = 4$ **only**

Square both sides of the equation. Be careful, on the left, to treat the product as the square of a binomial (don't forget the middle term).

$$\left(\sqrt{3x-3}+2\right)^2=\left(\sqrt{5x+5}\right)^2$$

$$\left(\sqrt{3x-3}\right)^2+4\sqrt{3x-3}+4=\left(\sqrt{5x+5}\right)^2$$

$$3x-3+4\sqrt{3x-3}+4=5x+5$$

Simplify the equation by combining like terms on the left.

$$3x+1+4\sqrt{3x-3}=5x+5$$

Now isolate the radical on the left by adding $-3x$ and -1 to each side of the equation.

$$\begin{array}{rcrcrcrcr} 3x & + & 1 & + & 4\sqrt{3x-3} & = & 5x & + & 5 \\ -3x & & -1 & & & & -3x & & -1 \\ \hline & & & & 4\sqrt{3x-3} & = & 2x & + & 4 \end{array}$$

Next, square both sides of the equation.

$$\left(4\sqrt{3x-3}\right)^2=(2x+4)^2$$

$$16(3x-3)=(2x+4)^2$$

$$48x-48=4x^2+16x+16$$

Now add $-48x$ and 48 to each side of the equation.

$$\begin{array}{rcrcrcrcr} 48x & - & 48 & = & 4x^2 & + & 16x & + & 16 \\ -48x & & +48 & & & & -48x & & +48 \\ \hline & & 0 & = & 4x^2 & - & 32x & + & 64 \end{array}$$

Factor the quadratic.

$$4x^2-32x+64=4(x^2-8x+16)=4(x-4)^2=0$$

Setting the binomial equal to 0, the solution is $x = 4$.

Check to be sure the solution isn't extraneous by substituting the answer back into the original equation.

$$\sqrt{3(4)-3}+2\stackrel{?}{=}\sqrt{5(4)+5}$$

$$\sqrt{12-3}+2\stackrel{?}{=}\sqrt{20+5}$$

$$\sqrt{9}+2\stackrel{?}{=}\sqrt{25}$$

$$3+2\stackrel{?}{=}5$$

$$5=5$$

It works so the solution is $x = 4$.

700. $x = -4$ only

Isolate the first radical by adding $-\sqrt{x+8}$ to each side.

$$\begin{array}{rcccl} 4\sqrt{x+5} & + & \sqrt{x+8} & = & 6 \\ & & -\sqrt{x+8} & & -\sqrt{x+8} \\ \hline 4\sqrt{x+5} & & & = & 6 - \sqrt{x+8} \end{array}$$

Now square both sides of the equation.

$$\left(4\sqrt{x+5}\right)^2 = \left(6 - \sqrt{x+8}\right)^2$$

$$16\left(\sqrt{x+5}\right)^2 = 36 - 12\sqrt{x+8} + \left(\sqrt{x+8}\right)^2$$

$$16(x+5) = 36 - 12\sqrt{x+8} + x + 8$$

Simplify the equation.

$$16x + 80 = 44 + x - 12\sqrt{x+8}$$

Now isolate the radical by adding –44 and $-x$ to each side of the equation.

$$\begin{array}{rcccccccl} 16x & + & 80 & = & 44 & + & x & - & 12\sqrt{x+8} \\ -x & & -44 & & -44 & & -x & & \\ \hline 15x & + & 36 & = & & & & - & 12\sqrt{x+8} \end{array}$$

Square both sides of the equation.

$$(15x+36)^2 = \left(-12\sqrt{x+8}\right)^2$$

$$225x^2 + 1080x + 1296 = 144(x+8)$$

$$225x^2 + 1080x + 1296 = 144x + 1152$$

Now add $-144x$ and –1152 to each side of the equation.

$$\begin{array}{rcccccccc} 225x^2 & + & 1080x & + & 1296 & = & 144x & + & 1152 \\ & & -144x & & -1152 & & -144x & & -1152 \\ \hline 225x^2 & + & 936x & + & 144 & = & 0 & & \end{array}$$

The quadratic does factor, but this would certainly be a case where the quadratic formula is warranted – as long as you have a calculator handy.

Factoring the quadratic will be easier after the GCF, 9, is factored out.

$$225x^2 + 936x + 144 = 9(25x^2 + 104x + 16) = 9(25x + 4)(x + 4) = 0$$

Setting the two binomials equal to zero, the solutions are $-\frac{4}{25}$ or –4.

Check to be sure neither solution is extraneous by substituting the answers back into the original equation. When $x = -\frac{4}{25}$:

$$4\sqrt{-\frac{4}{25}+5}+\sqrt{-\frac{4}{25}+8}\stackrel{?}{=}6$$

$$4\sqrt{-\frac{4}{25}+\frac{125}{25}}+\sqrt{-\frac{4}{25}+\frac{200}{25}}\stackrel{?}{=}6$$

$$4\sqrt{\frac{121}{25}}+\sqrt{\frac{196}{25}}\stackrel{?}{=}6$$

$$4\left(\frac{11}{5}\right)+\frac{14}{5}\stackrel{?}{=}6$$

$$\frac{44}{5}+\frac{14}{5}=\frac{58}{5}\neq 6$$

This solution is extraneous. Now check $x = -4$.

$$4\sqrt{-4+5}+\sqrt{-4+8}\stackrel{?}{=}6$$

$$4\sqrt{1}+\sqrt{4}\stackrel{?}{=}6$$

$$4+2\stackrel{?}{=}6$$

$$6=6$$

This solution works.

701.

$x = 10$ only

Isolate the first radical expression by adding $\sqrt{x+6}$ to each side.

$$\begin{array}{rcrcl} 3\sqrt{x-1} & - & \sqrt{x+6} & = & 5 \\ & & +\sqrt{x+6} & & +\sqrt{x+6} \\ \hline 3\sqrt{x-1} & & & = & 5+\sqrt{x+6} \end{array}$$

Now square both sides of the equation.

$$\left(3\sqrt{x-1}\right)^2=\left(5+\sqrt{x+6}\right)^2$$

$$9\left(\sqrt{x-1}\right)^2=25+10\sqrt{x+6}+\left(\sqrt{x+6}\right)^2$$

$$9(x-1)=25+10\sqrt{x+6}+x+6$$

Simplify the equation.

$$9x-9=31+x+10\sqrt{x+6}$$

Now isolate the radical by adding –31 and $-x$ to each side of the equation.

$$\begin{array}{rcrcrcrcl} 9x & - & 9 & = & 31 & + & x & + & 10\sqrt{x+6} \\ -x & & -31 & & -31 & & -x & & \\ \hline 8x & - & 40 & = & & & & & 10\sqrt{x+6} \end{array}$$

Before squaring both sides of the equation, factor out the GCF, 2, by dividing each term by 2.

$$4x-20=5\sqrt{x+6}$$

Now square both sides of the equation.

$$(4x-20)^2=\left(5\sqrt{x+6}\right)^2$$
$$16x^2-160x+400=25(x+6)$$
$$16x^2-160x+400=25x+150$$

Next add $-25x$ and -150 to each side of the equation.

$$\begin{array}{rcrcrcrcr} 16x^2 & - & 160x & + & 400 & = & 25x & + & 150 \\ & & -25x & & -150 & & -25x & & -150 \\ \hline 16x^2 & - & 185x & + & 250 & = & 0 & & \end{array}$$

The quadratic does factor, but this would certainly be a case where the quadratic formula is warranted – as long as you have a calculator handy.

Factoring the quadratic,

$$16x^2-185x+250=(16x-25)(x-10)=0.$$

Setting the two binomials equal to zero, the solutions are $\frac{25}{16}$ or 10.

Check to be sure neither solution is extraneous by substituting the answer back into the original equation. When $x=\frac{25}{16}$:

$$3\sqrt{\frac{25}{16}-1}-\sqrt{\frac{25}{16}+6}\stackrel{?}{=}5$$
$$3\sqrt{\frac{25}{16}-\frac{16}{16}}-\sqrt{\frac{25}{16}+\frac{96}{16}}\stackrel{?}{=}5$$
$$3\sqrt{\frac{9}{16}}-\sqrt{\frac{121}{16}}\stackrel{?}{=}5$$
$$3\left(\frac{3}{4}\right)-\frac{11}{4}\stackrel{?}{=}5$$
$$\frac{9}{4}-\frac{11}{4}=-\frac{2}{4}\neq 5$$

So this number is not a solution of the original equation. Now try $x = 10$.

$$3\sqrt{10-1}-\sqrt{10+6}\stackrel{?}{=}5$$
$$3\sqrt{9}-\sqrt{16}\stackrel{?}{=}5$$
$$3(3)-4\stackrel{?}{=}5$$
$$9-4\stackrel{?}{=}5$$
$$5=5$$

The solution is $x = 10$.

702. $x = 5$

First add -2 to each side of the equation.

$$\sqrt[3]{x-4}=1$$

Now raise each side of the equation to the third power.

$$\left(\sqrt[3]{x-4}\right)^3=1^3$$
$$x-4=1$$

Adding 4 to each side, you get $x = 5$.

703. $y = -240$

First add 2 to each side of the equation.

$$\sqrt[5]{3-y}=3$$

Now raise each side to the fifth power.

$$\left(\sqrt[5]{3-y}\right)^5=3^5$$

$$3-y=243$$

Now add –3 to each side.

$$-y=240$$

Multiplying each side by –1 gives you $y = -240$.

704. $x = 5$ or $x = -11$

The absolute value equation $|x + 3| = 8$ is equivalent to the two linear equations $x + 3 = 8$ and $x + 3 = -8$.

First solving $x + 3 = 8$, you add –3 to each side to get $x = 5$.

Then solving $x + 3 = -8$, you add –3 to each side to get $x = -11$.

The solution is $x = 5$ or $x = -11$.

705. $y = 7$ or $y = 1$

The absolute value equation $|y - 4| = 3$ is equivalent to the two linear equations $y - 4 = 3$ and $y - 4 = -3$.

First solving $y - 4 = 3$, you add 4 to each side to get $y = 7$.

Then solving $y - 4 = -3$, you add 4 to each side to get $y = 1$.

The solution is $y = 7$ or $y = 1$.

706. $z=-\frac{1}{5}$ or $z = -1$

The absolute value equation $|5z + 3| = 2$ is equivalent to the two linear equations $5z + 3 = 2$ and $5z + 3 = -2$.

First solving $5z + 3 = 2$, you add –3 to each side to get $5z = -1$. Dividing each side by 5, you find $z=-\frac{1}{5}$.

Then solving $5z + 3 = -2$, you add –3 to each side to get $5z = -5$. Dividing each side by 5, you find $z = -1$.

The solution is $z=-\frac{1}{5}$ or $z = -1$.

707. $x=-\frac{1}{2}$ or $x=\frac{7}{2}$

The absolute value equation $|3 - 2x| = 4$ is equivalent to the two linear equations $3 - 2x = 4$ and $3 - 2x = -4$.

First solving $3 - 2x = 4$, you add –3 to each side to get $-2x = 1$. Dividing each side of the equation by –2, you find $x = -\frac{1}{2}$.

Then solving $3 - 2x = -4$, you add –3 to each side to get $-2x = -7$. Dividing each side of the equation by –2, you find $x = \frac{7}{2}$.

The solution is $x = -\frac{1}{2}$ or $x = \frac{7}{2}$.

708. $w = 4$ or $w = -\frac{7}{2}$

The absolute value equation $|4w - 1| - 6 = 9$ needs to be rewritten before applying the equivalency involving linear equations.

First add 6 to each side to isolate the absolute value expression on the left. You then have $|4w - 1| = 15$.

This equation is equivalent to the two linear equations $4w - 1 = 15$ and $4w - 1 = -15$.

Solving $4w - 1 = 15$ for w, first add 1 to each side to get $4w = 16$. Then divide each side by 4 and you have $w = 4$.

Then solving $4w - 1 = -15$, first add 1 to each side to get $4w = -14$. Then divide each side by 4 and you have $w = -\frac{14}{4} = -\frac{7}{2}$.

The two solutions are $w = 4$ or $w = -\frac{7}{2}$.

709. $w = 0$ or $w = 4$

The absolute value equation $8 + |2 - w| = 10$ needs to be rewritten before applying the equivalency involving linear equations.

First add –8 to each side to isolate the absolute value expression on the left. You then have $|2 - w| = 2$. This equation is equivalent to the two linear equations $2 - w = 2$ and $2 - w = -2$.

Solving $2 - w = 2$ for w, first add –2 to each side to get $-w = 0$. Then divide each side by –1 and you have $w = 0$.

Then solving $2 - w = -2$, first add –2 to each side to get $-w = -4$. Then divide each side by –1 and you have $w = 4$.

So the solution is $w = 0$ or $w = 4$.

710. $x = \frac{1}{3}$ or $x = -1$

First divide each side of the equation by 5, giving you $|3x + 1| = 2$. The absolute value equation is equivalent to the two linear equations $3x + 1 = 2$ and $3x + 1 = -2$.

First solving $3x + 1 = 2$, you add –1 to each side to get $3x = 1$. Dividing each side of the equation by 3, you find $x = \frac{1}{3}$.

Then solving $3x + 1 = -2$, you add –1 to each side to get $3x = -3$. Dividing each side of the equation by 3, you find $x = -1$.

The solution is $x = \frac{1}{3}$ or $x = -1$.

711. $x = -1$ **or** $x = -7$

The absolute value equation $3|x + 4| - 2 = 7$ needs to be rewritten before applying the equivalency involving linear equations.

First add 2 to each side to isolate the absolute value expression on the left; you get $3|x + 4| = 9$. Next divide each side of the equation by 3 to get $|x + 4| = 3$.

The absolute value equation is equivalent to the two linear equations $x + 4 = 3$ and $x + 4 = -3$.

Adding –4 to each side of the two equations, you get $x = -1$ and $x = -7$, respectively.

712. $x = -\frac{4}{3}$ **or** $x = \frac{4}{3}$

The absolute value equation is equivalent to the two linear equations $-3x = 4$ and $-3x = -4$. Dividing each side of the two equations by –3, you get $x = -\frac{4}{3}$ or $x = \frac{4}{3}$, respectively.

713. $x = -9$ **or** $x = 6$

The absolute value equation is equivalent to the two linear equations $-2x - 3 = 15$ and $-2x - 3 = -15$.

Solving the first equation, you add 3 to each side of the equation to get $-2x = 18$. Dividing each side of the equation by –2, you have $x = -9$.

Solving the second equation, you add 3 to each side of the equation to get $-2x = -12$. Dividing each side by –2, you have $x = 6$.

714. **No solution**

The absolute value equation $|3x-2|+4=1$ needs to be rewritten before applying the equivalency involving linear equations.

Adding –4 to each side, you get $|3x-2|=-3$.

Right here, you have doubts. How can the absolute value of something have a negative result? Continuing on, though, and applying the equivalency, you get the two equations:

$$3x - 2 = -3 \text{ and } 3x - 2 = 3.$$

Solving the first equation, you add 2 to each side to get $3x = -1$.

Dividing each side by 3, you have $x = -\frac{1}{3}$.

Checking this answer in the original equation,

$$\left|3\left(-\frac{1}{3}\right)-2\right|+4\stackrel{?}{=}1$$

$$|-1-2|+4\stackrel{?}{=}1$$

$$|-3|+4\stackrel{?}{=}1$$

$$3+4\stackrel{?}{=}1$$

$$7\neq 1$$

This first answer doesn't work.

Solving the second linear equation, $3x - 2 = 3$, you add 2 to each side and get $3x = 5$. Dividing each side by 3 you get, $x = \frac{5}{3}$.

Check this answer in the original equation.

$$\left|3\left(\frac{5}{3}\right)-2\right|+4\stackrel{?}{=}1$$
$$|5-2|+4\stackrel{?}{=}1$$
$$|3|+4\stackrel{?}{=}1$$
$$3+4\stackrel{?}{=}1$$
$$7\neq 1$$

This doesn't work, either. The original problem has no solution.

715. No solution

The absolute value equation $3-|4-5x|=7$ needs to be rewritten before applying the equivalency involving linear equations.

Adding –3 to each side, you get $-|4-5x|=4$.

Then, multiplying each side by –1, you have $|4-5x|=-4$

Right here, you have doubts. How can the absolute value of something have a negative result? Continuing on, though, and applying the equivalency, you get the two equations:

$4 - 5x = -4$ and $4 - 5x = 4$.

Solving the first equation, you add –4 to each side to get $-5x = -8$.

Dividing each side by –5, you have $x = \frac{8}{5}$.

Check this answer in the original equation.

$$3-\left|4-5\left(\frac{8}{5}\right)\right|\stackrel{?}{=}7$$
$$3-|4-8|\stackrel{?}{=}7$$
$$3-|-4|\stackrel{?}{=}7$$
$$3-4\stackrel{?}{=}7$$
$$-1\neq 7$$

This first answer doesn't work.

Solving the second linear equation, $4 - 5x = 4$, you add –4 to each side and get $-5x = 0$.

Dividing each side by –5 you get, $x = 0$.

Check this answer in the original equation.

$$3-|4-5(0)|\stackrel{?}{=}7$$
$$3-|4-0|\stackrel{?}{=}7$$
$$3-|4|\stackrel{?}{=}7$$
$$3-4\stackrel{?}{=}7$$
$$-1\neq 7$$

This doesn't work, either. The original problem has no solution.

716. $-20 < -4$

When adding or subtracting values, the direction (sense) of the inequality stays the same.

$$\begin{array}{rcr} 7 & > & 3 \\ -2 & & -2 \\ \hline 5 & > & 1 \end{array}$$

When multiplying each side of an inequality by a negative number, the direction (sense) of the inequality is reversed.

$$\begin{array}{rcr} 5 & > & 1 \\ -4(5) & < & -4(1) \\ -20 & < & -4 \end{array}$$

717. $5 > -5$

When multiplying each side of an inequality by a negative number, the direction (sense) of the inequality is reversed.

$$\begin{array}{rcr} -4 & < & 1 \\ -2(-4) & > & -2(1) \\ 8 & > & -2 \end{array}$$

When adding or subtracting values, the direction (sense) of the inequality stays the same.

$$\begin{array}{rcr} 8 & > & -2 \\ -3 & & -3 \\ \hline 5 & > & -5 \end{array}$$

718. $5 \geq 1$

When dividing each side of an inequality by a negative number, the direction (sense) of the inequality is reversed.

$$\begin{array}{rcr} -6 & \leq & 6 \\ \frac{-6}{-3} & \geq & \frac{6}{-3} \\ 2 & \geq & -2 \end{array}$$

When adding or subtracting values, the direction (sense) of the inequality stays the same.

$$\begin{array}{rcr} 2 & \geq & -2 \\ +3 & & +3 \\ \hline 5 & \geq & 1 \end{array}$$

719. $-3 \leq 1$

When adding or subtracting values, the direction (sense) of the inequality stays the same.

$$\begin{array}{rcr} 0 & \geq & -4 \\ +3 & & +3 \\ \hline 3 & \geq & -1 \end{array}$$

When multiplying each side of an inequality by a negative number, the direction (sense) of the inequality is reversed.

$$\begin{array}{rcr} 3 & \geq & -1 \\ -1(3) & \leq & -1(-1) \\ -3 & \leq & 1 \end{array}$$

720. $[-3, 2)$

The *less-than-or-equal-to* symbol is represented with "[", and the *less-than* symbol is represented with ")". Write the numbers in order as they appear on the number line.

$[-3, 2)$

721. $[0, 4]$

The *less-than-or-equal-to* symbol is represented with "[" on the left and with "]" on the right. Write the numbers in order as they appear on the number line.

$[0, 4]$

722. $(-3, \infty)$

The *greater-than* symbol is represented by "(" on the left and is used with the –3. And then, because *infinity* has no endpoint, you use ")" on the right, also. Write the number and the *infinity* in order as they appear on the number line.

$(-3, \infty)$

723. $(-\infty, 7]$

The *less-than-or-equal-to* symbol is represented by "]" on the right with the 7. And then, because *infinity* has no endpoint, you use "(" on the left. Write the number and the *negative infinity* in order as they appear on the number line.

$(-\infty, 7]$

724. $x \geq -6$

The bracket, [, indicates that you use that endpoint and show that with *greater-than-or-equal-to*. The x represents all the numbers that are *greater-than-or-equal-to* the –6.

$$x \geq -6$$

725. $x < -2$

The parenthesis,) , indicates that the –2 is not included. The x represents all the numbers that are *less-than* the –2.

$$x < -2$$

726. $-4 \leq x < 7$

The bracket, [, indicates that you use that endpoint and show that with *greater-than-or-equal-to*. The parenthesis,) , indicates that the 7 is not included, so just use *less than*. The x indicates all those numbers that are *greater-than-or-equal-to* –4 while, at the same time, *less-than* 7.

$$-4 \leq x < 7$$

727. $2 < x < 3$

The parentheses indicate that you want all the numbers between 2 and 3 but not including either. You have to be careful not to confuse this with the coordinates (2, 3) in graphing. Use *less-than* for both inequalities.

$$2 < x < 3$$

728. $x < 4$

First add 5 to each side of the inequality.

$$\begin{array}{rcrcr} 2x & - & 5 & < & 3 \\ & & +5 & & +5 \\ \hline 2x & & & < & 8 \end{array}$$

Now divide each side by 2.

$$\frac{2x}{2} < \frac{8}{2}$$

$$x < 4$$

In interval notation, the answer is written $(-\infty, 4)$.

729. $x \leq -5$

Add $-4x$ to each side and add 2 to each side.

$$\begin{array}{rcrcrcr} 3x & - & 2 & \geq & 4x & + & 3 \\ -4x & & +2 & & -4x & & +2 \\ \hline -x & & & \geq & & & 5 \end{array}$$

Another choice would have been to add $-3x$ to each side and -3 to each side. It's just a matter of preference, and this method keeps the variable on the left.

Now divide each side by -1. Be sure to switch the sense/direction of the inequality symbol.

$$-x \geq 5$$
$$\frac{-x}{-1} \leq \frac{5}{-1}$$
$$x \leq -5$$

In interval notation, this solution is written $(-\infty, -5]$.

730. $x \geq -6$

First distribute the -3 over the two terms on the left.

$$-3x - 21 \leq 2x + 9$$

Now add $3x$ to each side and -9 to each side.

$$\begin{array}{rrcrr} -3x & -\ 21 & \leq & 2x & +\ 9 \\ +3x & -9 & & +3x & -9 \\ \hline & -\ 30 & \leq & 5x & \end{array}$$

Another choice would have been to add $-2x$ to each side and $+21$ to each side. It's just a matter of preference, and this method keeps the variable positive.

Divide each side of the inequality by 5.

$$-30 \leq 5x$$
$$\frac{-30}{5} \leq \frac{5x}{5}$$
$$-6 \leq x$$

A better way to describe the solution is to reverse the statement (switch the number and variable as well as the inequality symbol). It then reads $x \geq -6$, and the corresponding interval notation is $[-6, \infty)$.

731. $x < 1$

First multiply each side of the inequality by -2. The sense/direction of the inequality gets reversed because the multiplier is negative.

$$\frac{x-3}{-2} > x$$
$$\cancel{-2} \cdot \frac{x-3}{\cancel{-2}} < -2 \cdot x$$
$$x - 3 < -2x$$

Now add $-x$ to each side of the inequality.

$$\begin{array}{rrcr} x & -\ 3 & < & -2x \\ -x & & & -x \\ \hline & -\ 3 & < & -3x \end{array}$$

Another choice would have been to add $2x$ to each side and $+3$ to each side. It's just a matter of preference, and this method keeps it to just one operation.

Answers 701–800

Now divide each side by –3. Be sure to switch the sense/direction of the inequality symbol.

$$\begin{array}{ccc} -3 & < & -3x \\ \frac{-3}{-3} & > & \frac{-3x}{-3} \\ 1 & > & x \end{array}$$

A better way to describe the solution is to reverse the statement (switch the number and variable as well as the inequality symbol). It then reads $x < 1$, and the corresponding interval notation is $(-\infty, 1)$.

732. $x > 12$

First, clear the fractions by multiplying each term by 12, the common denominator of the two fractions.

$$\cancel{12}^{4} \cdot \frac{x}{\cancel{3}} + \cancel{12}^{3} \cdot \frac{x}{\cancel{4}} > 12 \cdot 7$$

$$4x + 3x > 84$$

Combine like terms on the left, and then divide each side by 7.

$$7x > 84$$

$$\frac{\cancel{7}x}{\cancel{7}} > \frac{84}{7}$$

$$x > 12$$

In interval notation, this is written $(12, \infty)$.

733. $x > 1$

First, clear the fractions by multiplying each term by 60, the common denominator of all the fractions.

$$\cancel{60}^{15} \cdot \frac{3}{\cancel{4}} - \cancel{60}^{12} \cdot \frac{x}{\cancel{5}} < \cancel{60}^{20} \cdot \frac{2x}{\cancel{3}} - \cancel{60} \cdot \frac{7}{\cancel{60}}$$

$$45 - 12x < 40x - 7$$

Now add –40 x to each side, and then add –45 to each side.

$$\begin{array}{ccccccc} 45 & - & 12x & < & 40x & - & 7 \\ & & -40x & & -40x & & \\ \hline 45 & - & 52x & < & & - & 7 \\ -45 & & & & & & -45 \\ \hline & - & 52x & < & & - & 52 \end{array}$$

Divide each side by –52, reversing the inequality sign.

$$\frac{-52x}{-52} > \frac{-52}{-52}$$

$$x > 1$$

In interval notation, this is written $(1, \infty)$.

734. $-2 \le x < 2$

First add –1 to each section of the inequality.

$$\begin{array}{rcrcrcr} -5 & \le & 3x & + & 1 & < & 7 \\ -1 & & & & -1 & & -1 \\ \hline -6 & \le & 3x & & & < & 6 \end{array}$$

Now divide each section by 3.

$$\frac{-6}{3} \le \frac{3x}{3} < \frac{6}{3}$$

$$-2 \le x < 2$$

In interval notation, this solution is written [–2, 2).

735. $-1 < x < 2$

First add –6 to each section of the inequality.

$$\begin{array}{rcrcrcr} -4 & < & 6 & - & 5x & < & 11 \\ -6 & & -6 & & & & -6 \\ \hline -10 & < & & - & 5x & < & 5 \end{array}$$

Now divide each section by –5. The senses/directions of the inequality need to be reversed.

$$-10 < -5x < 5$$

$$\frac{-10}{-5} > \frac{-5x}{-5} > \frac{5}{-5}$$

$$2 > x > -1$$

To rewrite the statement with the smaller number to the left as found on the number line, reverse the positions of the numbers and reverse the inequality symbols.

$-1 < x < 2$ is written (–1, 2) in interval notation.

736. $9 \le x \le 29$

First multiply each section of the inequality by 4.

$$4 \cdot 2 \le \not{4} \cdot \frac{x-1}{\not{4}} \le 4 \cdot 7$$

$$8 \le x - 1 \le 28$$

Now add 1 to each section.

$$\begin{array}{rcrcrcr} 8 & \le & x & - & 1 & \le & 28 \\ +1 & & & & +1 & & +1 \\ \hline 9 & \le & x & & & \le & 29 \end{array}$$

This solution is written [9, 29] in interval notation.

737. $-1 < x < 0$

You could first distribute the –3 in the middle section, but, since the numbers to the left and right of the inequality signs are divisible by 3, you can divide each section by –3 to simplify things. Be sure to reverse the inequality symbols.

Answers 701–800

$$-15 < -3(3-2x) < -9$$

$$\frac{-15}{-3} > \frac{\cancel{-3}(3-2x)}{\cancel{-3}} > \frac{-9}{-3}$$

$$5 > 3-2x > 3$$

Now add –3 to each section.

$$\begin{array}{rcl} 5 > 3 - 2x > 3 \\ -3 \quad -3 \qquad\quad -3 \\ \hline 2 > \quad - 2x > 0 \end{array}$$

Finally, divide each section by –2, reversing the senses/directions of the inequality signs again.

$$2 > -2x > 0$$

$$\frac{2}{-2} < \frac{-2x}{-2} < \frac{0}{-2}$$

$$-1 < x < 0$$

In interval notation, the solution is written (–1, 0).

738. $-4 < x < 3$

You want the product of the binomials to be a negative number. Find the *critical numbers* by setting each binomial equal to zero and solving for the variable.

The critical numbers are $x = 3$ and $x = -4$. Place those numbers on a number line. Use an open circle to mark those positions, because the inequality is *less-than* and doesn't include the endpoints.

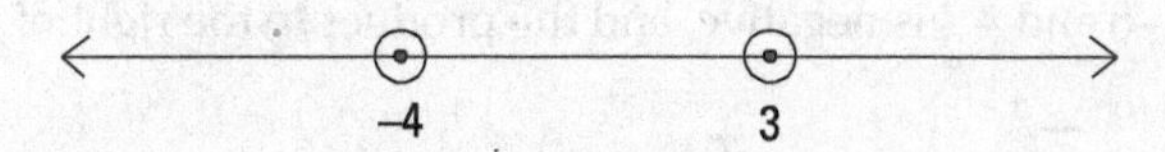

Now determine the sign of each binomial within each region determined by the critical numbers.

$$\begin{array}{llll} (x-3) & - & - & + \\ (x+4) & - & + & + \\ \hline & & -4 & 3 \end{array}$$

The above chart shows that the factor $(x-3)$ is negative to the left of –4, negative between –4 and 3, and positive to the right of 3. Those determinations are made by just choosing a number in each region and substituting it into the binomial. Likewise, the binomial $(x+4)$ is negative to the left of –4 and then positive in both regions to the right.

Since both factors are negative in the region to the left of –4, their product is positive. The product between –4 and 3 is negative, and the product to the right of 3 is positive.

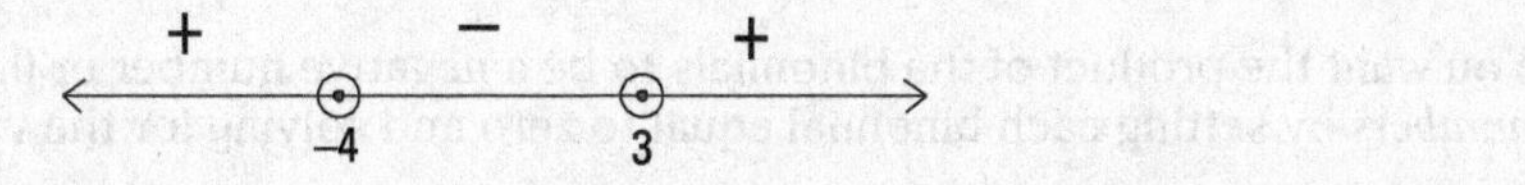

The problem asks for when the product is negative, so you want all values between –4 and 3, written $-4 < x < 3$ or $(-4, 3)$.

739.

$x \le -8$ or $x \ge -\frac{5}{2}$

You want the product of the binomials to be a positive number or 0. Find the *critical numbers* by setting each binomial equal to zero and solving for the variable.

The critical numbers are $x = -\frac{5}{2}$ and $x = -8$. Place those numbers on a number line. Use closed circles to mark those positions, because the inequality is *greater-than-or-equal to* and therefore includes the endpoints.

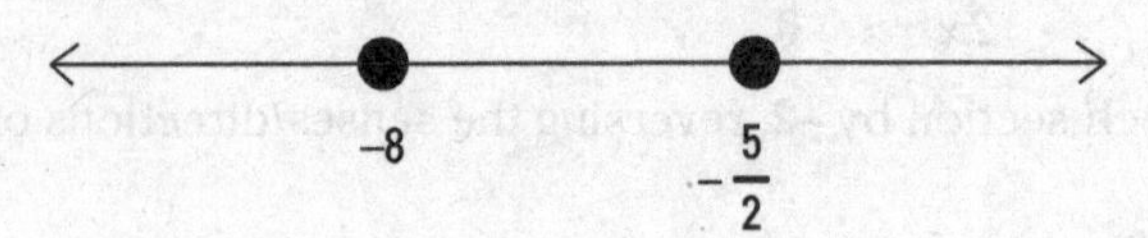

Now determine the sign of each binomial within each region determined by the critical numbers.

$(2x+5)$	–	–	+
$(x+8)$	–	+	+
	–8	$-\frac{5}{2}$	

The above chart shows that the factor $(2x+5)$ is negative to the left of –8, negative between –8 and $-\frac{5}{2}$, and positive to the right of $-\frac{5}{2}$. Those determinations are made by just choosing a number in each region and substituting it into the binomial. Likewise, the binomial $(x + 8)$ is negative to the left of –8 and then positive in both regions to the right.

Since both factors are negative in the region to the left of –8, their product is positive. The product between –8 and $-\frac{5}{2}$ is negative, and the product to the right of $-\frac{5}{2}$ is positive.

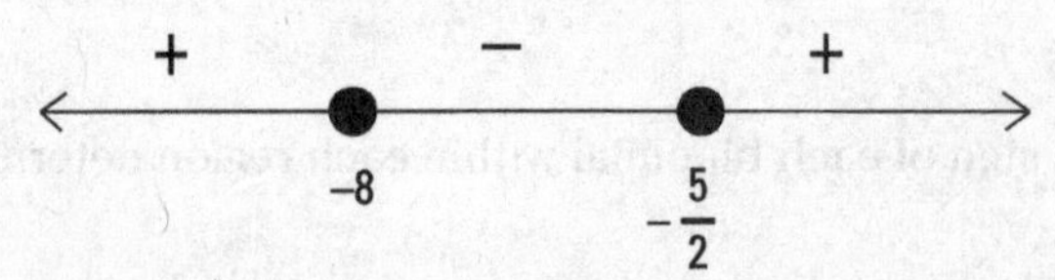

The problem asks for when the product is positive or 0, so you want all the numbers to the left of –8, including the –8, and all the numbers to the right of $-\frac{5}{2}$, including the $-\frac{5}{2}$. The solution is written $x \le -8$ or $x \ge -\frac{5}{2}$. In interval notation it's $(-\infty,-8] \cup \left[-\frac{5}{2},\infty\right)$. The U shape stands for "union" and means "or."

740.

$-1 \le x \le 9$

You first need to factor the quadratic into $(x-9)(x+1) \le 0$. This allows you to determine the *critical numbers*.

You want the product of the binomials to be a negative number or 0. Find the *critical numbers* by setting each binomial equal to zero and solving for the variable.

The critical numbers are $x = 9$ and $x = -1$. Place those numbers on a number line. Use closed circles to mark those positions, because the inequality is *less-than-or-equal to* and therefore includes the endpoints.

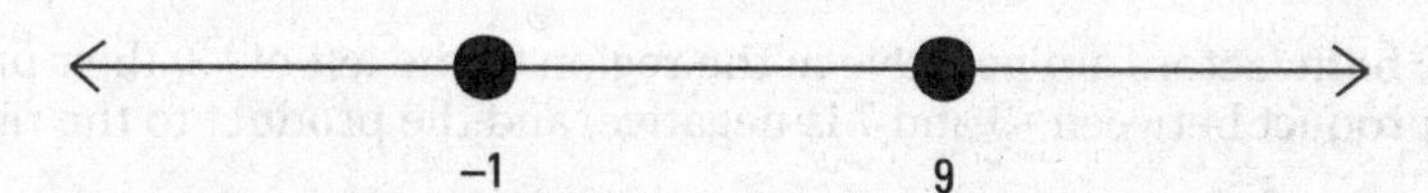

Now determine the sign of each binomial within each region determined by the critical numbers.

$(x-9)$	−	−	+
$(x+1)$	−	+	+
		−1	9

The above chart shows that the factor $(x-9)$ is negative to the left of −1, negative between −1 and 9, and positive to the right of 9. Those determinations are made by just choosing a number in each region and substituting it into the binomial. Likewise, the binomial $(x+1)$ is negative to the left of −1 and then positive in both regions to the right.

Since both factors are negative in the region to the left of −1, their product is positive. The product between −1 and 9 is negative, and the product to the right of 9 is positive.

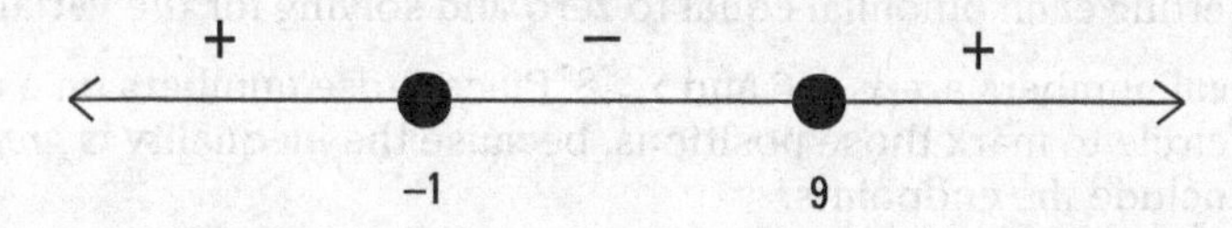

The problem asks for when the product is negative or 0, so you want all the numbers between −1 and 9, including both −1 and 9. The solution is written $-1 \le x \le 9$. In interval notation it's $[-1, 9]$.

741. $x < -3$ or $x > 7$

You first need to factor the quadratic into $(x+3)(x-7) > 0$. This allows you to determine the *critical numbers*.

You want the product of the binomials to be a positive number. Find the *critical numbers* by setting each binomial equal to zero and solving for the variable.

The critical numbers are $x = -3$ and $x = 7$. Place those numbers on a number line. Use an open circle to mark those positions, because the inequality is *greater-than* and doesn't include the endpoints.

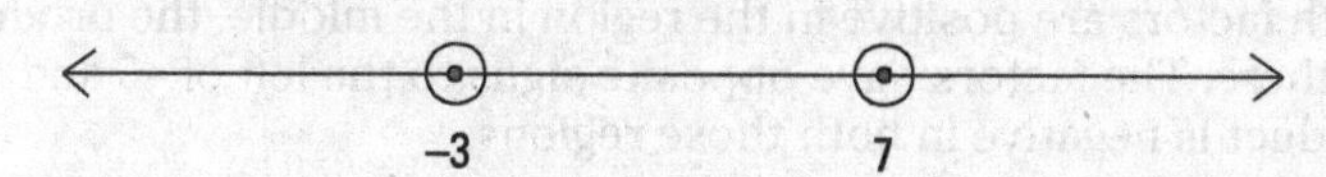

Now determine the sign of each binomial within each region determined by the critical numbers.

$(x+3)$	−	+	+
$(x-7)$	−	−	+
		−3	7

The above chart shows that the factor $(x+3)$ is negative to the left of −3 and then positive in both regions to the right. Those determinations are made by just choosing a number in each region and substituting it into the binomial. Likewise, the binomial $(x-7)$ is negative to the left of −3, negative between −3 and 7, and then positive to the right of 7.

Since both factors are negative in the region to the left of –3, their product is positive. The product between –3 and 7 is negative, and the product to the right of 7 is positive.

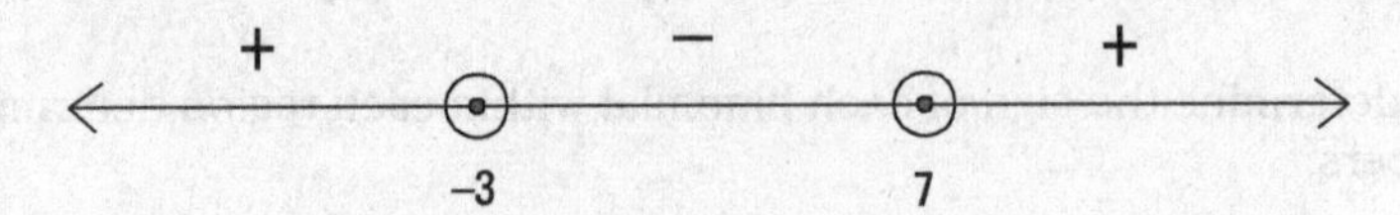

The problem asks for when the product is positive, so you want all the numbers to the left of –3 and to the right of 7. The solution is written $x < -3$ or $x > 7$, or, in interval notation, $(-\infty, -3) \cup (7, \infty)$. The "U" symbol stands for "union" meaning "or."

742. $-6 < x < 8$

First add $2x$ to each side of the equation, giving you $48 + 2x - x^2 > 0$. Now you need to factor the quadratic into $(6 + x)(8 - x) > 0$. This allows you to determine the *critical numbers*.

You want the product of the binomials to be a positive number. Find the *critical numbers* by setting each binomial equal to zero and solving for the variable.

The critical numbers are $x = -6$ and $x = 8$. Place those numbers on a number line. Use an open circle to mark those positions, because the inequality is *greater-than* and doesn't include the endpoints.

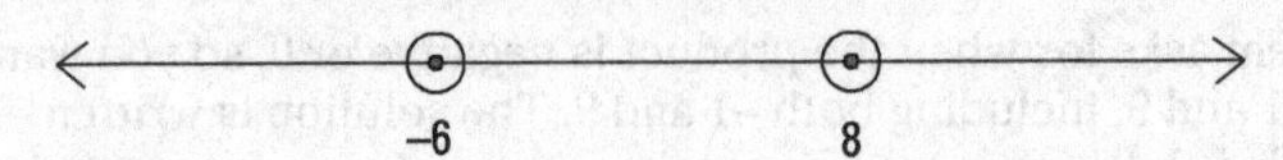

Now determine the sign of each binomial within each region determined by the critical numbers.

$(6+x)$	–	+	+
$(8-x)$	+	+	–
	–6	8	

The above chart shows that the factor $(6 + x)$ is negative to the left of –6 and then positive in both regions to the right. Those determinations are made by just choosing a number in each region and substituting it into the binomial. Likewise, the binomial $(8 - x)$ is positive in both regions to the left of 8 and negative to the right of 8.

Since both factors are positive in the region in the middle, the product of the factors is positive there. The factors have opposite signs to the left of –6 and to the right of 8, so their product is negative in both those regions.

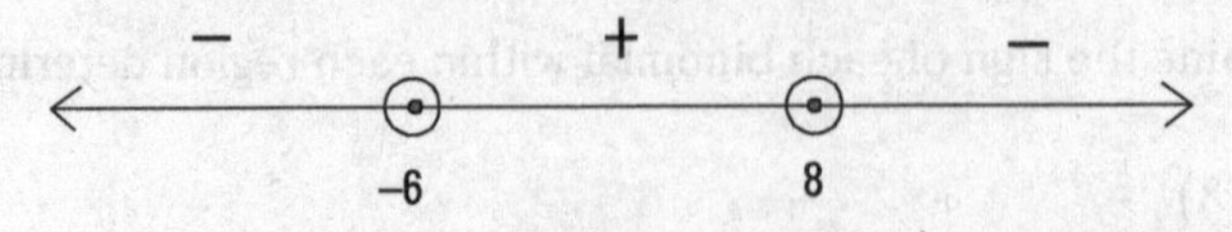

The problem asks for when the product is positive, so you want all the numbers between –6 and 8. The solution is written $-6 < x < 8$ or, in interval notation, $(-6, 8)$.

743. $x \leq -6$ or $x \geq 6$

You first need to factor the quadratic into $(6 + x)(6 - x) \leq 0$. This allows you to determine the *critical numbers*.

You want the product of the binomials to be a negative number or 0. Find the *critical numbers* by setting each binomial equal to zero and solving for the variable.

The critical numbers are $x = -6$ and $x = 6$. Place those numbers on a number line. Use closed circles to mark those positions, because the inequality is *less-than-or-equal to* and therefore includes the endpoints.

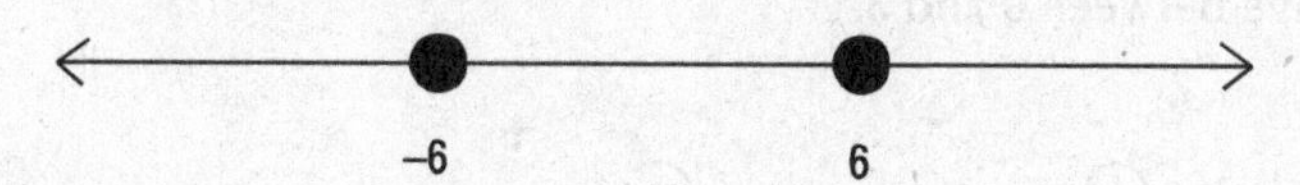

Now determine the sign of each binomial within each region determined by the critical numbers.

$(6+x)$	−	+	+
$(6-x)$	+	+	−
	−6	6	

The above chart shows that the factor $(6 + x)$ is negative to the left of −6 and positive in both regions to the right of −6. Those determinations are made by just choosing a number in each region and substituting it into the binomial. Likewise, the binomial is $(6 - x)$ is negative to the right of 6 and then positive in both regions to the left.

Since both factors are positive in the region between −6 and 6, their product is positive. The product to the left of −6 is negative, and the product to the right of 6 is negative.

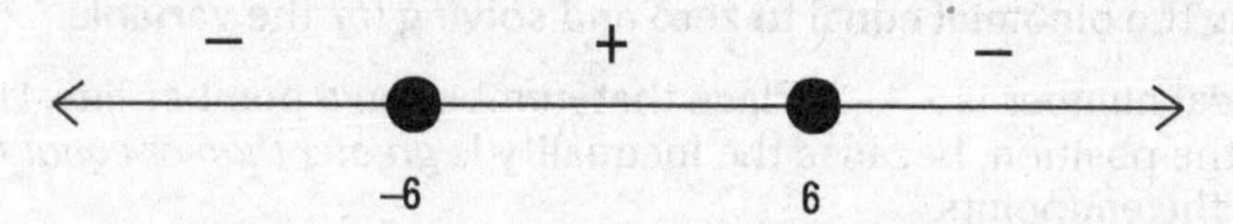

The problem asks for when the product is negative or 0, so you want all the numbers to the left of −6, including −6 as well as all the numbers to the right of 6, including the 6. The solution is written $x \le -6$ or $x \ge 6$. In interval notation it's $(-\infty, -6] \cup [6, \infty)$. The "U" symbol stands for "union" meaning "or."

744. $0 < x < 3$

First add $-15x$ to each side of the equation, giving you $5x^2 - 15x < 0$. Now you need to factor the quadratic into $5x(x - 3) < 0$. This allows you to determine the *critical numbers*. You want the product of the binomials to be a negative number. Find the *critical numbers* by setting each binomial equal to zero and solving for the variable.

The critical numbers are $x = 0$ and $x = 3$. Place those numbers on a number line. Use an open circle to mark those positions, because the inequality is *greater-than* and doesn't include the endpoints.

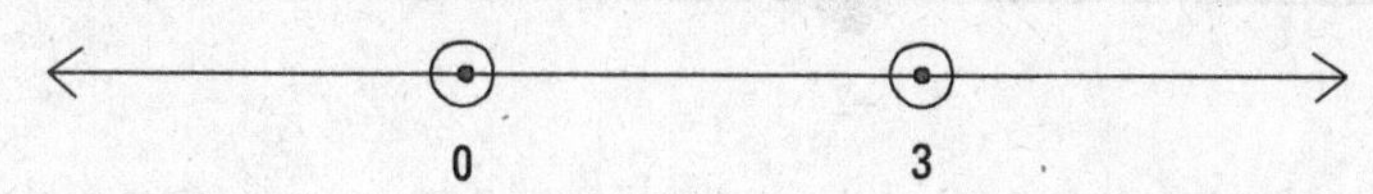

Now determine the sign of each binomial within each region determined by the critical numbers.

$5x$	−	+	+
$(x-3)$	−	−	+
	0	3	

The above chart shows that the factor $5x$ is negative to the left of 0 and then positive in both regions to the right. Those determinations are made by just choosing a number in each region and substituting it into the binomial. Likewise, the binomial $(x - 3)$ is negative to the left of 0 and between 0 and 3, and it's positive to the right of 3.

The product of the two factors is positive to the left of 0 and positive to the right of 3 and negative between 0 and 3.

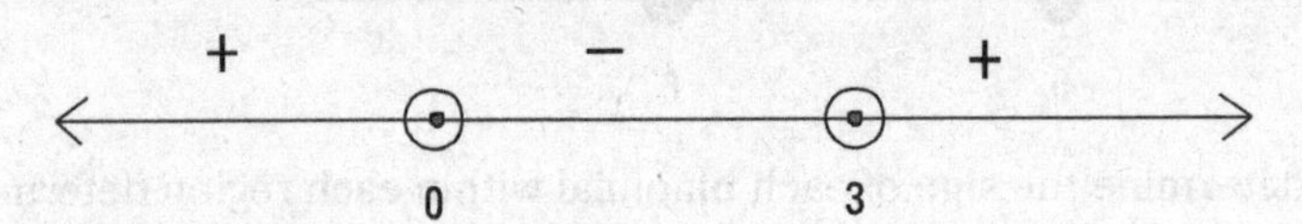

The revised equation asks for when the product is negative, so you want all the numbers between 0 and 3. The solution is written $0 < x < 3$ or, in interval notation, $(0, 3)$.

745. All real numbers

You first need to factor the quadratic into $(x + 2)^2 \geq 0$. Right at this point, you can see that the factored expression is *always* either positive or 0. The discussion continues, though, to confirm that conclusion.

The factored form allows you to determine the *critical number.*

You want the factored expression to be a positive number or 0. Find the *critical number* by setting the binomial equal to zero and solving for the variable.

The critical number is $x = -2$. Place that number on a number line. Use a closed circle to mark the position, because the inequality is *greater-than-or-equal to* and therefore includes the endpoints.

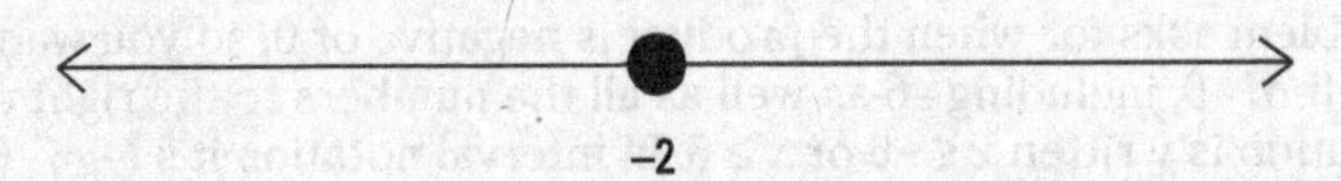

Now determine the sign of the binomial within each region determined by the critical number.

$(x+2)^2$ + +

–2

The above chart shows that the factor $(x + 2)^2$ is positive to the left of –2 and positive to the right of –2. Those determinations are made by just choosing a number in each region and substituting it into the binomial. The binomial is 0 at the critical number, so every real number makes the original statement true.

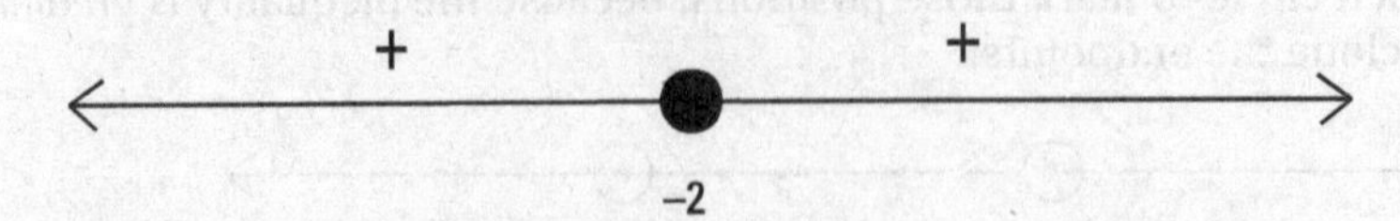

The solution is written *all real numbers* or, in interval notation, $(-\infty, \infty)$.

746. $-3 < x < 0$ or $x > 2$

You want the product of the factors to be a positive number. Find the *critical numbers* by setting each factor equal to zero and solving for the variable.

The critical numbers are $x = 0$, $x = -3$, and $x = 2$. Place those numbers on a number line. Use an open circle to mark those positions, because the inequality is *greater-than* and doesn't include the endpoints.

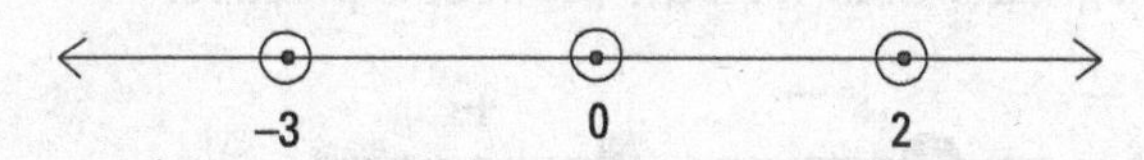

Now determine the sign of each factor within each region determined by the critical numbers.

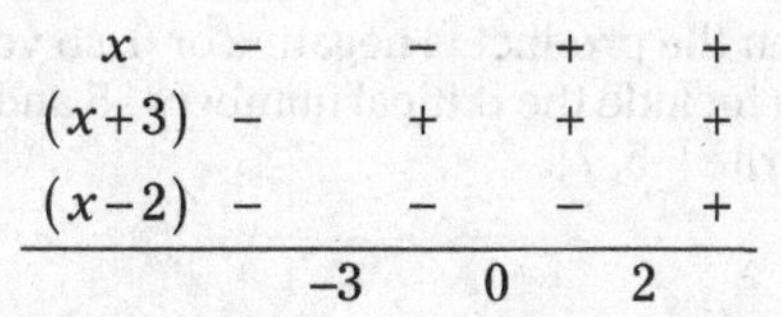

Consider each region formed by the critical numbers. Three factors are negative when x is less than -3, so their product is negative. Two factors are negative between -3 and 0, so the product of all three factors is positive. One factor is negative between 0 and 2, so the product of all three factors is negative. And all the factors are positive when x is greater than 2, so the product is positive.

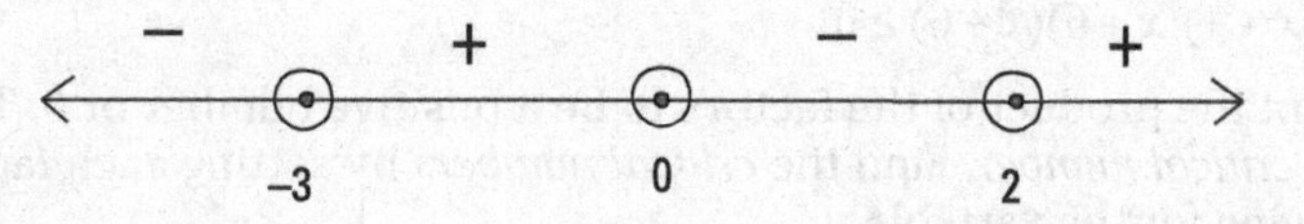

The problem asks for when the product is positive, so you want all the numbers between -3 and 0 and also all the numbers greater than 2. This is written $-3 < x < 0$ or $x > 2$. In interval notation you write $(-3, 0) \cup (2, \infty)$. The "U" symbol represents "union" which means "or."

747. $-5 \le x \le 7$

You want the product of the factors to be a negative number or 0. The product will be 0 at each *critical number*. Find the *critical numbers* by setting each factor equal to zero and solving for the variable.

The critical numbers are $x = -1$, $x = -5$, and $x = 7$. Place those numbers on a number line. Use closed circles to mark those positions, because the inequality is *less-than-or-equal to* and therefore includes the endpoints.

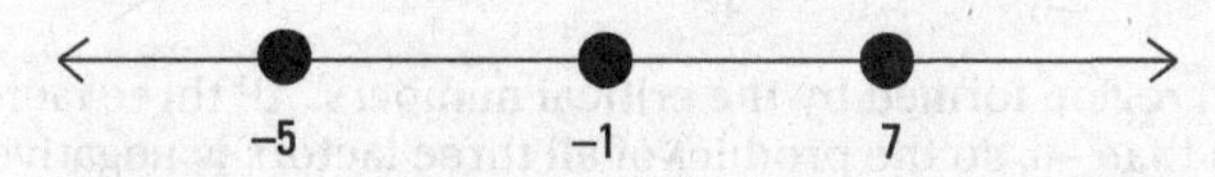

Now determine the sign of each factor within each region determined by the critical numbers.

$(x+1)^2$	+	+	+	+
$(x+5)$	−	+	+	+
$(x-7)$	−	−	−	+
		−5	−1	7

Notice that the squared factor is always positive (except at the critical number, $x = -1$). Consider each region formed by the critical numbers. Two factors are negative when x

is less than –5, so the product of all three factors is positive. One factor is negative between –5 and –1, so the product of all three factors is negative. One factor is negative between –1 and 7, so the product of all three factors is negative. And all the factors are positive when x is greater than 7, so the product is positive.

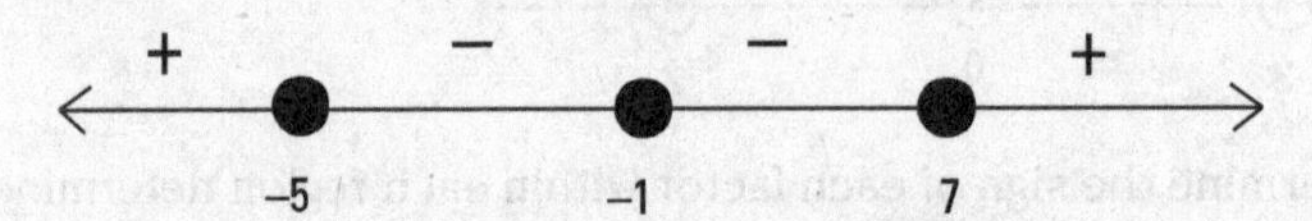

The problem asks for when the product is negative or 0, so you want all the numbers between –5 and 7 and you include the critical numbers –5 and 7. This is written $-5 \le x \le 7$. In interval notation you write [–5, 7].

748. $-6 \le x \le -1$ or $x \ge 6$

First factor the polynomial using grouping.

$$x^2(x+1) - 36(x+1) \ge 0$$

$$(x+1)(x^2-36) \ge 0$$

$$(x+1)(x+6)(x-6) \ge 0$$

You want the product of the factors to be a positive number or 0. The product will be 0 at each *critical number*. Find the *critical numbers* by setting each factor equal to zero and solving for the variable.

The critical numbers are $x = -1$, $x = -6$, and $x = 6$. Place those numbers on a number line. Use closed circles to mark those positions, because the inequality is *greater-than-or-equal to* and therefore includes the endpoints.

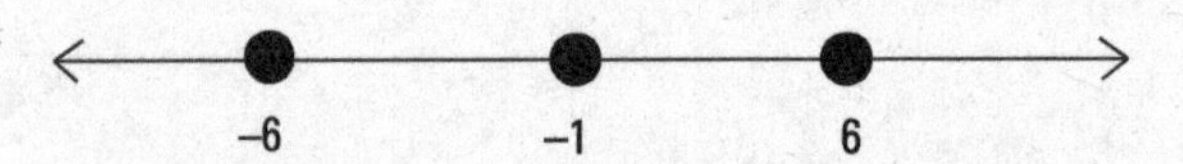

Now determine the sign of each factor within each region determined by the critical numbers.

$(x+1)$	–	–	+	+
$(x+6)$	–	+	+	+
$(x-6)$	–	–	–	+
	–6	–1	6	

Consider each region formed by the critical numbers. All three factors are negative when x is less than –6, so the product of all three factors is negative. Two factors are negative between –6 and –1, so the product of all three factors is positive. One factor is negative between –1 and 6, so the product of all three factors is negative. And all the factors are positive when x is greater than 6, so the product is positive.

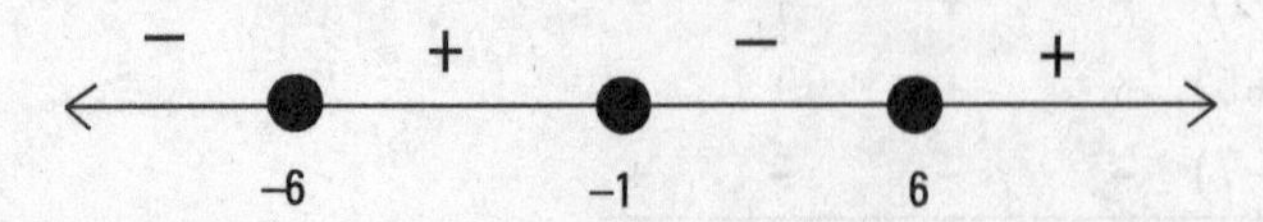

The problem asks for when the product is positive or 0, so you want all the numbers between –6 and –1 plus all the numbers greater than 6; include the critical numbers –6 and –1 and 6. This is written $-6 \le x \le -1$ or $x \ge 6$. In interval notation you write $[-6, -1] \cup [6, \infty)$.

749. $x < 0$

First factor the polynomial.

$$x(x^2 - 2x + 1) < 0$$

$$x(x - 1)^2 < 0$$

You want the product of the factors to be a negative number. Find the *critical numbers* by setting each factor equal to zero and solving for the variable.

The critical numbers are $x = 0$ and $x = 1$. Place those numbers on a number line. Use an open circle to mark those positions, because the inequality is *less-than* and doesn't include the endpoints.

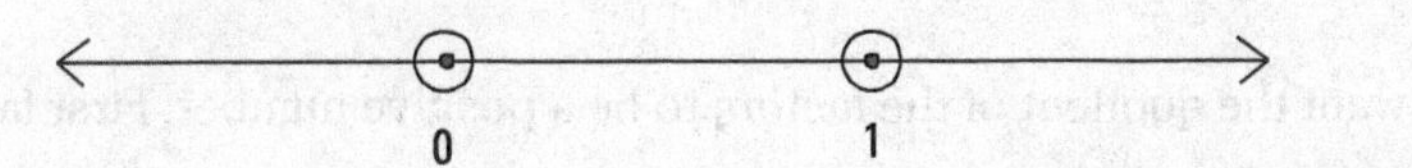

Now determine the sign of each factor within each region determined by the critical numbers.

		0		1	
x	−		+		+
$(x-1)^2$	+		+		+

Consider each region formed by the critical numbers. One factor is negative to the left of 0, so the product of the factors is negative. Both of the factors are positive between 0 and 1 and when x is greater than 1, so the product is positive.

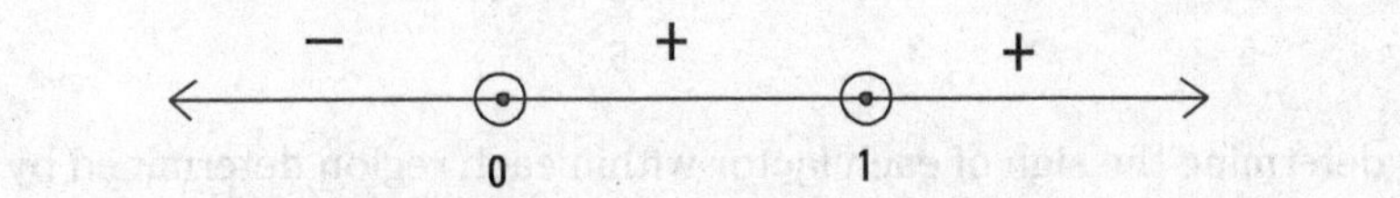

The problem asks for when the product is negative, so you want all the numbers less than 0. The solution is written $x < 0$. In interval notation you write $(-\infty, 0)$.

750. $-2 < x < -1$

You want the quotient of the factors to be a negative number. Find the *critical numbers* by setting each factor equal to zero and solving for the variable. Even though you can't use a number that puts a 0 in the denominator, you still have to use that critical number. The critical numbers are $x = -1$ and $x = -2$. Place those numbers on a number line. Use an open circle to mark those positions, because the inequality is *less-than* and doesn't include the endpoints and you can't include the value that causes division by 0.

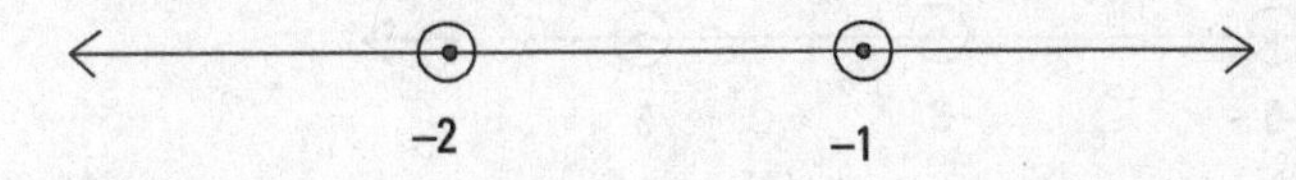

Now determine the sign of each factor within each region determined by the critical numbers.

		−2		−1	
$x+2$	−		+		+
$x+1$	−		−		+

Consider each region formed by the critical numbers. Both factors are negative to the left of –2, so the quotient of the factors is positive. One factor is negative between –2 and –1, so the quotient of the factors is negative. Both of the factors are positive when x is greater than –1, so the quotient is positive.

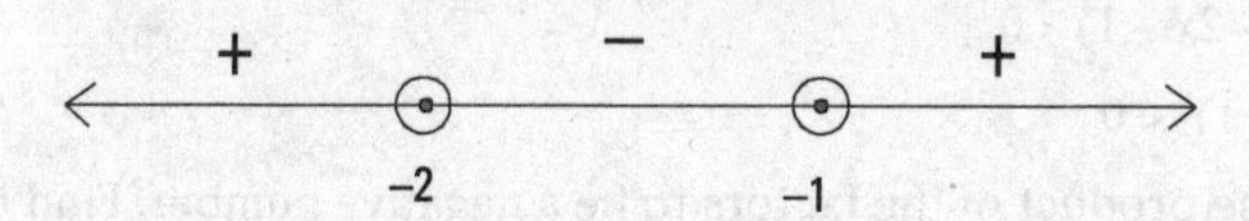

The problem asks for when the quotient is negative, so you want all the numbers between –2 and –1. The solution is written $-2 < x < -1$. In interval notation you write $(-2, -1)$.

751. $-5 < x < -3$ or $x > 5$

You want the quotient of the factors to be a positive number. First factor the numerator.

$$\frac{(x+5)(x-5)}{x+3} > 0$$

Find the *critical numbers* by setting each factor equal to zero and solving for the variable. Even though you can't use a number that puts a 0 in the denominator, you still have to use that critical number.

The critical numbers are $x = -5$, $x = 5$, and $x = -3$. Place those numbers on a number line. Use an open circle to mark those positions, because the inequality is *greater-than* and doesn't include the endpoints and you can't include the value that causes division by 0.

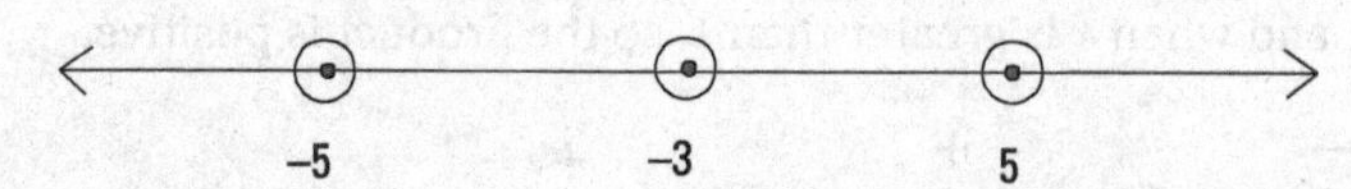

Now determine the sign of each factor within each region determined by the critical numbers.

$(x+5)$	–	+	+	+
$(x-5)$	–	–	–	+
$(x+3)$	–	–	+	+
	–5	–3	5	

Consider each region formed by the critical numbers. All three factors are negative to the left of –5, so the product/quotient is negative. Two factors are negative between –5 and –3, so the product/quotient of the factors is positive. One factor is negative between –3 and –5, so the product/quotient of the factors is negative. And all of the factors are positive when x is greater than 5, so the product/quotient is positive.

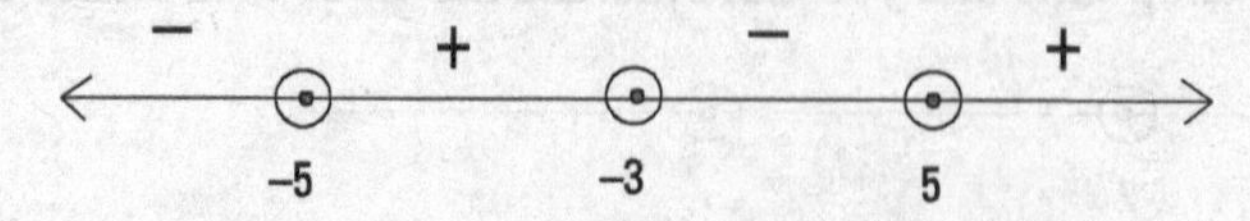

The problem asks for when the result is positive, so you want all the numbers between –5 and –3, and also the numbers greater than 5. The solution is written $-5 < x < -3$ or $x > 5$. In interval notation you write $(-5, -3) \cup (5, \infty)$.

752. $0 \le x < 4$

You want the quotient of the factors to be a negative number or zero. Find the *critical numbers* by setting each factor equal to zero and solving for the variable.

The critical numbers are $x = 0$, $x = 4$. Place those numbers on a number line. Use a closed circle for $x = 0$ and an open circle for $x = 4$. Even though 4 is a critical number, it cannot be used in the solution, because you cannot divide by 0.

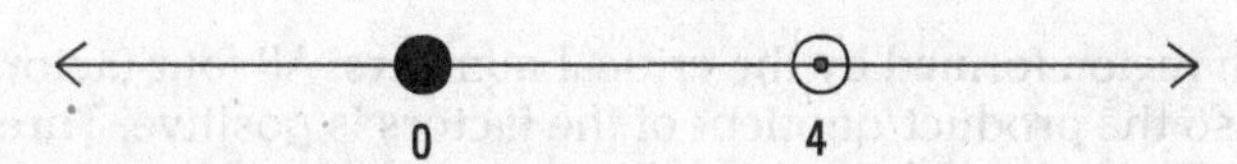

Now determine the sign of each factor within each region determined by the critical numbers.

x	−	+	+
$x-4$	−	−	+
	0	4	

Two factors are negative when x is less than 0, so the quotient of the factors is positive. One factor is negative between 0 and 4, so the quotient of the factors is negative. Both factors are positive when x is greater than 4, so the quotient is positive.

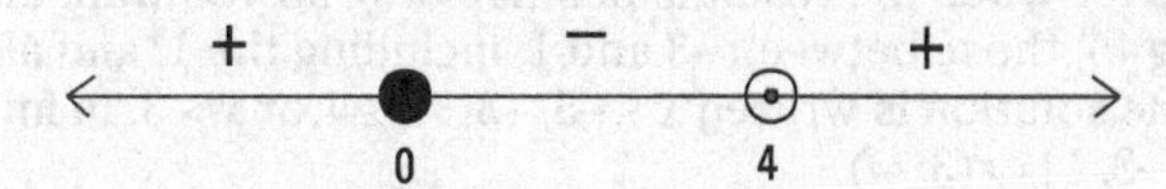

The problem asks for when the quotient is negative or 0, so you want all the numbers between 0 and 4; include the 0 but not the 4. This is written $0 \le x < 4$. In interval notation you write $[0, 4)$.

753. $x \le -5$ or $-3 < x \le 1$ or $x > 3$

You want the quotient of the expression to be a positive number or 0. A 0 in the numerator will result in a 0 for the value, but you cannot have a 0 in the denominator.

First factor the numerator and denominator.

$$\frac{(x+5)(x-1)}{(x+3)(x-3)} \ge 0$$

Find the *critical numbers* by setting each factor equal to zero and solving for the variable. Even though you can't use a number that puts a 0 in the denominator, you still have to use that critical number.

The critical numbers are $x = -5$, $x = 1$, $x = -3$, and $x = 3$. Place those numbers on a number line. Use an open circle to mark the positions of the critical numbers coming from the denominator and solid circles for those coming from the numerator.

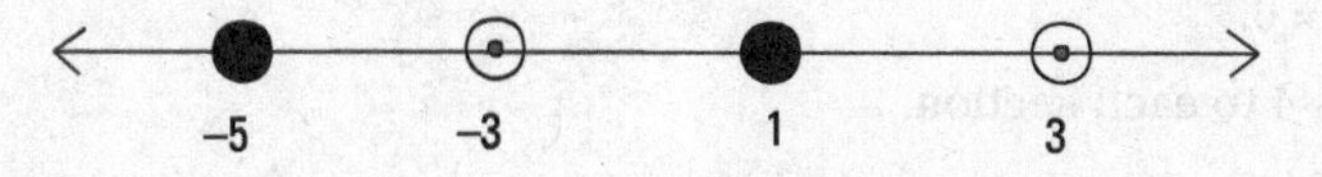

Now determine the sign of each factor within each region determined by the critical numbers.

$(x+5)$	−	+	+	+	+
$(x-1)$	−	−	−	+	+
$(x+3)$	−	−	+	+	+
$(x-3)$	−	−	−	−	+
		−5	−3	1	3

Consider each region formed by the critical numbers. All four factors are negative to the left of –5, so the product/quotient of the factors is positive. Three factors are negative between –5 and –3, so the product/quotient of the factors is negative. Two factors are negative between –3 and 1, so the product/quotient of the factors is positive. One factor is negative between 1 and 3, so the product/quotient of the factors is negative. And all of the factors are positive when x is greater than 3, so the product/quotient of the factors is positive.

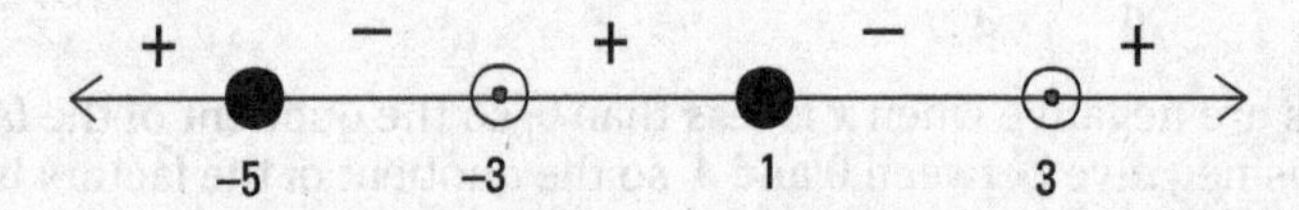

The problem asks for when the result is positive or 0, so you want all the numbers less than –5, including –5, those between –3 and 1, including the 1, and also the numbers greater than 3. The solution is written $x \leq -5$, $-3 < x \leq 1$ or $x > 3$. In interval notation you write $(-\infty, -5] \cup (-3, 1] \cup (3, \infty)$.

754. $x \geq \frac{5}{3}$ **or** $x \leq -3$

The absolute value inequality statement is equivalent to the solutions of the two inequalities $3x + 2 \geq 7$ and $3x + 2 \leq -7$.

First solving $3x + 2 \geq 7$, add –2 to each side. Then divide both sides by 3.

$$3x \geq 5$$

$$x \geq \frac{5}{3}$$

Next, solving $3x + 2 \leq -7$, add –2 to each side. Then divide both sides by 3.

$$3x \leq -9$$

$$x \leq -3$$

The solution consists of $x \geq \frac{5}{3}$ or $x \leq -3$. In interval notation, this is written $(-\infty, -3] \cup \left[\frac{5}{3}, \infty\right)$.

755. $-2 < x < 10$

The absolute value inequality statement is equivalent to the solution of the inequality $-6 < 4 - x < 6$.

First add –4 to each section.

$$-10 < -x < 2$$

Now divide each section by –1, which reverses all the senses.

$10 > x > -2$.

A more traditional statement has the numbers in order from left to right: $-2 < x < 10$. In interval notation, this is written (–2, 10).

756. $-\frac{5}{2} \le x \le \frac{3}{2}$

You first divide each side of the inequality by 5 to get $|2x+1| \le 4$.

This absolute value inequality is equivalent to the inequality

$-4 \le 2x + 1 \le 4$.

Adding –1 to each section you get:

$-5 \le 2x \le 3$.

Now divide each section by 2 and you get:

$-\frac{5}{2} \le x \le \frac{3}{2}$.

In interval notation, this is written $\left[-\frac{5}{2}, \frac{3}{2}\right]$.

757. $x \le -5$ **or** $x \ge \frac{19}{5}$

First, multiply each side of the inequality by 2.

$|5x+3| \ge 22$

This absolute value inequality is equivalent to the solutions of the two inequalities $5x + 3 \ge 22$ and $5x + 3 \le -22$.

Solving $5x + 3 \ge 22$, you first add –3 to each side and get $5x \ge 19$. Dividing each side by 5 you get:

$x \ge \frac{19}{5}$.

Now, solving $5x + 3 \le -22$, you add –3 to each side and get $5x \le -25$. Dividing each side by 5, you get $x \le -5$.

So the solution is $x \le -5$ or $x \ge \frac{19}{5}$. In interval notation, this is written $(-\infty, -5] \cup \left[\frac{19}{5}, \infty\right)$.

758. $x > 4$ **or** $x < -12$

First add 5 to each side of the inequality to isolate the absolute value portion on the left.

$|x+4| > 8$

This absolute value inequality is equivalent to the solutions of the two inequalities $x + 4 > 8$ and $x + 4 < -8$.

Solving $x + 4 > 8$ by adding –4 to each side, you have $x > 4$.

Solving $x + 4 < -8$ by adding –4 to each side, you have $x < -12$. So the solution is written:

$x > 4$ or $x < -12$. In interval notation, this is written $(-\infty, -12) \cup (4, \infty)$.

759. $1 \le x \le 4$

First add –4 to each side of the inequality to isolate the absolute value.

$$|5-2x| \le 3$$

Now write the equivalent compound inequality.

$$-3 \le 5-2x \le 3$$

Add –5 to each section of the inequality.

Then divide each section by –2. Be sure to reverse the senses/inequality signs.

$$-8 \le -2x \le -2$$

$$\frac{-8}{-2} \ge \frac{-2x}{-2} \ge \frac{-2}{-2}$$

$$4 \ge x \ge 1$$

A more conventional way to write this statement is to "flip" the statement, reversing the numbers and the inequality symbols. This puts the numbers in the order found on the number line.

$1 \le x \le 4$ is written [1, 4] in interval notation.

760. $x \ge 6$ or $x \le 4$

Before applying the equivalence relationship between the absolute value inequality and the inequality statements to be solved, the absolute value portion needs to be isolated. This is accomplished by adding 4 to each side and then dividing each side by 2.

$$2|x-5|-4 \ge -2$$

$$2|x-5| \ge 2$$

$$|x-5| \ge 1$$

Now, using the inequality equivalences, you have $x-5 \ge 1$ and $x-5 \le -1$.

Solving $x-5 \ge 1$, you add 5 to each side and have $x \ge 6$.

Solving $x-5 \le -1$, you add 5 to each side and have $x \le 4$.

The solution is $x \ge 6$ or $x \le 4$. In interval notation, this is $(-\infty, 4] \cup [6, \infty)$.

761. $\frac{11}{3} < x < \frac{13}{3}$

Before applying the equivalence relationship between the absolute value inequality and the inequality statements to be solved, the absolute value portion needs to be isolated. This is accomplished by adding –3 to each side and then multiplying each side by 6.

$$\frac{1}{6}|3x-1|+3<5$$

$$\frac{1}{6}|3x-1|<2$$

$$|3x-1|<12$$

Answers 701–800

Now, using the compound inequality equivalence, you have $-12 < 3x - 1 < 12$.

Solving $-12 < 3x - 1 < 12$, you add 1 to each section to get $-11 < 3x < 13$ and then divide each section by 3 to get $\frac{-11}{3} < x < \frac{13}{3}$.

Written in interval notation, it's $\left(\frac{-11}{3}, \frac{13}{3}\right)$.

762. $-2 < x \leq 1$

The inequality needs to be broken into two separate problems and the intersection (shared portions) of the two problems used as the solution to the original statement.

First solve $-4 < 3x + 2$.

Adding -2 to each side, you have $-6 < 3x$. Dividing each side by 3, you have $-2 < x$ or $x > -2$.

Solving the second inequality, $3x + 2 \leq 2x + 3$, you first add $-2x$ to each side to get $x + 2 \leq 3$ and then add -2 to each side to get $x \leq 1$.

Graph the two solutions, $x > -2$ and $x \leq 1$ on a number line.

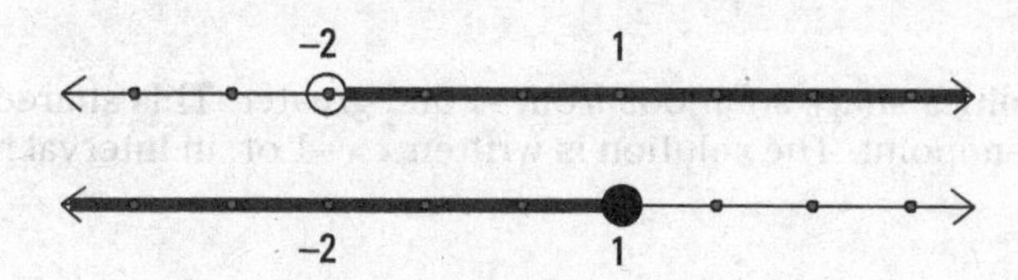

The two inequalities share solutions from -2 up to $+1$. This shared solution does not include -2, but it includes $+1$. The solution is written $-2 < x \leq 1$ or, in interval notation, $(-2, 1]$.

763. $3 \leq x \leq 9$

The inequality needs to be broken into two separate problems and the intersection (shared portions) of the two problems used as the solution to the original statement.

First solve $1 \leq 2x - 5$.

Adding 5 to each side, you have $6 \leq 2x$. Dividing each side by 2, you have $3 \leq x$ or $x \geq 3$.

Solving the second inequality, $2x - 5 \leq \frac{1}{3}x + 10$, you first add 5 to each side to get $2x \leq \frac{1}{3}x + 15$. Then add $-\frac{1}{3}x$ to each side to get $\frac{5}{3}x \leq 15$. Now multiply each side of the equation by $\frac{3}{5}$ to get $x \leq 9$.

Graph the two solutions, $x \geq 3$ and $x \leq 9$, on a number line.

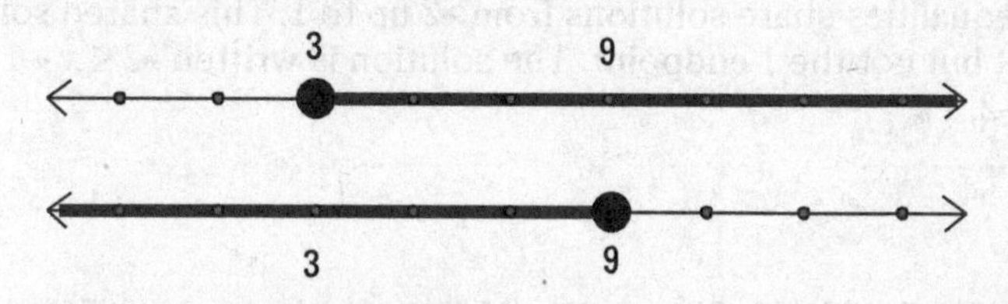

The two inequalities share solutions from 3 up to 9. This shared solution includes both endpoints. The solution is written $3 \leq x \leq 9$ or, in interval notation, $[3, 9]$.

764. $x > -1$

The inequality needs to be broken into two separate problems and the intersection (shared portions) of the two problems used as the solution to the original statement.

First solve $-5 < 4x - 1$.

Adding 1 to each side, you have $-4 < 4x$. Dividing each side by 4, you have $-1 < x$ or $x > -1$.

Solving the second inequality, $4x - 1 < 6x + 7$, you first add 1 to each side to get $4x < 6x + 8$. Then add $-6x$ to each side to get $-2x < 8$. Now divide each side of the equation by -2 to get $x > -4$.

Graph the two solutions, $x > -1$ and $x > -4$ on a number line.

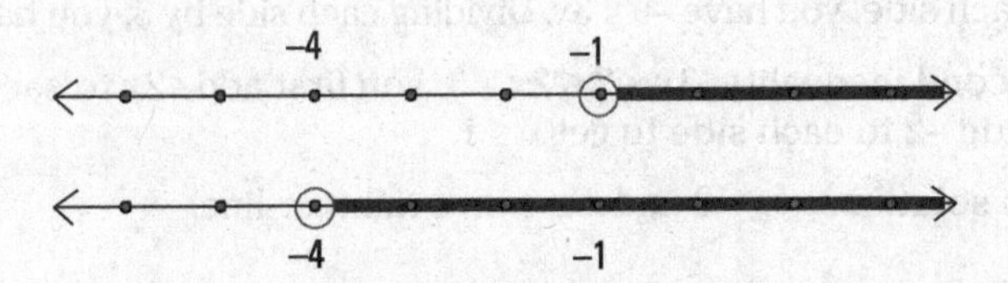

The two inequalities share solutions from –1 and greater. This shared solution doesn't include the –1 endpoint. The solution is written $x > -1$ or, in interval notation, $(-1, \infty)$.

765. $-2 \le x < 1$

The inequality needs to be broken into two separate problems and the intersection (shared portions) of the two problems used as the solution to the original statement.

First solve $x + 1 \le 3x + 5$.

Adding –1 to each side, you have $x \le 3x + 4$. Then adding $-3x$ to each side you have $-2x \le 4$. Dividing each side by –2, you have $x \ge -2$.

Solving the second inequality, $3x + 5 < 8$, you first add –5 to each side to get $3x < 3$. Then divide each side of the equation by 3, giving you $x < 1$.

Graph the two solutions, $x \ge -2$ and $x < 1$ on a number line.

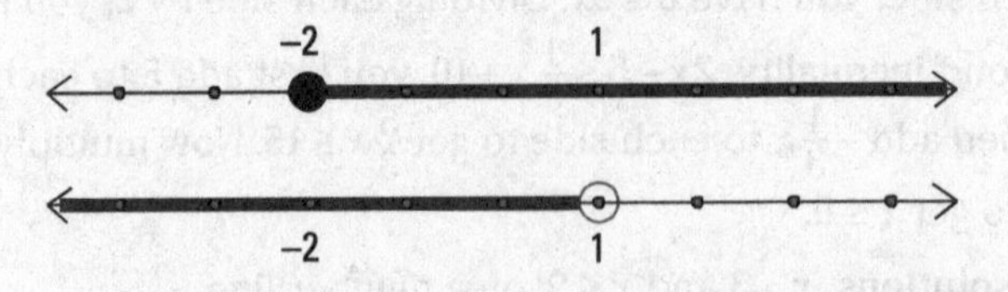

The two inequalities share solutions from –2 up to 1. This shared solution includes the –2 endpoint but not the 1 endpoint. The solution is written $-2 \le x < 1$ or, in interval notation, $[-2, 1)$.

766. $5,000

Using $I = Prt$, let $P = 20{,}000$. Change the 2.5% to the decimal 0.025 and let it represent r. The t is replaced by 10.

$$I = 20{,}000(0.025)(10) = 5{,}000$$

767. **$4,320**

Using $I = Prt$, let $P = 4{,}000$. Change the 4% to the decimal 0.04 and let it represent r. The t is replaced by 2.

$I = 4{,}000(0.04)(2) = 320$. This is the amount of interest. Add this to the amount loaned for a total payback of $4,320.

768. **$133**

Using $I = Prt$, let $P = 3{,}600$. Change the 11% to the decimal 0.11 and let it represent r. The t is replaced by 3.

$I = 3{,}600(0.11)(3) = 1{,}188$. This is just the amount of interest you owe. Add this to the cost of the television:

$$3{,}600 + 1{,}188 = 4{,}788$$

Now divide the total by the number of months in 3 years; $3(12) = 36$.

$$4{,}788 \div 36 = 133$$

769. **8 years**

If the total is to be $13,400, then you need to earn $3,400 in interest (subtract $10,000 from $13,400).

Using $I = Prt$, let $P = 10{,}000$. Change the $4\frac{1}{4}\%$ to the decimal 0.0425 and let it represent r. The $3,400 interest is I. The equation you work with is:

$$3{,}400 = 10{,}000(0.0425)t.$$

To solve for t, first multiply $10{,}000(0.0425) = 425$.

$$3{,}400 = 425t$$

Now divide each side by 425.

It will take 8 years. Will the cost of the boat be the same then?

$$\frac{3400}{425} = \frac{\cancel{425}t}{\cancel{425}}$$

$$8 = t$$

770. **98.6°**

Using the formula $°F = \frac{9}{5}°C + 32$, substitute the given value of 37 for $°C$ and simplify.

$$\begin{aligned} °F &= \frac{9}{5}(37) + 32 \\ &= \frac{333}{5} + 32 = 66\frac{3}{5} + 32 \\ &= 98\frac{3}{5} = 98.6 \end{aligned}$$

771. **15°**

Using the formula $°C = \frac{5}{9}(°F - 32)$, substitute the given value of 59 for $°F$ and simplify.

$$°C = \frac{5}{9}(59 - 32)$$
$$= \frac{5}{9}(27) = 15$$

772. **100°**

Using the formula $°C = \frac{5}{9}(°F - 32)$, substitute the given value of 212 for $°F$ and simplify.

$$°C = \frac{5}{9}(212 - 32)$$
$$= \frac{5}{9}(180) = 100$$

773. **–40°**

Using the formula $°F = \frac{9}{5}°C + 32$, substitute the given value of –40 for $°C$ and simplify.

$$°F = \frac{9}{5}(-40) + 32$$
$$= -72 + 32 = -40$$

This is the only temperature at which the two measures are the same.

774. **1,275**

Using the formula $S_n = \frac{n(n+1)}{2}$, let $n = 50$.

$$S_{50} = \frac{50(50+1)}{2} = \frac{{}^{25}\cancel{50}(51)}{\cancel{2}} = 25(51) = 1275$$

775. **1 through 100**

Using the formula $S_n = \frac{n(n+1)}{2}$, let $S_n = 5050$.

$$5050 = \frac{n(n+1)}{2}$$

Multiply each side of the equation by 2, and then distribute the n on the right.

$$2 \cdot 5050 = \cancel{2} \cdot \frac{n(n+1)}{\cancel{2}}$$
$$10{,}100 = n(n+1)$$
$$10{,}100 = n^2 + n$$

Add –10,100 to each side of the equation.

$$0 = n^2 + n - 10{,}100$$

Factor the quadratic.

$$0 = (n - 100)(n + 101)$$

Setting the two binomials equal to 0, you have $n = 100$ or $n = -101$. Discard the negative solution, and you have $n = 100$. The solution is the natural numbers from 1 through 100.

776. **1,050**

To do this problem, you find the sum of the natural numbers from 1 through 60 and then subtract the sum of the numbers from 1 through 39.

$$S_{60} = \frac{60(60+1)}{2} = 30(61) = 1830$$

$$S_{39} = \frac{39(39+1)}{2} = \frac{39(40)}{2} = 39(20) = 780$$

$$S_{60} - S_{39} = 1830 - 780 = 1050$$

777. **1,515**

To count the seats, you need the sum of the numbers from 36 through 65. If the first row has 36 seats, the second row has 37 seats, the third row has 38 seats, and so on, then you find that the row number has 35 more seats than that number (30th row has 65 seats). So first find the sum of the numbers from 1 through 65 and then subtract the sum of the numbers from 1 through 35.

$$S_{65} = \frac{65(65+1)}{2} = 65(33) = 2145$$

$$S_{35} = \frac{35(35+1)}{2} = 35(18) = 630$$

$$S_{65} - S_{35} = 2145 - 630 = 1{,}515$$

778. **330 miles**

Using $d = rt$, replace the r with 55 and the t with 6.

$$d = (55)(6) = 330$$

779. **60 mph**

Using $d = rt$, replace the d with 450 and the t with 7.5.

$$450 = r(7.5)$$

Divide each side of the equation by 7.5.

$$\frac{450}{7.5} = \frac{r(\cancel{7.5})}{\cancel{7.5}}$$

$$60 = r$$

780. **17.5 hours**

Using $d = rt$, replace the d with 1,050 and the r with 60.

$$1050 = 60t$$

Answers 701–800

Divide each side of the equation by 60.

$$\frac{1050}{60} = \frac{\cancel{60}t}{\cancel{60}}$$
$$17.5 = t$$

781. **50 mph**

Using $d = rt$, you have a total distance of 150 + 200 miles = 350 miles. The total amount of time spent driving is seven hours. You find this by determining that it's eight hours from 8:00 am until 4:00 pm and then subtracting the one hour you stopped.

Replacing d with 350 and t with 7,

$$350 = r(7)$$

Divide each side of the equation by 7.

$$\frac{350}{7} = \frac{r(\cancel{7})}{\cancel{7}}$$
$$50 = r$$

782. **11:00 am**

To solve this problem, you need the distance that Hank traveled to be equal to the distance that Helen traveled (since Helen eventually catches Hank). Using $d = rt$, distance Hank traveled = distance Helen traveled:

$$d_{Hank} = d_{Helen}$$
$$r_{Hank}t_{Hank} = r_{Helen}t_{Helen}$$

Let t be the amount of time Hank traveled until Helen caught up with him.

And let $t - 0.5$ be the amount of time that Helen traveled (since Helen started driving 0.5 hour after Hank).

$$r_{Hank}t_{Hank} = r_{Helen}t_{Helen}$$
$$50t = 60(t - 0.5)$$

Distribute the 60 on the right.

$$50t = 60t - 30$$

Add $-60t$ to each side of the equation.

$$-10t = -30$$

Now divide each side of the equation by -10.

$$\frac{\cancel{-10}t}{\cancel{-10}} = \frac{-30}{-10}$$
$$t = 3$$

Since t represents the amount of time that Hank traveled, and Hank left at 8:00, then it must have been 11:00 when Helen caught up with him.

783. **3:00 pm**

To solve this problem, you need the distance that the first bus traveled and the distance that the second bus traveled to have a sum of 800 miles.

Using $d = rt$:

distance first bus + distance second bus = 800

$$d_{First} + d_{Second} = 800$$

$$r_{First}t_{First} + r_{Second}t_{Second} = 800$$

Let t represent the amount of time that the first bus traveled, and let 40 represent its rate.

And let $t - 1$ be the amount of time that the second bus traveled (since it left one hour after the first bus), and let 55 represent its rate.

$$r_{First}t_{First} + r_{Second}t_{Second} = 800$$

$$40t + 55(t-1) = 800$$

Distribute the 55 and combine like terms.

$$40t + 55t - 55 = 800$$

$$95t - 55 = 800$$

Add 55 to each side of the equation.

$$95t = 855$$

Now divide each side of the equation by 95.

$$\frac{\cancel{95}t}{\cancel{95}} = \frac{855}{95}$$

$$t = 9$$

The first bus traveled 9 hours. Since it left at 6:00 am, 9 hours later is 3:00 pm.

784. **$1\frac{1}{4}$ hour**

To solve this problem, you need the total distance that Claire walked and Charlie walked to have a sum of 10 miles.

Using $d = rt$:

distance Claire + distance Charlie = 10

$$d_{Claire} + d_{Charlie} = 10$$

$$r_{Claire}t_{Claire} + r_{Charlie}t_{Charlie} = 10$$

Let t represent the amount of time that the Claire walked, and let 4 represent her rate.

Then let $t - \frac{1}{4}$ be the amount of time that Charlie walked (15 minutes is $\frac{1}{4}$ hour, and he started 15 minutes after Claire), and let 5 represent his rate.

$$r_{Claire}t_{Claire} + r_{Charlie}t_{Charlie} = 10$$

$$4t + 5\left(t - \frac{1}{4}\right) = 10$$

Distribute the 5 and combine like terms.

$$4t+5t-\frac{5}{4}=10$$
$$9t-\frac{5}{4}=10$$

Add $\frac{5}{4}$ to each side of the equation, and then multiply each side by $\frac{1}{9}$.

$$9t=\frac{5}{4}+10=\frac{45}{4}$$
$$\frac{1}{\not{9}}\cdot\not{9}t=\frac{1}{\not{9}}\cdot\frac{\not{45}^{5}}{4}$$
$$t=\frac{5}{4}=1\frac{1}{4}$$

It took 1 hour and 15 minutes for them to meet up.

785. **200 miles**

To solve this problem, you need the distance that Bill traveled to be equal to the distance that Will traveled.

Using $d = rt$:

distance Bill traveled = distance Will traveled

$$d_{Bill}=d_{Will}$$
$$r_{Bill}t_{Bill}=r_{Will}t_{Will}$$

You know that the amount of time Bill traveled is 5, so replace his time with 5, and let r represent his rate.

Then let $5 - 1 = 4$ be the amount of time that Will traveled (Bill's time was one hour more than Will's) and his rate be represented by $r + 10$.

$$r_{Bill}t_{Bill}=r_{Will}t_{Will}$$
$$(r)5=(r+10)4$$

Simplify both sides.

$$5r = 4r + 40$$

Add $-4r$ to each side.

$$r = 40$$

This doesn't answer the question about how far the trip is. But if 40 is Bill's rate, and if he takes 5 hours, then $d = rt$ gives you $d = 40(5) = 200$ miles.

786. **720°**

Using the formula $A = 180(n - 2)$, replace the n with 6.

$$A = 180(6-2) = 180(4) = 720$$

787. **1440°**

A decagon has 10 sides. Using the formula $A = 180(n-2)$, replace the n with 10.

$$A = 180(10-2) = 180(8) = 1440$$

788. **octagon**

Using the formula $A = 180(n-2)$, replace the A with 1080 and solve for n.

$$1080 = 180(n-2) = 180n - 360$$

Add 360 to both sides of the equation.

$$1440 = 180n$$

Now divide each side of the equation by 180.

$$\frac{1440}{180} = \frac{\cancel{180}n}{\cancel{180}}$$
$$8 = n$$

The polygon is an octagon.

789. **150°**

Using the formula $A = 180(n-2)$, replace the A with 1800.

$$1800 = 180(n-2) = 180n - 360$$

Add 360 to both sides of the equation.

$$2160 = 180n$$

Now divide each side of the equation by 180.

$$\frac{2160}{180} = \frac{\cancel{180}n}{\cancel{180}}$$
$$12 = n$$

The polygon is a dodecahedron. To find the measure of one angle, divide the total number of degrees by 12.

$$\frac{1800}{12} = 150$$

Each angle measures 150°.

790. **85**

Using $A = \frac{x_1 + x_2 + x_3 + \cdots + x_n}{n}$, put the scores in the numerator of the fraction, and replace the n with 4 (the total number of scores).

$$A = \frac{81+67+93+99}{4}$$
$$= \frac{340}{4} = 85$$

791. 8.3

Using $A=\frac{x_1+x_2+x_3+\cdots+x_n}{n}$, put the scores in the numerator of the fraction. Be careful with all the multiple scores. The total number of quizzes is 10, so you replace the n with 10.

$$A=\frac{3(9)+2(10)+4(7)+1(8)}{10}$$
$$=\frac{27+20+28+8}{10}=\frac{83}{10}=8.3$$

792. At least 85

Using $A=\frac{x_1+x_2+x_3+\cdots+x_n}{n}$, put the given scores in the numerator, and use x to represent the needed score. There will be a total of 5 exams, so replace the n with 5. Joel wants the average to come out to be at least 90, so replace the A with 90.

$$90=\frac{85+87+93+100+x}{5}$$

Multiply each side of the equation by 5, and add the scores in the numerator.

$$5\cdot 90=\cancel{5}\cdot\frac{85+87+93+100+x}{\cancel{5}}$$
$$450=365+x$$

Subtract 365 from both sides.

$$85=x$$

Joel needs at least an 85 to have an average of 90 for all five exams.

793. At least 99

Using $A=\frac{x_1+x_2+x_3+\cdots+x_n}{n}$, put 3(91) for the sum of the first three exams and let x represent what he needs on the last exam. Replace the n with 4, and use 93 for the average, A.

$$93=\frac{3(91)+x}{4}$$

Multiply both sides of the equation by 4.

$$4\cdot 93=\cancel{4}\cdot\frac{3(91)+x}{\cancel{4}}$$
$$372=273+x$$

Subtract 273 from both sides of the equation.

$$99=x$$

Tomas will have to be practically perfect.

794. 91

The numbers are the squares of the first six natural numbers. Using the formula $S = \frac{n(n+1)(2n+1)}{6}$, replace the n with 6.

$$S = \frac{\cancel{6}(6+1)(2 \cdot 6+1)}{\cancel{6}} = 7(13) = 91$$

795. 1,240

You want the sum of the squares of the first fifteen natural numbers. Using the formula $S = \frac{n(n+1)(2n+1)}{6}$, replace the n with 15.

$$S = \frac{15(15+1)(2 \cdot 15+1)}{6}$$

$$= \frac{15(16)(31)}{6} = \frac{7440}{6} = 1240$$

796. 37

Using $a_n = a_1 + (n-1)d$, replace the a_1 with 1, the n with 10, and the d with 4.

$$a_{10} = 1 + (10-1) \cdot 4 = 1 + 9 \cdot 4 = 37$$

797. 192

Using $a_n = a_1 + (n-1)d$, replace the a_1 with –6, the n with 100, and the d with 2.

$$a_{100} = -6 + (100-1) \cdot 2 = -6 + 99 \cdot 2 = 192$$

798. 159

Using $a_n = a_1 + (n-1)d$, replace the a_1 with 3, the n with 40, and the d with 4. You find the value of d by subtracting any term from the very next term in the sequence.

$$a_{40} = 3 + (40-1) \cdot 4 = 3 + 39 \cdot 4 = 159$$

799. 43

Using $a_n = a_1 + (n-1)d$, replace the a_1 with 100, the n with 20, and the d with –3. You find the value of d by subtracting any term from the very next term in the sequence.

$$a_{20} = 100 + (20-1)(-3) = 100 + (19)(-3) = 43$$

800. 550

Using $S_n = \frac{n}{2}(a_1 + a_n)$, you have to identify n. You already have the first and last terms, 40 and 60. One way to count the number of even numbers between 40 and 60 (including both those numbers) is to come up with a system. There are five even numbers from 40 through 48, five even numbers from 50 through 58, and then just one more for the 60. That's 11 numbers.

$$S_{11} = \frac{11}{2}(40+60) = \frac{11}{2}(100) = 11(50) = 550$$

801. 2,420

Using $S_n = \frac{n}{2}(a_1 + a_n)$, you have to identify the last number in the sequence. You do this with the formula for finding the *n*th term, $a_n = a_1 + (n-1)d$.

The 40th term is 2 + (40 – 1)3 = 2 + 39(3) = 119.

Now you find the sum.

$$S_{40} = \frac{40}{2}(2+119) = 20(121) = 2420$$

802. 784

The list of cubes includes the cubes of the numbers 1 through 7. Using $S_n = \frac{n^2(n+1)^2}{4}$, replace the *n* with 7.

$$S_7 = \frac{7^2(7+1)^2}{4}$$
$$= \frac{49(64)}{4} = 49(16) = 784$$

803. 44,000

The list of cubes includes the cubes of the numbers 5 through 20, and the formula, $S_n = \frac{n^2(n+1)^2}{4}$, gives you the sum starting with the number 1. You need to find the sum of the cubes from 1 through 20 and subtract the sum of the cubes from 1 through 4.

$$S_{20} = \frac{20^2(20+1)^2}{4}$$
$$= \frac{400(441)}{4} = 100(441) = 44{,}100$$

$$S_4 = \frac{4^2(4+1)^2}{4}$$
$$= \frac{16(25)}{4} = 4(25) = 100$$

$$S_{20} - S_4 = 44{,}100 - 100 = 44{,}000$$

804. \$6,741.74

Using the formula $A = P\left(1+\frac{r}{n}\right)^{nt}$, replace the *P* with 5000, the *r* with 0.03, the *n* with 4, and the *t* with 10.

$$A = 5000\left(1+\frac{0.03}{4}\right)^{4(10)}$$

Using a scientific calculator, you need to combine the terms in the parentheses first, then raise the answer to the 40th power, and then multiply by 5000.

$$\begin{aligned} A &= 5000\left(1+\frac{0.03}{4}\right)^{4(10)} \\ &= 5000(1.0075)^{40} \\ &\approx 5000(1.34835) \\ &\approx 6741.74 \end{aligned}$$

805. **$11,977.29**

Using the formula $A = P\left(1+\frac{r}{n}\right)^{nt}$, replace the P with 40,000, the r with 0.0525, the n with 12, and the t with 5.

$$A = 40{,}000\left(1+\frac{0.0525}{12}\right)^{12(5)}$$

Using a scientific calculator, you need to combine the terms in the parentheses first, then raise the answer to the 60th power, and then multiply by 40,000.

$$\begin{aligned} A &= 40{,}000\left(1+\frac{0.0525}{12}\right)^{12(5)} \\ &= 40{,}000(1.004375)^{60} \\ &\approx 40{,}000(1.29943) \\ &\approx 51{,}977.29 \end{aligned}$$

The interest earned is the difference between the total amount and the principal: 51,977.29 – 40,000 = 11,977.29.

806. **$49,188.42**

Using the formula $A = P\left(1+\frac{r}{n}\right)^{nt}$, replace the P with 20,000, the r with 0.06, the n with 365, and the t with 15.

$$A = 20{,}000\left(1+\frac{0.06}{365}\right)^{365(15)}$$

Using a scientific calculator, you need to combine the terms in the parentheses first, then raise the answer to the 5475th power, and then multiply by 20,000.

$$\begin{aligned} A &= 20{,}000\left(1+\frac{0.06}{365}\right)^{365(15)} \\ &\approx 20{,}000(1.000164)^{5475} \\ &\approx 20{,}000(2.459421) \\ &\approx 49{,}188.42 \end{aligned}$$

807. **$32,019.05**

Using the formula $A = P\left(1+\frac{r}{n}\right)^{nt}$, replace the P with 1, the r with 0.02, the n with 4, and the t with 520 (subtracting 2012 – 1492).

$$A = 1\left(1+\frac{0.02}{4}\right)^{4(520)}$$

Using a scientific calculator, you need to combine the terms in the parentheses first, then raise the answer to the 2080th power, and then multiply by 1 (if you think that's necessary).

$$\begin{aligned}A &= 1\left(1+\frac{0.02}{4}\right)^{4(520)}\\ &= 1(1.005)^{2080}\\ &\approx 1(32019.05083)\\ &\approx 32{,}019.05\end{aligned}$$

808. **28 yards**

Using $P = 2(l + w)$, replace the l with 8 and the w with 6.

$$P = 2(8 + 6) = 2(14) = 28$$

809. **150 feet**

Using $P = 2(l + w)$, replace the P with 400 and the w with 50.

$$400 = 2(l + 50)$$

Divide each side of the equation by 2.

$$200 = l + 50$$

Now add –50 to each side.

$$150 = l$$

810. **180 feet**

If the length is three times the width, then you can write $3w$ for the length. Replacing l with $3w$ in the formula $P = 2(l + w)$, you have $P = 2(3w + w) = 2(4w) = 8w$.

Now replace the P with 480, and you have $480 = 8w$.

Divide each side of the equation by 8, and $w = 60$.

The question asks for the length.

$$l = 3w = 3(60) = 180$$

811. **52 feet**

If the length is 4 feet greater than the width, then you can write the length as being $4 + w$.

Replacing l with $4 + w$ in the formula $P = 2(l + w)$, you have $P = 2(4 + w + w) = 2(4 + 2w) = 8 + 4w$.

Now replace P with 200, and the equation is $200 = 8 + 4w$.

Subtract 8 from each side of the equation.

$$192 = 4w$$

Now divide each side of the equation by 4.

$$48 = w$$

You want the length of the pool. Since $l = 4 + w$, then $l = 4 + 48 = 52$.

812. **135 feet**

If the width is 30 more than one-third the length, then $w = 30 + \frac{1}{3}l$. Writing the perimeter formula using this value for w, you have $P = 2\left(l + 30 + \frac{1}{3}l\right) = 2\left(30 + \frac{4}{3}l\right)$. Replace the P with 420.

$$420 = 2\left(30 + \frac{4}{3}l\right)$$

Now divide each side by 2.

$$210 = 30 + \frac{4}{3}l$$

Add –30 to each side of the equation.

$$180 = \frac{4}{3}l$$

And now multiply each side by the reciprocal of $\frac{4}{3}$.

$$\frac{3}{\cancel{4}} \cdot \cancel{180}^{45} = \frac{\cancel{3}}{\cancel{4}} \cdot \frac{\cancel{4}}{\cancel{3}}l$$

$$135 = l$$

813. **180 feet by 120 feet**

First write the ratio of length to width as a proportion.

$$\frac{l}{w} = \frac{3}{2}$$

Now multiply each side of the equation by w to solve for l.

$$\cancel{w} \cdot \frac{l}{\cancel{w}} = \frac{3}{2} \cdot w$$

$$l = \frac{3}{2}w$$

Replace the l in $P = 2(l + w)$ with its equivalent, and replace the P with 600.

$$600 = 2\left(\frac{3}{2}w + w\right)$$

Simplify on the right.

$$600 = 2\left(\frac{5}{2}w\right) = 5w$$

Divide each side of the equation by 5 to get $120 = w$.

Since $l = \frac{3}{2}w$, then, if $w = 120$,

$$l = \frac{3}{2}(120) = 180$$

814. **26 square feet**

Using the formula $A = \frac{1}{2}h(b_1 + b_2)$, replace the h with 4, and replace b_1 and b_2 with 5 and 8.

$$A = \frac{1}{2}(4)(5+8) = 2(13) = 26$$

815. **52 feet**

Using the formula $A = \frac{1}{2}h(b_1 + b_2)$, replace the A with 170, the h with 5, and the b_1 with 16.

$$170 = \frac{1}{2}(5)(16 + b_2)$$

Multiply each side of the equation by $\frac{2}{5}$.

$$\frac{2}{5} \cdot 170 = \frac{\cancel{2}}{\cancel{5}} \cdot \frac{1}{\cancel{2}}(\cancel{5})(16 + b_2)$$

$$68 = 16 + b_2$$

Subtract 16 from each side, and you find that $b_2 = 52$.

816. **32 feet**

The area of a square is equal to the side measure squared.

$$64 = s^2$$

Take the square root of each side of the equation, and you get $8 = s$.

To find the perimeter, multiply the length of the side times 4.

$$P = 4(8) = 32.$$

817. **256 square feet**

The perimeter of a square is four times the measure of a side.

$$64 = 4s$$

Divide each side of the equation by 4 to get $16 = s$.

Find the area by squaring the measure of the side.

$$A = 16^2 = 256$$

818. **9, 15, and 18 inches**

The two comparisons are both to the shortest side, so let x represent the measure of the shortest side. Then one side is twice the shortest side, $2x$. And another side is 6 inches greater than shortest side, $6 + x$.

The total perimeter, 42, is the sum of the measures of the sides.

$$42 = x + (2x) + (6 + x)$$

Simplify on the right by adding like terms.

$$42 = 4x + 6$$

Add –6 to each side.

$$36 = 4x$$

Divide by 4.

$$9 = x$$

The shortest side is 9 inches.

Twice the shortest is 2(9) = 18 inches.

Six greater than the shortest is 6 + 9 = 15 inches.

819. **16 feet**

First determine the lengths of the sides of the current triangle.

First side: 8 feet

Second side: three times 8 feet, 3(8) = 24 feet

Third side: 8 feet shorter than second side, 24 – 8 = 16 feet

Total perimeter: 8 + 24 + 16 = 48 feet.

An equilateral triangle has three sides all the same length. So divide the perimeter by 3.

$$\frac{48}{3} = 16$$

820. **10 feet**

Let w represent the width. Then the length is $l = 2w - 2$. Using the area formula, $A = lw$, replace the A with 180 and substitute in the length equivalence.

$$180 = (2w - 2)w$$

Simplify on the right.

$$180 = 2w^2 - 2w$$

Divide each term by 2.

$$90 = w^2 - w$$

Now add –90 to each side and factor the quadratic.

$$0 = w^2 - w - 90$$

$$0 = (w - 10)(w + 9)$$

Setting the binomials equal to 0, you get $w = 10$ or $w = -9$.

The width measure can't be negative, so just use $w = 10$.

821. **3,520 square feet**

First let the length of the rectangle be represented by l. Then the width is $w = \frac{1}{2}l + 4$. Using the perimeter formula, $P = 2(l + w)$, let P be 248 and substitute in the width equivalence.

$$248 = 2\left(l + \frac{1}{2}l + 4\right)$$

Simplify on the right. Then divide each side of the equation by 2.

$$248 = 2\left(\frac{3}{2}l + 4\right)$$

$$124 = \frac{3}{2}l + 4$$

Subtract 4 from each side.

$$120 = \frac{3}{2}l$$

Now multiply each side by $\frac{2}{3}$.

$$\frac{2}{3} \cdot 120 = \frac{\cancel{2}}{\cancel{3}} \cdot \frac{\cancel{3}}{\cancel{2}}l$$

$$80 = l$$

The length of the rectangle is 80. Use this to determine the width.

$$w = \frac{1}{2}l + 4 = \frac{1}{2}(80) + 4 = 40 + 4 = 44$$

The area of a rectangle is found by multiplying the length times the width.

$$A = lw = (80)(44) = 3520$$

822. **$5400\sqrt{3}$ sq. cm**

A regular hexagon has six sides – all the same length. Divide the perimeter, 360, by 6 to get 60 for the length of one side. Let 60 replace the x in the formula for the area of a regular hexagon.

$$A = \frac{3\sqrt{3}}{2}(60)^2 = \frac{3\sqrt{3}}{\cancel{2}} \cdot \cancel{3600}^{1800} = 5400\sqrt{3}$$

823. **72 feet**

Using the formula for the area of a regular hexagon, $A = \frac{3\sqrt{3}}{2}x^2$, replace the A with $216\sqrt{3}$ and solve for x, the length of one side.

$$216\sqrt{3} = \frac{3\sqrt{3}}{2}x^2$$

To solve for x^2, first multiply each side of the equation by the reciprocal of its coefficient.

$$\frac{2}{\cancel{3}\cancel{\sqrt{3}}} \cdot \cancel{216}^{72}\cancel{\sqrt{3}} = \frac{\cancel{2}}{\cancel{3\sqrt{3}}} \cdot \frac{\cancel{3\sqrt{3}}}{\cancel{2}}x^2$$

$$2 \cdot 72 = 144 = x^2$$

Taking the square root of each side, you find that $12 = x$.

Now, to find the perimeter, multiply the length of the side times 6 (the total number of sides in a hexagon). $P = 6(12) = 72$

824. **12 cm**

Using $A = \frac{1}{2}bh$, replace the A with 60 and the h with 10. Then solve for b.

$$60 = \frac{1}{2}b(10)$$

First simplify on the right.

$$60 = 5b$$

Now divide each side by 5.

$$12 = b$$

825. **16 inches**

Using the area formula, $A = \frac{1}{2}bh$, replace the A with 2(144) = 288 sq in and the b with 36. Then solve for h.

$$288 = \frac{1}{2}(36)h$$

Simplify on the right.

$$288 = 18h$$

Now divide each side of the equation by 18.

$$16 = h$$

826. **6 square feet**

Using the Pythagorean theorem to solve for the measure of the other leg, replace the c with 5 and the a with 3.

$$3^2 + b^2 = 5^2$$

$$9 + b^2 = 25$$

Subtracting 9 from each side,

$$b^2 = 16$$

Taking the square root of each side, $b = 4$. The two legs of a right triangle are perpendicular to one another, so one can be the base and the other the height of the triangle. Using the formula for the area of a triangle, $A = \frac{1}{2}bh$, replace the b with 4 and the h with 3. (The other way around would have worked, too.)

$$A = \frac{1}{2}(4)(3) = 6$$

827. **13 feet**

First write the measure of one leg in terms of the other. If the first leg has measure a, and if the other, b, is two more than twice a, then the measure of $b = 2a + 2$.

The two legs are perpendicular to one another, so they can be used for the base and height in the area formula.

$$A = \frac{1}{2}bh = \frac{1}{2}(a)(2a+2)$$

Replace the area, A, with 30.

$$30 = \frac{1}{2}(a)(2a+2)$$

Distribute the $\frac{1}{2}$ over the two terms in the binomial on the right side.

$$30 = (a)(a + 1)$$

Distribute the a on the right, and then add –30 to each side of the equation.

$$30 = a^2 + a$$

$$0 = a^2 + a - 30$$

Now factor the quadratic.

$$0 = (a + 6)(a - 5)$$

Setting the two binomials equal to 0, you get $a = -6$ or $a = 5$.

The leg of the triangle won't have a negative measure, so $a = 5$.

You find the measure of b using $2a + 2$, so $b = 2(5) + 2 = 10 + 2 = 12$.

The two legs of the right triangle are 5 and 12.

Use the Pythagorean theorem to find the measure of the hypotenuse.

$$a^2 + b^2 = c^2$$

$$5^2 + 12^2 = c^2$$

$$25 + 144 = 169 = c^2$$

Take the square root of each side, and $13 = c$.

828. **8 inches**

Using the volume formula, $V = lwh$, replace the V with 48, the l with $4w$, and the h with 3.

$$48 = (4w)(w)(3)$$

Simplify on the right.

$$48 = 12w^2$$

Divide each side by 12.

$$4 = w^2$$

Taking the square root of each side, you get $w = 2$.

Since the length is 4 times the width, $l = 4(2) = 8$

829. **4 cm**

A cube is a right rectangular prism in which all the faces are congruent squares and all the edges are the same measure. So $l = w = h$.

Replace all three dimensions with x. Then the formula for the volume is $V = x^3$.

Letting $V = 64$, you have $64 = x^3$. Taking the cube root of each side, you have $4 = x$.

830. $\frac{6}{\pi}$ **inches**

Using the formula for the volume of a cylinder, $V = \pi r^2 h$, replace the V with 54 and the r with 3.

$$54 = \pi(3)^2 h$$

Simplify on the right.

$$54 = 9\pi h$$

Solve for h by dividing each side of the equation by 9π.

$$\frac{\cancel{54}^{6}}{\cancel{9}\pi} = \frac{\cancel{9\pi}h}{\cancel{9\pi}}$$

$$\frac{6}{\pi} = h$$

The height is a little less than 2 inches.

831. **4 feet**

Using the formula for the volume of a cylinder, $V = \pi r^2 h$, replace the V with 192π and the h with 12.

$$192\pi = \pi r^2(12)$$

Divide each side of the equation by 12π.

$$\frac{192\cancel{\pi}}{12\cancel{\pi}} = \frac{\cancel{\pi}r^2(\cancel{12})}{\cancel{12}\cancel{\pi}}$$

$$\frac{192}{12} = 16 = r^2$$

Take the square root of both sides to get $4 = r$.

832. **45°**

Using the formula $\frac{\theta^D}{180} = \frac{\theta^R}{\pi}$, place the $\frac{\pi}{4}$ in the numerator above the π.

$$\frac{\theta^D}{180} = \frac{\frac{\pi}{4}}{\pi}$$

Simplify the complex fraction on the right by multiplying the numerator by the reciprocal of the denominator.

$$\frac{\theta^D}{180} = \frac{\cancel{\pi}}{4} \cdot \frac{1}{\cancel{\pi}} = \frac{1}{4}$$

Now solve for the degree measure by multiplying each side of the equation by 180.

$$180 \cdot \frac{\theta^D}{180} = 180^{45} \cdot \frac{1}{4}$$

$$\theta^D = 45$$

833. $\frac{5\pi}{6}$

Using the formula $\frac{\theta^D}{180} = \frac{\theta^R}{\pi}$, place the 150 in the numerator above the 180.

$$\frac{150}{180} = \frac{\theta^R}{\pi}$$

Reduce the fraction on the left.

$$\frac{150^5}{180^6} = \frac{\theta^R}{\pi}$$

Now solve for the radian measure by multiplying each side of the equation by π.

$$\pi \cdot \frac{5}{6} = \pi \cdot \frac{\theta^R}{\pi}$$

$$\frac{5\pi}{6} = \theta^R$$

834. $\frac{5\pi}{3}$

Using the formula $\frac{\theta^D}{180} = \frac{\theta^R}{\pi}$, place the 300 in the numerator above the 180.

$$\frac{300}{180} = \frac{\theta^R}{\pi}$$

Reduce the fraction on the left.

$$\frac{300^5}{180^3} = \frac{\theta^R}{\pi}$$

Now solve for the radian measure by multiplying each side of the equation by π.

$$\pi \cdot \frac{5}{3} = \pi \cdot \frac{\theta^R}{\pi}$$

$$\frac{5\pi}{3} = \theta^R$$

835. 165°

Using the formula $\frac{\theta^D}{180} = \frac{\theta^R}{\pi}$, place the $\frac{11\pi}{12}$ in the numerator above the π.

$$\frac{\theta^D}{180} = \frac{\frac{11\pi}{12}}{\pi}$$

Simplify the complex fraction on the right by multiplying the numerator by the reciprocal of the denominator.

$$\frac{\theta^D}{180} = \frac{11\pi}{12} \cdot \frac{1}{\pi} = \frac{11}{12}$$

Answers 801–900

Now solve for the degree measure by multiplying each side of the equation by 180.

$$\not{180}\cdot\frac{\theta^D}{\not{180}}=\not{180}^{15}\cdot\frac{11}{\not{12}}$$
$$\theta^D=165$$

836. $-\frac{\pi}{6}$

Using the formula $\frac{\theta^D}{180}=\frac{\theta^R}{\pi}$, place the –30 in the numerator above the 180.

$$\frac{-30}{180}=\frac{\theta^R}{\pi}$$

Reduce the fraction on the left.

$$\frac{\not{-30}^{1}}{\not{180}^{6}}=\frac{\theta^R}{\pi}$$

Now solve for the radian measure by multiplying each side of the equation by π.

$$\pi\cdot\frac{-1}{6}=\not{\pi}\cdot\frac{\theta^R}{\not{\pi}}$$
$$-\frac{\pi}{6}=\theta^R$$

837. About 115°

Using the formula $\frac{\theta^D}{180}=\frac{\theta^R}{\pi}$, place a 2 in the numerator above the π.

$$\frac{\theta^D}{180}=\frac{2}{\pi}$$

Solve for the degree measure by multiplying each side of the equation by 180.

$$\not{180}\cdot\frac{\theta^D}{\not{180}}=180\cdot\frac{2}{\pi}$$
$$\theta^D=\frac{360}{\pi}$$

If you want a decimal approximation, divide 360 by 3.1416 to get about 115°.

838. 32 feet

Using $h=-16t^2+48t$, replace the t's with 2.

$$h=-16(2)^2+48(2)=-64+96=32$$

839. 20 feet

Using $h=-16t^2+48t$, replace the t's with 2.5.

$$h=-16(2.5)^2+48(2.5)=-100+120=20$$

This height is lower at 2.5 seconds than at 2 seconds, because the ball is coming back down toward the ground.

840. **0.5 and 2.5 seconds**

Using $h = -16t^2 + 48t$, replace the h with 20 and solve for t.

$$20 = -16t^2 + 48t$$

Add –20 to each side of the equation, and then factor on the right.

$$0 = -16t^2 + 48t - 20$$
$$0 = -4(4t^2 - 12t + 5)$$
$$0 = -4(2t - 1)(2t - 5)$$

Setting the two binomials equal to 0, you get $t = \frac{1}{2}$ or $t = \frac{5}{2}$.

Both answers are correct. After 0.5 seconds, the ball is 20 feet in the air. It goes higher and then starts down, reaching 20 feet again after 2.5 seconds.

841. **3 seconds**

Using $h = -16t^2 + 48t$, replace the h with 0 (the height of the ball is 0 feet when it hits the ground) and solve for t.

$$0 = -16t^2 + 48t$$

Factor on the right.

$$0 = -16t(t - 3)$$

Setting the two factors equal to 0, you get $t = 0$ or $t = 3$. The $t = 0$ represents the beginning – right before the launching of the ball. The $t = 3$ represents when the ball comes back down and hits the ground.

842. **30 square inches**

Using $A = \sqrt{s(s-a)(s-b)(s-c)}$, you first find the *semi-perimeter* (half the perimeter) by adding up the measures of the sides and finding half of that measure: 5 + 12 + 13 = 30. Half of 30 is 15. Now use the 15 for s in the formula, and replace the a, b, and c with the measures of the sides and simplify.

$$A = \sqrt{15(15-5)(15-12)(15-13)}$$
$$= \sqrt{15(10)(3)(2)} = \sqrt{900} = 30$$

843. **24 square feet**

Using $A = \sqrt{s(s-a)(s-b)(s-c)}$, you first find the *semi-perimeter* (half the perimeter) by adding up the measures of the sides and finding half of that measure: 6 + 8 + 10 = 24. Half of 24 is 12. Now use the 12 for s in the formula, and replace the a, b, and c with the measures of the sides and simplify.

$$A = \sqrt{12(12-6)(12-8)(12-10)}$$
$$= \sqrt{12(6)(4)(2)} = \sqrt{576} = 24$$

844. **4 or $2\sqrt{13}$**

Using $A=\sqrt{s(s-a)(s-b)(s-c)}$, you have two of the sides; let $a = 3$ and $b = 5$. You need c. The semi-perimeter is half the sum of the measures of the sides, so $s=\frac{1}{2}(3+5+c)=\frac{1}{2}(8+c)=4+\frac{1}{2}c$. Fill in the values for a, b, s, and the area, A.

$$6=\sqrt{\left(4+\frac{1}{2}c\right)\left(4+\frac{1}{2}c-3\right)\left(4+\frac{1}{2}c-5\right)\left(4+\frac{1}{2}c-c\right)}$$

Simplify what's in the parentheses.

$$6=\sqrt{\left(4+\frac{1}{2}c\right)\left(1+\frac{1}{2}c\right)\left(-1+\frac{1}{2}c\right)\left(4-\frac{1}{2}c\right)}$$

Rearrange the orders of the binomials, and you find that you have two pairs of *sum-and-difference* binomials. They multiply together rather nicely.

$$6=\sqrt{\left(1+\frac{1}{2}c\right)\left(-1+\frac{1}{2}c\right)\left(4+\frac{1}{2}c\right)\left(4-\frac{1}{2}c\right)}$$

$$6=\sqrt{\left(-1+\frac{1}{4}c^2\right)\left(16-\frac{1}{4}c^2\right)}$$

Continue by multiplying the two binomials together.

$$6=\sqrt{\left(-16+\frac{17}{4}c^2-\frac{1}{16}c^4\right)}$$

Now square both sides of the equation to get rid of the radical.

$$6^2=\left(\sqrt{\left(-16+\frac{17}{4}c^2-\frac{1}{16}c^4\right)}\right)^2$$

$$36=-16+\frac{17}{4}c^2-\frac{1}{16}c^4$$

Add –36 to each side of the equation. Then reverse all the signs (same as multiplying both sides by –1) to get the polynomial ready to be factored.

$$0=-52+\frac{17}{4}c^2-\frac{1}{16}c^4$$

$$\frac{1}{16}c^4-\frac{17}{4}c^2+52=0$$

Even though it will make the constant pretty big, you should multiply each side of the equation by 16 to get rid of the fractions. You then have a *quadratic-like* trinomial that can be factored.

$$16\left(\frac{1}{16}c^4-\frac{17}{4}c^2+52\right)=16(0)$$

$$c^4-68c^2+832=0$$

$$\left(c^2-52\right)\left(c^2-16\right)=0$$

Setting the two binomials equal to zero, you get the four solutions: $c=\sqrt{52}=2\sqrt{13}$, $c=-\sqrt{52}=-2\sqrt{13}$, $c = 4$, or $c = -4$. You disregard the two negative solutions; the problem involves the measures of the sides of a triangle. The other two solutions work fine.

845. **30 inches**

Using $A=\sqrt{s(s-a)(s-b)(s-c)}$, you can call all three sides the same thing. Let them all be represented by a. The semi-perimeter is half the sum of the measures of the sides, so $s=\frac{1}{2}(a+a+a)=\frac{1}{2}(3a)=\frac{3}{2}a$.

Replace the area, A, with $25\sqrt{3}$, the s with $\frac{3}{2}a$, and the measures of the sides with a.

$$25\sqrt{3}=\sqrt{\frac{3}{2}a\left(\frac{3}{2}a-a\right)\left(\frac{3}{2}a-a\right)\left(\frac{3}{2}a-a\right)}$$

Simplify under the radical.

$$25\sqrt{3}=\sqrt{\frac{3}{2}a\left(\frac{1}{2}a\right)\left(\frac{1}{2}a\right)\left(\frac{1}{2}a\right)}$$

$$25\sqrt{3}=\sqrt{\frac{3a^4}{16}}$$

You could simplify the radical on the right, but you'll be squaring both sides, anyway, so just skip the simplifying for now and square both sides of the equation.

$$\left(25\sqrt{3}\right)^2=\left(\sqrt{\frac{3a^4}{16}}\right)^2$$

$$625(3)=\frac{3a^4}{16}$$

Now multiply each side by $\frac{16}{3}$.

$$\frac{16}{\cancel{3}}\cdot 625(\cancel{3})=\frac{\cancel{16}}{\cancel{3}}\cdot\frac{\cancel{3}a^4}{\cancel{16}}$$

$$10{,}000=a^4$$

The number 10 raised to the fourth power is 10,000, so the fourth root of 10,000 is 10. You don't bother with the negative solution.

The question asks for the perimeter of the triangle, so you add up the three sides, all measuring 10, and get 30.

846. **Jim: 10; Jon: 20**

Jim's age: n

Since Jon is twice as old as Jim, then *Jon's age*: $2n$.

Five years ago, each was 5 years younger, so subtract 5 from their current ages.

Jim's age five years ago: $n-5$

Jon's age five years ago: $2n-5$

The statement needed to solve the problem says that, five years ago, Jon *was* (*equals*) three times as old as Jim.

Jon's age five years ago = three times Jim's age five years ago

Plug in the expressions for Jim and Jon's relative ages five years ago.

$$2n-5=3(n-5)$$

Distribute on the right.

$2n - 5 = 3n - 15$

Add $-2n$ to each side and add 15 to each side.

$10 = n$

Jim is 10 years old, and Jon is twice as old, 20 years old.

847. **Greta: 6; Grace: 18**

Greta's age: g

Since Grace is three times as old as Greta, then

Grace's age: $3g$.

In six years, each will be six years older, so add 6 to their current ages.

Greta's age in six years: $g + 6$

Grace's age in six years: $3g + 6$

The statement needed to solve the problem says that, in six years, Grace *will be* (*equals*) twice as old as Greta.

Grace's age in six years = two times Greta's age in six years

Plug in the expressions for Grace and Greta's relative ages in six years.

$3g + 6 = 2(g + 6)$

Distribute on the right.

$3g + 6 = 2g + 12$

Add $-2g$ to each side and add -6 to each side.

$g = 6$

Greta is 6 years old, and Grace is three times that, or 18 years old.

848. **Stefanie: 12; Amanda: 36**

Stefanie's age: n

Since Amanda is three times as old as Stefanie, then *Amanda's age:* $3n$.

Eight years from now, both will be eight years older, so add eight to each age.

Stefanie in eight years: $n + 8$

Amanda in eight years: $3n + 8$

The statement needed to solve the problem says that, in eight years, Amanda *will be* (*equals*) four more than twice Stefanie's age.

Amanda's age in eight years = four more than twice Stefanie's age in eight years

Plug in the expressions for Amanda and Stefanie's relative ages in eight years.

$3n + 8 = 4 + 2(n + 8)$

Distribute on the right.

$3n + 8 = 4 + 2n + 16$

Simplify on the right.

$3n + 8 = 2n + 20$

Add $-2n$ to each side, and add -8 to each side.

$n = 12$

Stefanie is 12 and Amanda is three times that, or 36.

849. Hal: 18: Hank: 20

Hal's age: h

Since Hank is two years older than Hal:

Hank's age: $h + 2$.

Twelve years ago means to subtract 12 from their ages:

Hal's age 12 years ago: $h - 12$

Hank's age 12 years ago: $(h + 2) - 12 = h - 10$

The statement needed to solve the problem is that Hank's age 12 years ago *was* (*equals*) 4 years less than twice Hal's age 12 years ago.

Hank's age 12 years ago = twice Hal's age 12 years ago minus 4.

Plug in the expressions for Hank and Hal's relative ages 12 years ago.

$h - 10 = 2(h - 12) - 4$

Simplify on the right.

$h - 10 = 2h - 24 - 4$

$h - 10 = 2h - 28$

Add $-h$ to each side and add 28 to each side.

$18 = h$

Hal is 18, and Hank is two years older, or 20.

850. Bart: 18; Betty: 22

Bart's age: b

Betty is four years older than Bart.

Betty's age: $b + 4$

In three years, they will be older, so add 3 to each age.

Bart's age in three years: $b + 3$

Betty's age in three years: $b + 4 + 3 = b + 7$

The statement needed to solve the problem is Betty's age in three years plus Bart's age in three years *will be* (*equals*) 46.

Plug in Betty and Bart's respective ages in three years.

$(b + 3) + (b + 7) = 46$

Simplify on the left.

$2b + 10 = 46$

Add –10 to each side.

$2b = 36$

Divide each side by 2.

$b = 18$

Bart is 18 and Betty, four years older, is 22.

851. **Maura: 10; Les: 20**

Maura's age: m

Les is ten years older than Maura.

Les's age: $m + 10$

Five years ago they were both younger; subtract 5 from each age.

Maura's age five years ago: $m - 5$

Les's age five years ago: $m + 10 - 5 = m + 5$

The statement needed to solve the problem says that Maura's age five years ago plus Les's age five years ago has a sum of 20.

$(m - 5) + (m + 5) = 20$

Simplify on the left.

$2m = 20$

Divide each side by 2.

$m = 10$

Maura is 10, and Les, who is ten years older, is 20.

852. **Moe: 5; Joe: 10; Louie: 12**

Moe's age: e

Moe is twice as old as Joe.

Joe's age: $2e$

Louie is two years older than Joe.

Louie's age: $2e + 2$

In two years, they'll each be two years older.

Moe's age in two years: $e + 2$

Joe's age in two years: $2e + 2$

Louie's age in two years: $2e + 2 + 2 = 2e + 4$

The statement needed to solve the problem says that Joe's age in two years plus Moe's age in two years plus Louie's age in two years *will be* (*equals*) 33.

$(2e + 2) + (e + 2) + (2e + 4) = 33$

Simplify on the left.

$5e + 8 = 33$

Add –8 to each side.

$5e = 25$

Divide each side by 5.

$e = 5$

Moe is 5. Joe is twice as old, or 10. Louie is two years older than Joe, or 12.

853. **Karen: 10; Barb: 13; Mary: 26**

Karen's age: k

Barb is three years older than Karen.

Barb's age: $k + 3$

Mary is twice Barb's age.

Mary's age: $2(k + 3) = 2k + 6$

Five years ago, each girl was five years younger, so subtract five from their ages.

Karen five years ago: $k - 5$

Barb five years ago: $k + 3 - 5 = k - 2$

Mary five years ago: $2k + 6 - 5 = 2k + 1$

The statement needed to solve the problem says that the sum of Karen's age five years ago plus Barb's age five years ago plus Mary's age five years ago *was* (*equals*) 34.

$(k - 5) + (k - 2) + (2k + 1) = 34$

Simplify on the left.

$4k - 6 = 34$

Add 6 to each side.

$4k = 40$

Divide each side by 4.

$k = 10$

Karen is 10. Barb is three years older than Karen, or 13. Mary is twice Barb's age, or 26.

854. **29, 30, and 31**

First integer: n

Second integer: $n + 1$

Third integer: $n + 2$

The statement is that the sum of the integers *is* (*equals*) 90.

$n + (n + 1) + (n + 2) = 90$

Simplify on the left.

$3n + 3 = 90$

Add –3 to each side.

$3n = 87$

Divide each side by 3.

$n = 29$

The three integers are 29, 30, and 31.

855. **–1, 0, 1, 2, and 3**

First integer: n

Second integer: n + 1

Third integer: n + 2

Fourth integer: n + 3

Fifth integer: n + 4

The statement is that the sum of the integers *is* (*equals*) 5.

$n + (n + 1) + (n + 2) + (n + 3) + (n + 4) = 5$

Simplify on the left.

$5n + 10 = 5$

Add –10 to each side.

$5n = -5$

Divide each side by 5.

$n = -1$

The five consecutive integers are: –1, 0, 1, 2, and 3.

856. **48**

First integer: n

Second integer: n + 2

Third integer: n + 4

The statement is that the sum of the even integers is 138.

$n + (n + 2) + (n + 4) = 138$

Simplify on the left.

$3n + 6 = 138$

Add –6 to each side.

$3n = 132$

Divide each side by 3.

$n = 44$

The three even integers are 44, 46, and 48, so the largest is 48.

857. 100

First integer: n

Second integer: $n + 2$

Third integer: $n + 4$

Fourth integer: $n + 6$

The statement is that the sum of the even integers is 412.

$$n + (n + 2) + (n + 4) + (n + 6) = 412$$

Simplify on the left.

$$4n + 12 = 412$$

Add −12 to each side.

$$4n = 400$$

Divide each side by 4.

$$n = 100$$

The four even integers are 100, 102, 104, and 106. The smallest is 100.

858. 47

First integer: n

Second integer: $n + 2$

Third integer: $n + 4$

Fourth integer: $n + 6$

The statement is that the sum of the odd integers is 176.

$$n + (n + 2) + (n + 4) + (n + 6) = 176$$

Simplify on the left.

$$4n + 12 = 176$$

Add −12 to each side.

$$4n = 164$$

Divide each side by 4.

$$n = 41$$

The four odd integers are 41, 43, 45, and 47. The largest is 47.

859. 151

First integer: n

Second integer: $n + 2$

Third integer: $n + 4$

Fourth integer: $n + 6$

Fifth integer: $n + 8$

The statement is that the sum of the odd integers is 755.

$$n + (n + 2) + (n + 4) + (n + 6) + (n + 8) = 755$$

Simplify on the left.

$$5n + 20 = 755$$

Add –20 to each side.

$$5n = 735$$

Divide each side by 5.

$$n = 147$$

The five odd integers are 147, 149, 151, 153, and 155. The middle number in the list is 151.

860. **36, 40, and 44**

First integer: n

Second integer: $n + 4$

Third integer: $n + 8$

The statement is that the sum of the multiples of 4 is 120.

$$n + (n + 4) + (n + 8) = 120$$

Simplify on the left.

$$3n + 12 = 120$$

Add –12 to each side.

$$3n = 108$$

Divide each side by 3.

$$n = 36$$

The numbers are 36, 40, and 44.

861. **36**

First integer: n

Second integer: $n + 6$

Third integer: $n + 12$

Fourth integer: $n + 18$

The statement is that the sum of the multiples of 6 is 108.

$$n + (n + 6) + (n + 12) + (n + 18) = 108$$

Simplify on the left.

$$4n + 36 = 108$$

Add –36 to each side.

$$4n = 72$$

Divide each side by 4.

$n = 18$

The numbers are 18, 24, 30, and 36. The largest is 36.

862. **105**

First integer: n

Second integer: $n + 7$

Third integer: $n + 14$

Fourth integer: $n + 21$

Fifth integer: $n + 28$

The statement is that the sum of the multiples of 7 is 525.

$n + (n + 7) + (n + 14) + (n + 21) + (n + 28) = 525$

Simplify on the left.

$5n + 70 = 525$

Add –70 to each side.

$5n = 455$

Divide each side by 5.

$n = 91$

The numbers are 91, 98, 105, 112, and 119. The middle number is 105.

863. **54**

First integer: n

Second integer: $n + 2$

Third integer: $n + 4$

Fourth integer: $n + 6$

Fifth integer: $n + 8$

The statement is that the sum of the first and fifth numbers in the list is 108.

$n + (n + 8) = 108$

Simplify on the left.

$2n + 8 = 108$

Add –8 to each side.

$2n = 100$

Divide each side by 2.

$n = 50$

The numbers are 50, 52, 54, 56, and 58. The middle number is 54.

864. 16

First integer: n

Second integer: $n + 4$

Third integer: $n + 8$

Fourth integer: $n + 12$

Fifth integer: $n + 16$

Sixth integer: $n + 20$

The statement is that the sum of the first and sixth numbers is 12.

$$n + (n + 20) = 12$$

Simplify on the left.

$$2n + 20 = 12$$

Add –20 to each side.

$$2n = -8$$

Divide each side by 2.

$$n = -4$$

The numbers are –4, 0, 4, 8, 12, and 16. The last number is 16.

865. 48

First integer: n

Second integer: $n + 3$

Third integer: $n + 6$

Fourth integer: $n + 9$

Fifth integer: $n + 12$

The statement is that the sum of the first and fifth numbers is 108.

$$n + (n + 12) = 108$$

Simplify on the left.

$$2n + 12 = 108$$

Add –12 to each side.

$$2n = 96$$

Divide each side by 2.

$$n = 48$$

The numbers are 48, 51, 54, 57, and 60. The first number is 48.

866. **12 and 14**

First integer: n

Second integer: $n + 2$

The statement is that the product of the two even integers *is* (*equals*) 142 more than the sum of the integers.

$$n(n+2) = 142 + n + (n+2)$$

Distribute on the left and simplify on the right.

$$n^2 + 2n = 144 + 2n$$

Add $-2n$ to each side.

$$n^2 = 144$$

Find the square root of each side, and you have $n = 12$ or $n = -12$.

The question asked for the positive integers so the numbers are 12 and 14.

867. **–4, 0, and 4**

First integer: n

Second integer: $n + 4$

Third integer: $n + 8$

The statement is that the product of the first and third multiples *is* (*equals*) 16 less than the sum of the two integers.

$$n(n+8) = n + (n+8) - 16$$

Distribute on the left and simplify on the right.

$$n^2 + 8n = 2n - 8$$

Add $-2n$ to each side, and add 8 to each side.

$$n^2 + 6n + 8 = 0$$

Factor the quadratic.

$$(n+2)(n+4) = 0$$

Setting the binomials equal to 0, you have $n = -2$ and $n = -4$. The number –2 isn't a multiple of 4, so the only answer is –4. The three numbers are –4, 0, and 4.

868. $2\frac{2}{5}$ **hours**

Let x represent the total amount of time it will take to weed the garden working together. Shirley can do $\frac{1}{4}$ of the job in one hour, and John can do $\frac{1}{6}$ of the job in one hour. So $\frac{x}{4} + \frac{x}{6} = 1$, where 1 represents completing the whole job.

Find a common denominator and multiply each term by that number to eliminate the fractions.

$$\cancel{24}^{6}\cdot\frac{x}{\cancel{4}}+\cancel{24}^{4}\cdot\frac{x}{\cancel{6}}=24\cdot 1$$

$$6x+4x=24$$

Simplify on the left.

$10x = 24$

Divide each side by 10.

$$\frac{\cancel{10}x}{\cancel{10}}=\frac{24}{10}$$

$$x=\frac{24}{10}=\frac{12}{5}=2\frac{2}{5}$$

Working together, it will take them $2\frac{2}{5}$ hours to weed the garden. That's 2 hours and 24 minutes.

869. 4 hours

Let x represent the total amount of time it will take to paint the room working together. Ken can do $\frac{1}{5}$ of the job in one hour, and Paula can do $\frac{1}{20}$ of the job in one hour. So $\frac{x}{5}+\frac{x}{20}=1$, where 1 represents completing the whole job.

Find a common denominator and multiply each term by that number to eliminate the fractions.

$$\cancel{20}^{4}\cdot\frac{x}{\cancel{5}}+\cancel{20}\cdot\frac{x}{\cancel{20}}=20\cdot 1$$

$$4x+x=20$$

Simplify on the left.

$5x = 20$

Divide each side by 5 to get $x = 4$.

Working together, it will take them 4 hours to paint the room.

870. $2\frac{2}{3}$ hours

Let x represent the total amount of time it will take to clean the house working together. Madeline can do $\frac{1}{4}$ of the job in one hour, and Katie can do $\frac{1}{8}$ of the job in one hour. So $\frac{x}{4}+\frac{x}{8}=1$, where 1 represents completing the whole job.

Find a common denominator and multiply each term by that number to eliminate the fractions.

$$\cancel{8}^{2}\cdot\frac{x}{\cancel{4}}+\cancel{8}\cdot\frac{x}{\cancel{8}}=8\cdot 1$$

$$2x+x=8$$

Simplify on the left.

$3x = 8$

Divide each side by 3.

$$\frac{\cancel{3}x}{\cancel{3}}=\frac{8}{3}$$

$$x=\frac{8}{3}=2\frac{2}{3}$$

Working together, it will take them $2\frac{2}{3}$ hours to clean the house. That's 2 hours and 40 minutes.

871. $1\frac{7}{11}$ **hours**

Let x represent the total amount of time it will take to clean the garage working together. Larry can do $\frac{1}{3}$ of the job in one hour, Moe can do $\frac{1}{6}$ of the job in one hour, and Curly can do $\frac{1}{9}$ of the job in one hour. So $\frac{x}{3}+\frac{x}{6}+\frac{x}{9}=1$, where 1 represents completing the whole job.

Find a common denominator and multiply each term by that number to eliminate the fractions.

$$\cancel{18}^{6}\cdot\frac{x}{\cancel{3}}+\cancel{18}^{3}\cdot\frac{x}{\cancel{6}}+\cancel{18}^{2}\cdot\frac{x}{\cancel{9}}=18\cdot 1$$

$$6x+3x+2x=18$$

Simplify on the left.

$$11x=18$$

Divide each side by 11.

$$\frac{\cancel{11}x}{\cancel{11}}=\frac{18}{11}$$

$$x=\frac{18}{11}=1\frac{7}{11}$$

Working together, it will take them $1\frac{7}{11}$ hours to clean the garage. That's about 1 hour and 38 minutes.

872. $2\frac{10}{31}$ **hours**

Let x represent the total amount of time it will take to grade the calculus papers working together. Dan can do $\frac{1}{4}$ of the job in one hour, Don can do $\frac{1}{8}$ of the job in one hour, and Duane can do $\frac{1}{18}$ of the job in one hour. So $\frac{x}{4}+\frac{x}{8}+\frac{x}{18}=1$, where 1 represents completing the whole job.

Find a common denominator and multiply each term by that number to eliminate the fractions.

$$\cancel{72}^{18}\cdot\frac{x}{\cancel{4}}+\cancel{72}^{9}\cdot\frac{x}{\cancel{8}}+\cancel{72}^{4}\cdot\frac{x}{\cancel{18}}=72\cdot 1$$

$$18x+9x+4x=72$$

Simplify on the left.

$$31x=72$$

Divide each side by 31.

$$\frac{\cancel{31}x}{\cancel{31}} = \frac{72}{31}$$

$$x = \frac{72}{31} = 2\frac{10}{31}$$

Working together, it will take them $2\frac{10}{31}$ hours to grade the papers. That's about 2 hour and 19 minutes.

873. **2 hours**

First change the 1 hour and 20 minutes to hours. You get $1\frac{20}{60} = 1\frac{1}{3} = \frac{4}{3}$ of an hour. Fred can do $\frac{1}{4}$ of the job in one hour. Let the amount of time that Ted needs be represented by n, so Ted can do $\frac{1}{n}$ of the job in one hour. Use $\frac{4/3}{4} + \frac{4/3}{n} = 1$, where 1 represents completing the whole job.

Find a common denominator and multiply each term by that number to simplify the fractions.

$$\cancel{4}n \cdot \frac{4/3}{\cancel{4}} + 4\cancel{n} \cdot \frac{4/3}{\cancel{n}} = 4n \cdot 1$$

$$\frac{4}{3}n + 4\left(\frac{4}{3}\right) = 4n$$

$$\frac{4}{3}n + \frac{16}{3} = 4n$$

Simplify and solve for n.

$$\cancel{3} \cdot \frac{4}{\cancel{3}}n + \cancel{3} \cdot \frac{16}{\cancel{3}} = 3 \cdot 4n$$

$$4n + 16 = 12n$$

$$16 = 8n$$

$$2 = n$$

It would take Ted 2 hours to power wash the deck by himself.

874. **15 hours**

First change the 2 and a half hours to 2.5 hours. Jake can do $\frac{1}{3}$ of the job in one hour. Let the amount of time that Blake needs be represented by n, so Blake can do $\frac{1}{n}$ of the job in one hour. Use $\frac{2.5}{3} + \frac{2.5}{n} = 1$, where 1 represents completing the whole job.

Find a common denominator and multiply each term by that number to eliminate the fractions.

$$\cancel{3}n \cdot \frac{2.5}{\cancel{3}} + 3\cancel{n} \cdot \frac{2.5}{\cancel{n}} = 3n \cdot 1$$

$$2.5n + 3(2.5) = 3n$$

$$2.5n + 7.5 = 3n$$

Simplify and solve for n.

$$2.5n + 7.5 = 3n$$

$$7.5 = 3n - 2.5n$$

$$7.5 = 0.5n$$

Dividing each side by 0.5,

$$\frac{7.5}{0.5} = \frac{\cancel{0.5}n}{\cancel{0.5}}$$

$$15 = n$$

It would take Blake 15 hours to wash the dishes by himself.

875. **18 hours**

Dasher can do $\frac{1}{12}$ of the job in one hour, and Dancer can do $\frac{1}{9}$ of the job in one hour. Let the amount of time that Prancer needs be represented by n, so Prancer can do $\frac{1}{n}$ of the job in one hour. Use $\frac{4}{12} + \frac{4}{9} + \frac{4}{n} = 1$, where 1 represents completing the whole job.

Find a common denominator and multiply each term by that number to eliminate the fractions.

$$\cancel{36}^{3} n \cdot \frac{4}{\cancel{12}} + \cancel{36}^{4} n \cdot \frac{4}{\cancel{9}} + 36\cancel{n} \cdot \frac{4}{\cancel{n}} = 36n \cdot 1$$

$$12n + 16n + 144 = 36n$$

Simplify on the left.

$$28n + 144 = 36n$$

Now subtract $28n$ from each side.

$$144 = 8n$$

Dividing each side by 8, you find that $n = 18$.

It would take Prancer 18 hours to trim the tree by himself.

876. **15 quarts**

You have three separate containers, each with a different quantity and concentration (quality).

Container 1 + Container 2 = Container 3

$$x \text{ qt } (70\%) + 10 \text{ qt } (20\%) = (x + 10)\text{qt } (50\%)$$

$$x(0.70) + 10(0.20) = (x + 10)(0.50)$$

$$0.7x + 2 = 0.5x + 5$$

Add $-0.5x$ and -2 to each side.

$$0.2x = 3$$

Divide each side by 0.2, and you get $x = 15$.

So you need to add 15 quarts of the 70% juice and you will have a total of 25 quarts of the 50% solution.

877. **24 ounces**

You have three separate containers, each with a different quantity and concentration (quality).

Container 1 + Container 2 = Container 3

x oz (60%) + 6 oz (10%) = (x + 6) oz (50%)

$x(0.60) + 6(0.10) = (x + 6)(0.50)$

$0.6x + .6 = 0.5x + 3$

Add $-0.5x$ and -0.6 to each side.

$0.1x = 2.4$

Divide each side by 0.1, and you get $x = 24$.

So you need 24 ounces of the 60% solution and you will have a total of 30 ounces of the 50% solution.

878. **2 ounces**

You have three separate containers, each with a different quantity and concentration (quality). The concentrations are in terms of chocolate, so the syrup is 100% concentrate, and the milk is 0% concentrate.

Container 1 + Container 2 = Container 3

x oz (100%) + 8 oz (0%) = (x + 8) oz (20%)

$x(1) + 8(0) = (x + 8)(0.20)$

$x + 0 = 0.2x + 1.6$

Add $-0.2x$ to each side.

$0.8x = 1.6$

Divide each side by 0.8, and you get $x = 2$.

So you need 2 ounces of the chocolate syrup to produce the 20% solution – now a total of 10 ounces of chocolate milk.

879. **6.4 quarts**

You have a radiator and will be removing a container's worth of solution and adding another container's worth of solution (100% pure antifreeze).

Radiator – Container + Container = Radiator

16 qt(50%) – x qt(50%) + x qt(100%) = 16qt (70%)

$16(0.50) - x(0.50) + x(1) = 16(0.70)$

Simplify.

$8 - 0.5x + x = 11.2$

$8 + 0.5x = 11.2$

$0.5x = 3.2$

Divide each side by 0.5, and you get $x = 6.4$. Drain 6.4 quarts of the current contents and replace with 6.4 quarts of antifreeze.

880. **160 quarters**

Let d represent the number of dimes.

Then $d + 60$ represents the number of quarters. Multiply how many of each coin times its value and set the sum of the results equal to $50.

$d(0.10) + (d + 60)(0.25) = \$50$

Simplify and solve for d.

$0.10d + 0.25d + 15 = 50$

$.35d = 35$

Divide each side of the equation by 0.35.

$d = 100$

Janie has 100 dimes and 160 quarters.

881. **60 quarters**

Let q represent the number of quarters. Then $100 - q$ represents the number of nickels. Multiply how many of each coin times its value and set the sum of the results equal to $17.

$q(0.25) + (100 - q)(0.05) = 17$

Simplify and solve for q.

$0.25q + 5 - 0.05q = 17$

$0.20q = 12$

Divide each side by 0.02, and $q = 60$. Stan has 60 quarters and 40 nickels.

882. **80 cents**

Let q represent the number of quarters. Twice as many nickels would be $2q$ nickels. Four more dimes than quarters would be $q + 4$.

Multiply how many of each coin times its value and set the sum of the results equal to $4.

$q(0.25) + 2q(0.05) + (q + 4)(0.10) = 4$

Simplify and solve for q.

$$0.25q + 0.10q + 0.10q + 0.40 = 4$$

$$0.45q + 0.40 = 4$$

$$0.45q = 3.60$$

Divide each side by 0.45, and $q = 8$. He has 8 quarters. Twice as many nickels would make 16 nickels, which is 80 cents.

883. **12 quarters, 36 dimes, and 30 nickels**

Let q represent the number of quarters. Then $3q$ represents the number of dimes.

The number of nickels is 78 minus the total number of quarters and dimes, which is $4q$. So the number of nickels is represented with $78 - 4q$.

Multiply how many of each coin times its value and set the sum of the results equal to $8.10.

$$q(0.25) + 3q(0.10) + (78 - 4q)(0.05) = 8.10$$

Simplify and solve for q.

$$0.25q + 0.30q + 3.90 - 0.2q = 8.10$$

$$0.35q + 3.90 = 8.10$$

$$0.35q = 4.20$$

Divide each side by 0.35, and $q = 12$. That's 12 quarters. Three times as many gives you 36 dimes, and $78 - 48 = 30$ nickels.

884. **$12.50**

Let h represent the number of half-dollars. Then $h + 5$ represents the number of quarters.

If there are 110 coins total, then the number of silver dollars is 110 less the number of quarters and half-dollars.

Silver dollars: $110 - (h + h + 5) = 105 - 2h$.

Multiply how many of each coin times its value and set the sum of the results equal to $50.00.

$$h(0.50) + (h + 5)(0.25) + (105 - 2h)(1) = 50$$

$$0.50h + 0.25h + 1.25 + 105 - 2h = 50$$

Simplify and solve for h.

$$-1.25h + 106.25 = 50$$

$$-1.25h = -56.25$$

$$h = 45$$

There are 45 half-dollars and $45 + 5$ quarters = 50 quarters.

$$50(0.25) = 12.50$$

885. **10**

Let n represent the number of \$10 bills.

Then $4n$ represents the number of \$5 bills.

And $6 + 4n$ represents the number of \$2 bills.

And $5(6 + 4n) - 10$ represents the number of \$1 bills.

Multiply how many of each bill times its value and set the sum of the results equal to \$90.00.

$$n(10) + 4n(5) + (6+4n)(2) + [5(6+4n) - 10](1) = 90$$

Simplify on the left,

$$10n + 20n + 12 + 8n + 30 + 20n - 10 = 90$$

$$58n + 32 = 90$$

Add –32 to each side.

$$58n = 58$$

Dividing each side by 58, you have $n = 1$.

The number of \$2 bills is represented by $6 + 4n$, so there are $6 + 4(1) = 10$ of them.

886. **A**

Starting from the origin, you move 2 units to the left and then 3 units up. The point is in QII.

887. **D**

Starting from the origin, you move 4 units to the right and then 1 unit down. The point is in QIV.

888. **A**

Starting from the origin, you don't move left or right, but you move 2 units up.

889. **C**

Starting from the origin, you move 4 units to the left. The point stays on the axis and doesn't move up or down from there.

890. ***x*-axis**

The point is not in QI or QIV, but it's on the positive x-axis, between those two quadrants.

891. **QIII**

The point is in QIII. Both coordinates have negative values.

892. **QII**

The point is in QII. The x-coordinate is negative, and the y-coordinate is positive.

893. ***y*-axis**

The point is not in QIII or QIV, but it's on the negative y-axis, between those two quadrants.

894. **(2, 0) and (0, 3)**

To find the x-intercept, let $y = 0$.

$3x + 2(0) = 6$

$3x = 6$

Divide each side by 3, and $x = 2$.

The x-intercept is (2, 0).

To find the y-intercept, let $x = 0$.

$3(0) + 2y = 6$

$2y = 6$

Divide each side by 2, and $y = 3$.

The y-intercept is (0, 3).

895. **(3, 0) and (0, –4)**

To find the x-intercept, let $y = 0$.

$4x - 3(0) = 12$

$4x = 12$

Divide each side by 4, and $x = 3$.

The x-intercept is (3, 0).

To find the y-intercept, let $x = 0$.

$4(0) - 3y = 12$

$-3y = 12$

Divide each side by –3, and $y = -4$.

The y-intercept is (0, –4).

896. **(0, 0) for both**

To find the x-intercept, let $y = 0$.

$5x + 2(0) = 0$

$5x = 0$

Divide each side by 5, and $x = 0$.

The x-intercept is (0, 0).

To find the y-intercept, let $x = 0$.

$5(0) + 2y = 0$

$2y = 0$

Divide each side by 2, and $y = 0$.

The y-intercept is (0, 0).

This line goes through the origin.

897. **(0, 0) for both**

To find the x-intercept, let $y = 0$.

$6x - 0 = 0$

$6x = 0$

Divide each side by 6, and $x = 0$.

The x-intercept is (0, 0).

To find the y-intercept, let $x = 0$.

$6(0) - y = 0$

$-y = 0$

Divide each side by –1, and $y = 0$.

The y-intercept is (0, 0).

This line goes through the origin.

898. $\mathbf{(0,-3)}$ **and** $\mathbf{\left(\frac{3}{4},0\right)}$

The equation is in the *slope-intercept* form, $y = mx + b$. The y-intercept is $(0, b)$.

So the y-intercept of this line is (0, –3).

To find the x-intercept, let $y = 0$.

$0 = 4x - 3$

$3 = 4x$

Divide each side by 4, and $x = \frac{3}{4}$.

The x-intercept is $\left(\frac{3}{4},0\right)$.

899. **(0, 2) and (2, 0)**

The equation is in the *slope-intercept* form, $y = mx + b$. The y-intercept is $(0, b)$.

So the y-intercept of this line is (0, 2).

To find the x-intercept, let $y = 0$.

$$0 = -x + 2$$

$$x = 2$$

The x-intercept is (2, 0).

900. **(0, 2) and (–6, 0)**

The equation is in the *slope-intercept* form, $y = mx + b$. The y-intercept is $(0, b)$.

So the y-intercept of this line is (0, 2).

To find the x-intercept, let $y = 0$.

$$0 = \frac{1}{3}x + 2$$

$$-\frac{1}{3}x = 2$$

Multiply each side by –3, and $x = -6$.

The x-intercept is (–6, 0).

901. **(0, –12) and (–16, 0)**

The equation is in the *slope-intercept* form, $y = mx + b$. The y-intercept is $(0, b)$.

So the y-intercept of this line is (0, –12).

To find the x-intercept, let $y = 0$.

$$0 = -\frac{3}{4}x - 12$$

$$\frac{3}{4}x = -12$$

Multiply each side by $\frac{4}{3}$, and $x = -16$.

The x-intercept is (–16, 0).

902. **(0, 8) only**

The graph of the line is horizontal, so there is no x-intercept. The y-intercept is at (0, 8).

903. **(–3, 0) only**

The graph of the line is vertical, so there is no y-intercept. The x-intercept is at (–3, 0).

904. $m = -1$

Using the formula for the slope through two points, $m = \frac{y_2 - y_1}{x_2 - x_1}$, substitute the values carefully, keeping the same order (coordinates from the same point above and below each another).

$$m = \frac{3-6}{2-(-1)} = \frac{-3}{3} = -1$$

Answers 901–1001

905. $m = -\frac{13}{5}$

Using the formula for the slope through two points, $m = \frac{y_2 - y_1}{x_2 - x_1}$, substitute the values carefully, keeping the same order (coordinates from the same point above and below each another).

$$m = \frac{4-(-9)}{0-5} = \frac{13}{-5} = -\frac{13}{5}$$

906. $m = \frac{1}{9}$

Using the formula for the slope through two points, $m = \frac{y_2 - y_1}{x_2 - x_1}$, substitute the values carefully, keeping the same order (coordinates from the same point above and below each another).

$$m = \frac{-3-(-2)}{-4-5} = \frac{-1}{-9} = \frac{1}{9}$$

907. $m = \frac{5}{4}$

Using the formula for the slope through two points, $m = \frac{y_2 - y_1}{x_2 - x_1}$, substitute the values carefully, keeping the same order (coordinates from the same point above and below each another).

$$m = \frac{5-0}{0-(-4)} = \frac{5}{4}$$

908. $m = 0$

Using the formula for the slope through two points, $m = \frac{y_2 - y_1}{x_2 - x_1}$, substitute the values carefully, keeping the same order (coordinates from the same point above and below each another).

$$m = \frac{5-5}{6-(-3)} = \frac{0}{9} = 0$$

909. **Undefined slope**

Using the formula for the slope through two points, $m = \frac{y_2 - y_1}{x_2 - x_1}$, substitute the values carefully, keeping the same order (coordinates from the same point above and below each another).

The slope is undefined.

$$m = \frac{2-(-4)}{-4-(-4)} = \frac{6}{0}$$

910. $m = -4$

The equation is in slope-intercept form, $y = mx + b$. The slope is the value of m, the coefficient of the variable x. So, for $y = -4x + 3$, the slope is $m = -4$.

911. $m = 2$

The equation is in slope-intercept form, $y = mx + b$. The slope is the value of m, the coefficient of the variable x. So, for $y = 2x - 1$, the slope is $m = 2$.

912. $m = -\frac{1}{2}$

First rewrite the equation in slope-intercept form, $y = mx + b$. The slope is the value of m, the coefficient of the variable x.

$$3x + 6y = 11$$

Add $-3x$ to each side.

$$6y = -3x + 11$$

Divide each term by 6.

$$\frac{6y}{6} = \frac{-3x}{6} + \frac{11}{6}$$

$$y = -\frac{1}{2}x + \frac{11}{6}$$

The slope is $m = -\frac{1}{2}$.

913. $m = \frac{4}{3}$

First rewrite the equation in slope-intercept form, $y = mx + b$. The slope is the value of m, the coefficient of the variable x.

$$4x - 3y = 7$$

Add $-4x$ to each side.

$$-3y = -4x + 7$$

Divide each term by -3.

$$\frac{-3y}{-3} = \frac{-4x}{-3} + \frac{7}{-3}$$

$$y = \frac{4}{3}x - \frac{7}{3}$$

The slope is $m = \frac{4}{3}$.

914. $m = 0$

The equation is in slope-intercept form, $y = mx + b$. The slope is the value of m, the coefficient of the variable x. Since you find no x variable, its coefficient must be 0; you could write the equation as $y = 0x - 6$. So the slope is $m = 0$.

915. **Undefined slope**

If you try to write the equation in slope-intercept form, $y = mx + b$, you have $0 = x - 3$ or $0 = -x + 3$. In either case, you have no y variable. This equation represents a vertical line, so its slope is undefined.

916. Also goes through (1, 2)

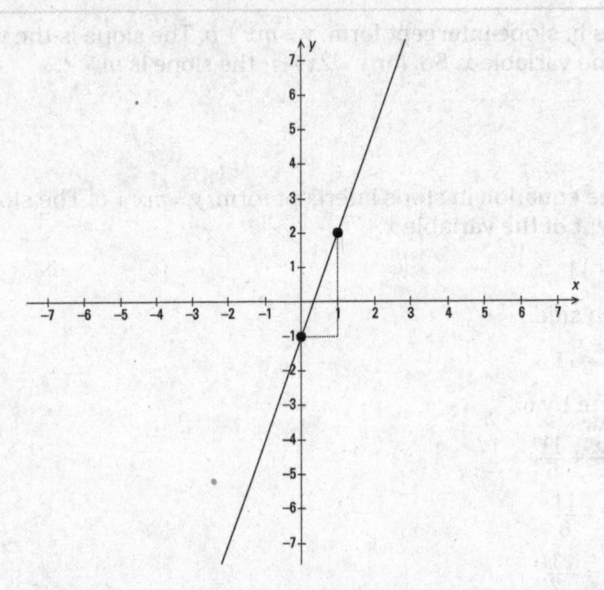

First plot the y-intercept, (0, –1). The slope is 3, which you think of as $\frac{3}{1}$, so you start at the y-intercept and move one unit to the right. From that point, move 3 units up. You're now at another point on the line (1, 2). Draw a line through this point and the y-intercept.

917. Also goes through (1, 1)

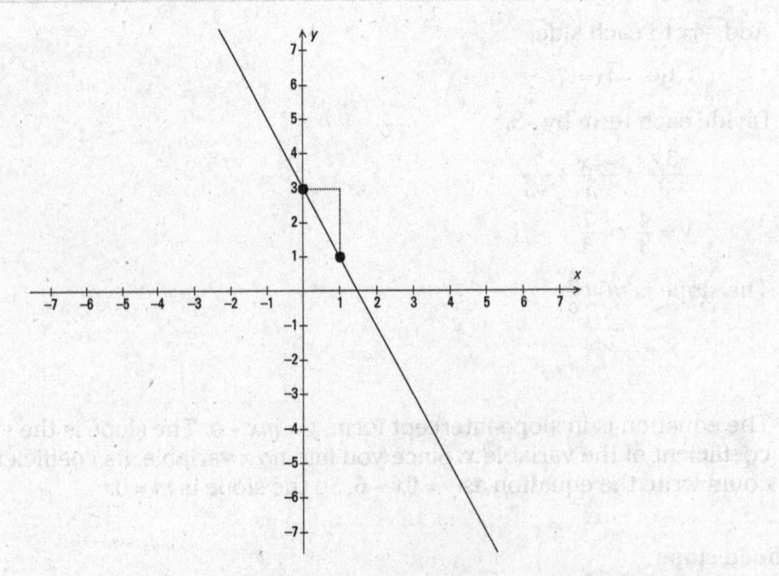

First plot the y-intercept, (0, 3). The slope is –2, which you think of as $\frac{-2}{1}$, so you start at the y-intercept and move 1 unit to the right. From that point, move 2 units down. You're now at another point on the line (1, 1). Draw a line through this point and the y-intercept.

918. Also goes through (3, –6)

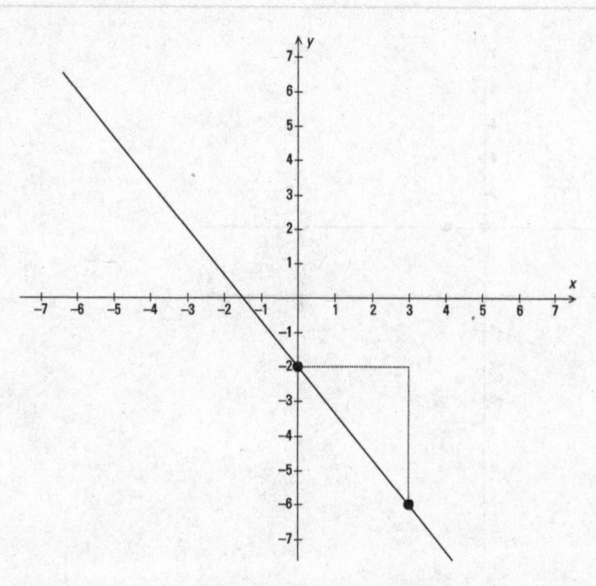

First plot the *y*-intercept, (0, –2). The slope is $-\frac{4}{3}$, so you start at the *y*-intercept and move 3 units to the right. From that point, move 4 units down. You're now at another point on the line (3, –6). Draw a line through this point and the *y*-intercept.

919. Also goes through (4, –2)

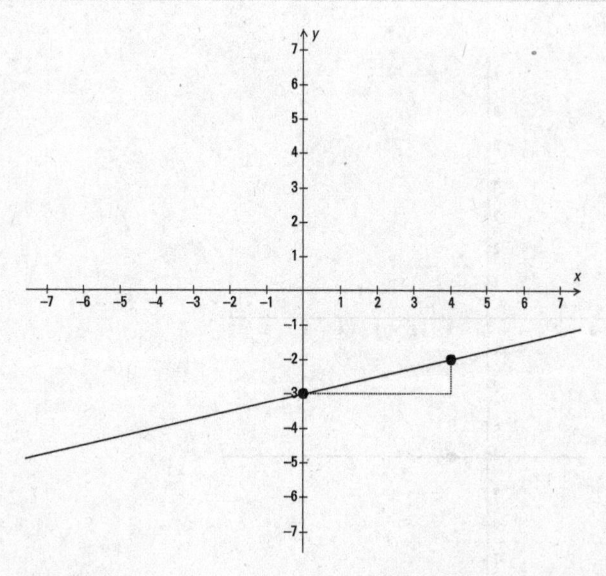

First plot the *y*-intercept, (0, –3). The slope is $\frac{1}{4}$. You start at the *y*-intercept and move 4 units to the right. From that point, move 1 unit up. You're now at another point on the line (4, –2). Draw a line through this point and the *y*-intercept.

920. Also goes through (1, 2)

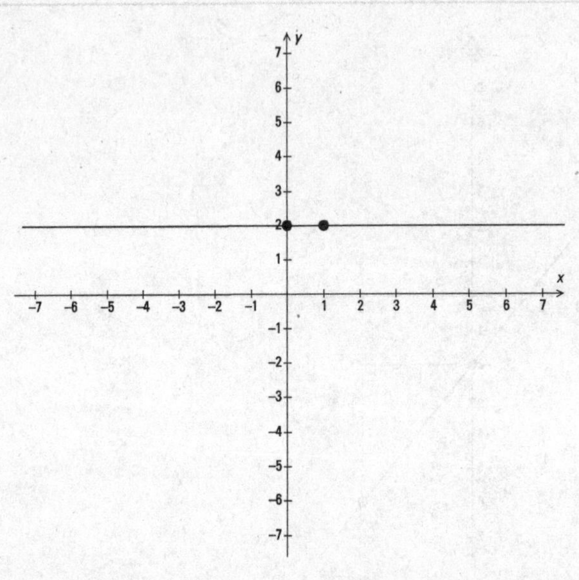

First plot the *y*-intercept, (0, 2). The slope is 0, because the coefficient of the *x* variable is 0. You can think of the slope as being $\frac{0}{1}$. Start at the *y*-intercept and move one unit to the right. From that point, you don't move up or down. You're now at another point on the line (1, 2). Draw a line through this point and the *y*-intercept. You have a horizontal line.

921. Also goes through (1, –4)

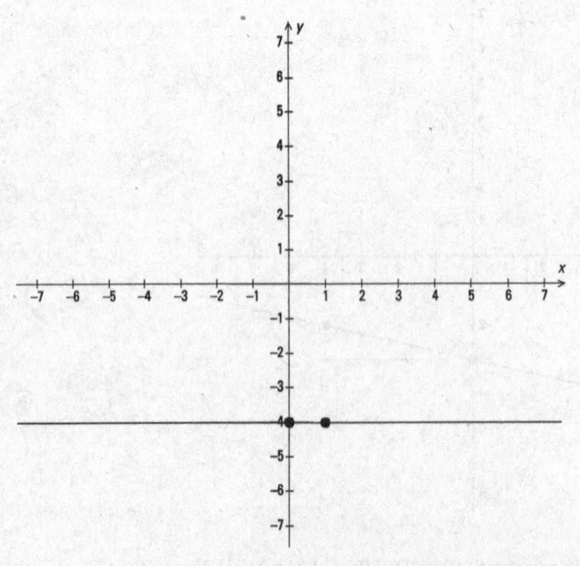

First plot the *y*-intercept, (0, –4). The slope is 0, because the coefficient of the *x* variable is 0. You can think of the slope as being $\frac{0}{1}$. Start at the *y*-intercept and move 1 unit to the right. From that point, you don't move up or down. You're now at another point on the line (1, –4). Draw a line through this point and the *y*-intercept. You have a horizontal line.

922. **Slope is $-\frac{5}{3}$**

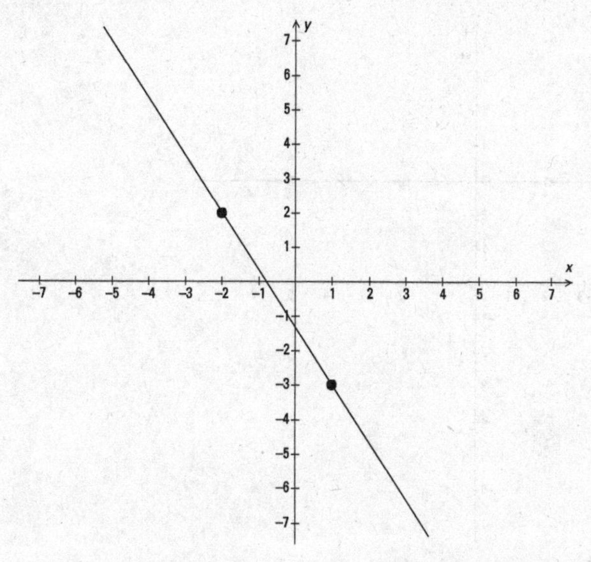

The slope of the line is $m=\frac{2-(-3)}{-2-1}=\frac{5}{-3}=-\frac{5}{3}$, so you are expecting the line to fall steeply from left to right.

923. **Line rises from left to right**

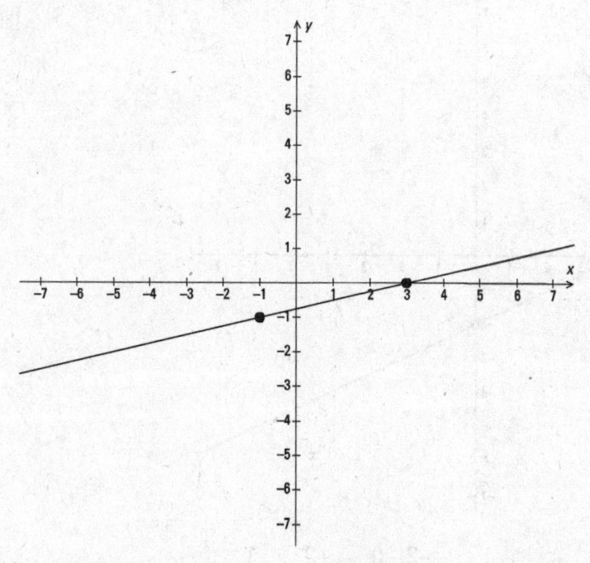

The slope of the line is $m=\frac{0-(-1)}{3-(-1)}=\frac{1}{4}$, so you are expecting the line to rise slowly from left to right.

924. **Line is horizontal**

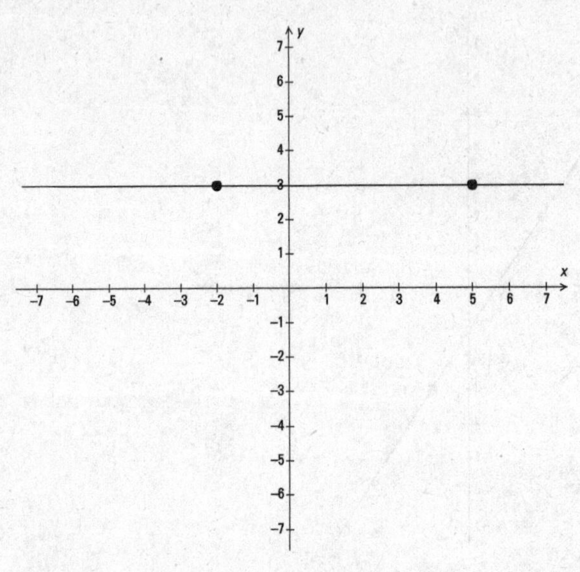

The slope of the line is $m = \frac{3-3}{-2-5} = \frac{0}{-7} = 0$, so you are expecting the line to be horizontal.

925. **Line falls from left to right**

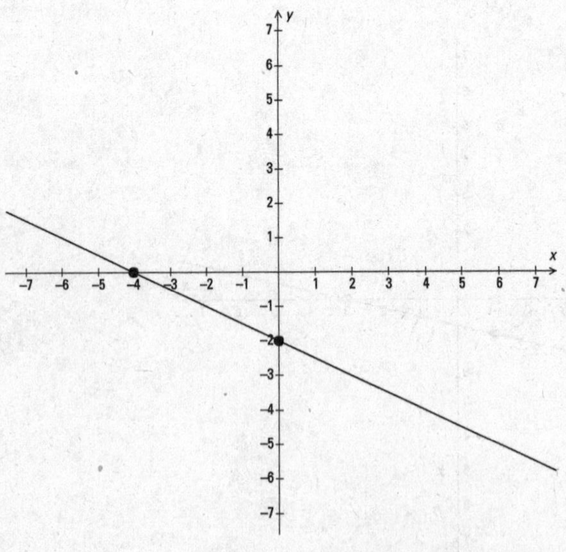

The slope of the line is $m = \frac{-2-0}{0-(-4)} = \frac{-2}{4} = -\frac{1}{2}$, so you are expecting the line to fall slowly from left to right.

926. $y = 2x + 10$

Using the point-slope form, $y - y_1 = m(x - x_1)$, replace the m with the given slope and the coordinates (x_1, y_1) with the respective coordinates of the given point.

$$y - 4 = 2(x - (-3))$$

Simplify and solve for y.

$$y - 4 = 2x + 6$$

$$y = 2x + 10$$

This is the slope-intercept form of the equation of the line.

927. $y = -\frac{1}{2}x - \frac{1}{2}$

Using the point-slope form, $y - y_1 = m(x - x_1)$, replace the m with the given slope and the coordinates (x_1, y_1) with the respective coordinates of the given point.

$$y - (-1) = -\frac{1}{2}(x - 1)$$

Simplify and solve for y.

$$y + 1 = -\frac{1}{2}x + \frac{1}{2}$$

$$y = -\frac{1}{2}x - \frac{1}{2}$$

This is the slope-intercept form. You can clear the fractions and write it in the standard form as: $x + 2y = -1$.

928. $y = -3x + 12$

Using the point-slope form, $y - y_1 = m(x - x_1)$, replace the m with the given slope and the coordinates (x_1, y_1) with the respective coordinates of the given point.

$$y - 0 = -3(x - 4)$$

Simplifying, you have:

$$y = -3x + 12.$$

This is the slope-intercept form.

929. $y = \frac{6}{7}x + \frac{17}{7}$

Using the point-slope form, $y - y_1 = m(x - x_1)$, replace the m with the given slope and the coordinates (x_1, y_1) with the respective coordinates of the given point.

$$y - 5 = \frac{6}{7}(x - 3)$$

Simplify and solve for y.

$$y-5=\frac{6}{7}x-\frac{18}{7}$$

$$y=\frac{6}{7}x+\frac{17}{7}$$

This is the slope-intercept form.

You can clear the fractions, and write it in the standard form as: $6x-7y=-17$.

930. $y=3$

Using the point-slope form, $y-y_1=m(x-x_1)$, replace the m with the given slope and the coordinates (x_1, y_1) with the respective coordinates of the given point.

$$y-3=0(x-4)$$

Simplify and solve for y.

$$y-3=0$$

$$y=3$$

This is the slope-intercept form.

931. $x=-3$

You can't really use the point-slope form, $y-y_1=m(x-x_1)$, because you have no number to replace the m with. Just recognize that you have a vertical line, and the equation will involve the x-coordinate.

$$x=-3$$

932. $y=-2x+7$

First find the slope of the line passing through the given points.

$$m=\frac{-1-(-5)}{4-6}=\frac{4}{-2}=-2$$

Now use the point-slope form and the coordinates of one of the points to determine the equation. Using (4, –1):

$$y-(-1)=-2(x-4)$$

$$y+1=-2x+8$$

$$y=-2x+7.$$

933. $y=\frac{5}{4}x-\frac{3}{4}$

First find the slope of the line passing through the given points.

$$m=\frac{3-(-7)}{3-(-5)}=\frac{10}{8}=\frac{5}{4}$$

Now use the point-slope form and the coordinates of one of the points to determine the equation. Using (3, 3):

$$y-3=\frac{5}{4}(x-3)$$

$$y-3=\frac{5}{4}x-\frac{15}{4}$$

$$y=\frac{5}{4}x-\frac{3}{4}$$

Rewriting the equation in the standard form eliminates the fractions: $5x-4y=3$.

934. $y=6$

First find the slope of the line passing through the given points.

$$m=\frac{6-6}{1-3}=\frac{0}{-2}=0$$

Now use the point-slope form and the coordinates of one of the points to determine the equation. Using (1, 6):

$$y-6=0(x-1)$$

$$y-6=0$$

$$y=6$$

This is a horizontal line.

935. $y=-\frac{5}{4}x+2$

First find the slope of the line passing through the given points.

$$m=\frac{-3-2}{4-0}=\frac{-5}{4}=-\frac{5}{4}$$

Now use the point-slope form and the coordinates of one of the points to determine the equation. Using (0,2):

$$y-2=-\frac{5}{4}(x-0)$$

$$y-2=-\frac{5}{4}x$$

$$y=-\frac{5}{4}x+2$$

Since the point (0, 2) is the y-intercept, the equation could have been written directly using the slope-intercept form. Also, the fraction can be eliminated when writing this in the standard form: $5x+4y=8$.

936. $x=-4$

First find the slope of the line passing through the given points.

$$m=\frac{5-(-5)}{-4-(-4)}=\frac{10}{0}$$

The slope is undefined. The line through the points is a vertical line, and its equation uses the x-coordinates of the points: $x=-4$.

937. $y = -\frac{8}{3}x$

First find the slope of the line passing through the given points.

$$m = \frac{0-(-8)}{0-3} = \frac{8}{-3} = -\frac{8}{3}$$

Now use the point-slope form and the coordinates of one of the points to determine the equation. Using (0, 0):

$$y - 0 = -\frac{8}{3}(x - 0)$$

$$y = -\frac{8}{3}x$$

Written in the standard form: $8x + 3y = 0$.

938. $\| m = 4$; $\perp m = -\frac{1}{4}$

Parallel lines have the same slope, so a line parallel to $y = 4x - 3$ has a slope of $m = 4$.

Perpendicular lines have slopes that are negative reciprocals of one another, so a line perpendicular to $y = 4x - 3$ has a slope of $m = -\frac{1}{4}$.

939. $\| m = -\frac{5}{3}$; $\perp m = \frac{3}{5}$

Parallel lines have the same slope, so a line parallel to $y = -\frac{5}{3}x + 7$ has a slope of $m = -\frac{5}{3}$.

Perpendicular lines have slopes that are negative reciprocals of one another, so a line perpendicular to $y = -\frac{5}{3}x + 7$ has a slope of $m = \frac{3}{5}$.

940. $\| m = \frac{2}{3}$; $\perp m = -\frac{3}{2}$

First write the equation in slope-intercept form.

$$2x - 3y = 7$$

$$-3y = -2x + 7$$

$$y = \frac{2}{3}x - \frac{7}{3}$$

Parallel lines have the same slope, so a line parallel to $y = \frac{2}{3}x - \frac{7}{3}$ has a slope of $m = \frac{2}{3}$.

Perpendicular lines have slopes that are negative reciprocals of one another, so a line perpendicular to $y = \frac{2}{3}x - \frac{7}{3}$ has a slope of $m = -\frac{3}{2}$.

941. $\| m = \frac{1}{4}$; $\perp m = -4$

First write the equation in slope-intercept form.

$$x - 4y = 8$$

$$-4y = -x + 8$$

$$y = \frac{1}{4}x - 2$$

Parallel lines have the same slope, so a line parallel to $y=\frac{1}{4}x-2$ has a slope of $m=\frac{1}{4}$. Perpendicular lines have slopes that are negative reciprocals of one another, so a line perpendicular to $y=\frac{1}{4}x-2$ has a slope of $m = -4$.

942. $\parallel m$ **has no slope;** $\perp m = 0$

The line corresponding to this equation is vertical and has no slope (undefined).

Parallel lines have the same slope, so a line parallel to $x = 5$ also has no slope.

A line perpendicular to a vertical line is a horizontal line, so its slope is $m = 0$.

943. $\parallel m = 0$; $\perp m$ **has no slope**

The line corresponding to this equation is horizontal and has a slope $m = 0$.

Parallel lines have the same slope, so a line parallel to $y = -6$ also has slope $m = 0$.

A line perpendicular to a horizontal line is a vertical line; vertical lines have no slope (undefined).

944. $\parallel y=-2x+3;\ \perp y=\frac{1}{2}x+3$

The slope of the line $y = -2x + 1$ is $m = -2$.

The slope of the parallel line is also $m = -2$. The parallel line through the point (0, 3) can be written using the slope-intercept form and inserting the slope and the y-intercept, which is the given point.

$$y=-2x+3$$

The slope of the perpendicular line is $m=\frac{1}{2}$. The perpendicular line through the point (0, 3) can be written using the slope-intercept form and inserting the slope and the y-intercept, which is the given point.

$$y=\frac{1}{2}x+3$$

945. $\parallel y=\frac{4}{3}x-3;\ \perp y=-\frac{3}{4}x-\frac{37}{4}$

The slope of the line $y=\frac{4}{3}x-3$ is $m=\frac{4}{3}$. A line parallel to the given line also has $m=\frac{4}{3}$. The parallel line through the point (–3, –7) can be written using the point-slope form and inserting the slope and the coordinates of the point.

$$y-(-7)=\frac{4}{3}(x-(-3))$$

$$y+7=\frac{4}{3}(x+3)$$

$$y+7=\frac{4}{3}x+4$$

$$y=\frac{4}{3}x-3$$

A line perpendicular to the given line has $m=-\frac{3}{4}$. The perpendicular line through the point (–3, –7) can be written using the point-slope form and inserting the slope and the coordinates of the point.

$$y-(-7)=-\frac{3}{4}(x-(-3))$$
$$y+7=-\frac{3}{4}(x+3)$$
$$y+7=-\frac{3}{4}x-\frac{9}{4}$$
$$y=-\frac{3}{4}x-\frac{37}{4}$$

946. $\| y=4x;\ \perp y=-\frac{1}{4}x$

First, rewrite the equation of the line in slope-intercept form.

$$4x-y=3$$
$$-y=-4x+3$$
$$y=4x-3$$

The slope of the line $4x-y=3$ is $m=4$.

A line parallel to the given line also has $m=4$. The parallel line through the point (0, 0) can be written using the slope-intercept form and inserting the slope and the y-intercept, which is the given point.

$$y=4x+0 \text{ or } y=4x$$

A line perpendicular to the given line has $m=-\frac{1}{4}$. The perpendicular line through the point (0, 0) can be written using the slope-intercept form and inserting the slope and the y-intercept, which is the given point.

$$y=-\frac{1}{4}x+0 \text{ or } y=-\frac{1}{4}x$$

947. $\| y=-2x-1;\ \perp y=\frac{1}{2}x+\frac{3}{2}$

First, rewrite the equation of the line in slope-intercept form.

$$3y=-6x+7$$
$$y=-2x+\frac{7}{3}$$

The slope of the line $6x+3y=7$ is $m=-2$.

A line parallel to the given line also has $m=-2$. The parallel line through the point (–1, 1) can be written using the point-slope form and inserting the slope and coordinates of the point.

$$y-1=-2(x-(-1))$$
$$y-1=-2x-2$$
$$y=-2x-1$$

A line perpendicular to the given line has $m = \frac{1}{2}$. The perpendicular line through the point (–1, 1) can be written using the point-slope form.

$$y - 1 = \frac{1}{2}(x - (-1))$$
$$y - 1 = \frac{1}{2}x + \frac{1}{2}$$
$$y = \frac{1}{2}x + \frac{3}{2}$$

948. **13 units**

Using the distance formula, $d = \sqrt{(x_2 - x_1)^2 + (y_2 - y_1)^2}$, plug in the coordinates of the points.

$$d = \sqrt{(-2-3)^2 + (-8-4)^2}$$
$$= \sqrt{(-5)^2 + (-12)^2} = \sqrt{25+144}$$
$$= \sqrt{169} = 13$$

949. **10 units**

Using the distance formula, $d = \sqrt{(x_2 - x_1)^2 + (y_2 - y_1)^2}$, plug in the coordinates of the points.

$$d = \sqrt{(-5-1)^2 + (5-(-3))^2}$$
$$= \sqrt{(-6)^2 + (8)^2} = \sqrt{36+64}$$
$$= \sqrt{100} = 10$$

950. **5 units**

Using the distance formula, $d = \sqrt{(x_2 - x_1)^2 + (y_2 - y_1)^2}$, plug in the coordinates of the points.

$$d = \sqrt{(0-(-4))^2 + (0-(-3))^2}$$
$$= \sqrt{(4)^2 + (3)^2} = \sqrt{16+9}$$
$$= \sqrt{25} = 5$$

951. **25 units**

Using the distance formula, $d = \sqrt{(x_2 - x_1)^2 + (y_2 - y_1)^2}$, plug in the coordinates of the points.

$$d = \sqrt{(-2-5)^2 + (22-(-2))^2}$$
$$= \sqrt{(-7)^2 + (24)^2} = \sqrt{49+576}$$
$$= \sqrt{625} = 25$$

952. **$5\sqrt{2}$ units**

Using the distance formula, $d=\sqrt{(x_2-x_1)^2+(y_2-y_1)^2}$, plug in the coordinates of the points.

$$\begin{aligned} d &= \sqrt{(-2-3)^2+(-2-3)^2} \\ &= \sqrt{(-5)^2+(-5)^2} = \sqrt{25+25} \\ &= \sqrt{50} = \sqrt{25}\sqrt{2} = 5\sqrt{2} \end{aligned}$$

953. **$2\sqrt{41}$ units**

Using the distance formula, $d=\sqrt{(x_2-x_1)^2+(y_2-y_1)^2}$, plug in the coordinates of the points.

$$\begin{aligned} d &= \sqrt{(6-(-4))^2+(9-1)^2} \\ &= \sqrt{(10)^2+(8)^2} = \sqrt{100+64} \\ &= \sqrt{164} = \sqrt{4}\sqrt{41} = 2\sqrt{41} \end{aligned}$$

954. **$3\sqrt{10}$ units**

Using the distance formula, $d=\sqrt{(x_2-x_1)^2+(y_2-y_1)^2}$, plug in the coordinates of the points.

$$\begin{aligned} d &= \sqrt{(0-(-3))^2+(-2-7)^2} \\ &= \sqrt{(3)^2+(-9)^2} = \sqrt{9+81} \\ &= \sqrt{90} = \sqrt{9}\sqrt{10} = 3\sqrt{10} \end{aligned}$$

955. **$4\sqrt{2}$ units**

Using the distance formula, $d=\sqrt{(x_2-x_1)^2+(y_2-y_1)^2}$, plug in the coordinates of the points.

$$\begin{aligned} d &= \sqrt{(4-0)^2+(0-4)^2} \\ &= \sqrt{(4)^2+(-4)^2} = \sqrt{16+16} \\ &= \sqrt{32} = \sqrt{16}\sqrt{2} = 4\sqrt{2} \end{aligned}$$

956. **2 units**

Using the distance formula, $d=\sqrt{(x_2-x_1)^2+(y_2-y_1)^2}$, plug in the coordinates of the points.

$$\begin{aligned} d &= \sqrt{(-3-(-3))^2+(8-6)^2} \\ &= \sqrt{(0)^2+(2)^2} = \sqrt{0+4} \\ &= \sqrt{4} = 2 \end{aligned}$$

The points are on the same vertical line and are 2 units apart.

957. **$2\sqrt{41}$ units**

Using the distance formula, $d=\sqrt{(x_2-x_1)^2+(y_2-y_1)^2}$, plug in the coordinates of the points.

$$\begin{aligned}d&=\sqrt{(5-(-5))^2+(-4-4)^2}\\&=\sqrt{(10)^2+(-8)^2}=\sqrt{100+64}\\&=\sqrt{164}=\sqrt{4}\sqrt{41}=2\sqrt{41}\end{aligned}$$

958. **(3, 1)**

Using the midpoint formula, $M=\left(\frac{x_1+x_2}{2},\frac{y_1+y_2}{2}\right)$, plug in the coordinates of the points.

$$M=\left(\frac{4+2}{2},\frac{7+(-5)}{2}\right)=\left(\frac{6}{2},\frac{2}{2}\right)=(3,1)$$

959. **(–4, 1)**

Using the midpoint formula, $M=\left(\frac{x_1+x_2}{2},\frac{y_1+y_2}{2}\right)$, plug in the coordinates of the points.

$$M=\left(\frac{-3+(-5)}{2},\frac{6+(-4)}{2}\right)=\left(\frac{-8}{2},\frac{2}{2}\right)=(-4,1)$$

960. **(–1, 2)**

Using the midpoint formula, $M=\left(\frac{x_1+x_2}{2},\frac{y_1+y_2}{2}\right)$, plug in the coordinates of the points.

$$M=\left(\frac{1+(-3)}{2},\frac{6+(-2)}{2}\right)=\left(\frac{-2}{2},\frac{4}{2}\right)=(-1,2)$$

961. **(–7, –3)**

Using the midpoint formula, $M=\left(\frac{x_1+x_2}{2},\frac{y_1+y_2}{2}\right)$, plug in the coordinates of the points.

$$M=\left(\frac{-6+(-8)}{2},\frac{-6+0}{2}\right)=\left(\frac{-14}{2},\frac{-6}{2}\right)=(-7,-3)$$

962. $\left(\frac{9}{2},3\right)$

Using the midpoint formula, $M=\left(\frac{x_1+x_2}{2},\frac{y_1+y_2}{2}\right)$, plug in the coordinates of the points.

$$M=\left(\frac{4+5}{2},\frac{-3+9}{2}\right)=\left(\frac{9}{2},\frac{6}{2}\right)=\left(\frac{9}{2},3\right)$$

963. (–3, 9)

Using the midpoint formula, $M=\left(\frac{x_1+x_2}{2},\frac{y_1+y_2}{2}\right)$, plug in the coordinates of the points.

$$M=\left(\frac{-3+(-3)}{2},\frac{8+10}{2}\right)=\left(\frac{-6}{2},\frac{18}{2}\right)=(-3,9)$$

The points are on the same vertical line.

964. (2, 0)

Using the midpoint formula, $M=\left(\frac{x_1+x_2}{2},\frac{y_1+y_2}{2}\right)$, plug in the coordinates of the points.

$$M=\left(\frac{4+0}{2},\frac{0+0}{2}\right)=\left(\frac{4}{2},\frac{0}{2}\right)=(2,0)$$

The points are on the same horizontal line and are 4 units apart, so the midpoint is 2 units away from each of them.

965. $\left(-\frac{1}{2},-\frac{3}{2}\right)$

Using the midpoint formula, $M=\left(\frac{x_1+x_2}{2},\frac{y_1+y_2}{2}\right)$, plug in the coordinates of the points.

$$M=\left(\frac{\frac{5}{2}+\left(-\frac{7}{2}\right)}{2},\frac{\frac{3}{2}+\left(-\frac{9}{2}\right)}{2}\right)$$

$$=\left(\frac{-1}{2},\frac{-3}{2}\right)=\left(-\frac{1}{2},-\frac{3}{2}\right)$$

966. (1, 5)

The line $y = x + 4$ has a *y*-intercept of 4 and a slope of 1. The line $y = -x + 6$ has a *y*-intercept of 6 and a slope of –1. The two lines intersect at the point (1, 5).

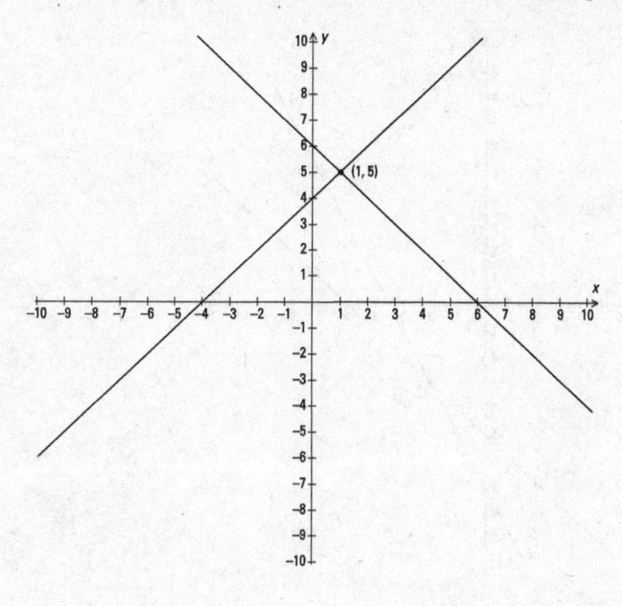

967. (1, –1)

The line $y = 2x - 3$ has a *y*-intercept of –3 and a slope of 2. The line $y = -2x + 1$ has a *y*-intercept of 1 and a slope of –2. The two lines intersect at the point (1, –1).

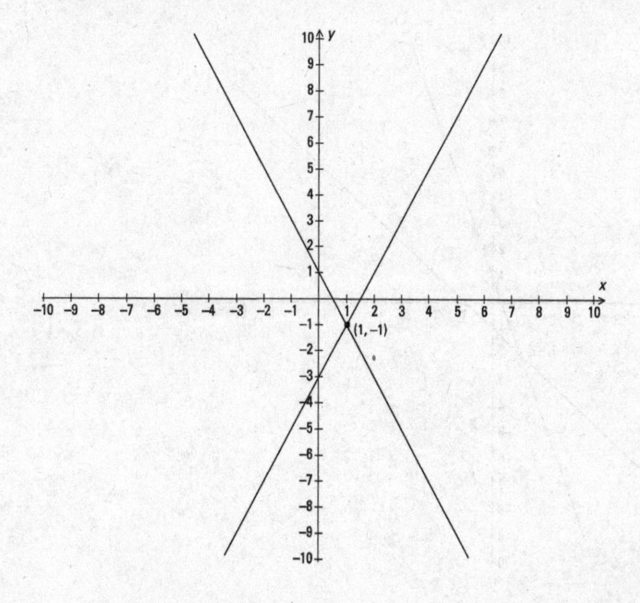

968. (2, –3)

The line $y = x - 5$ has a y-intercept of –5 and a slope of 1. The line $y = 2x - 7$ has a y-intercept of –7 and a slope of 2. The two lines intersect at the point (2, –3).

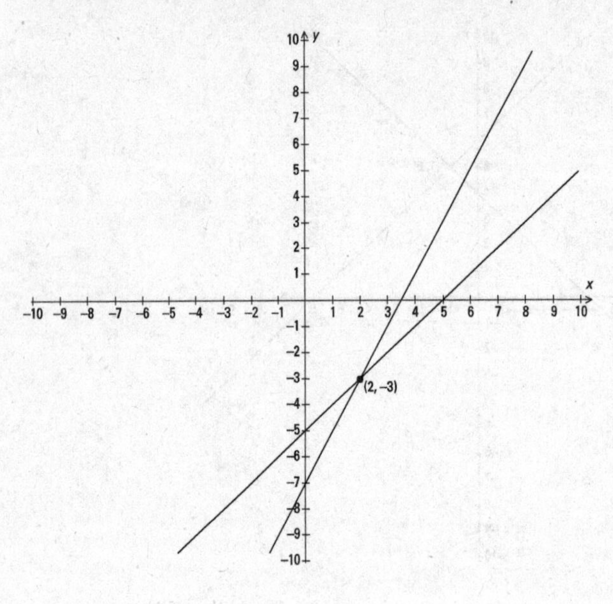

969. (–1, –2)

The line $y = 3x + 1$ has a y-intercept of 1 and a slope of 3. The line $y = x - 1$ has a y-intercept of –1 and a slope of 1. The two lines intersect at the point (–1, –2).

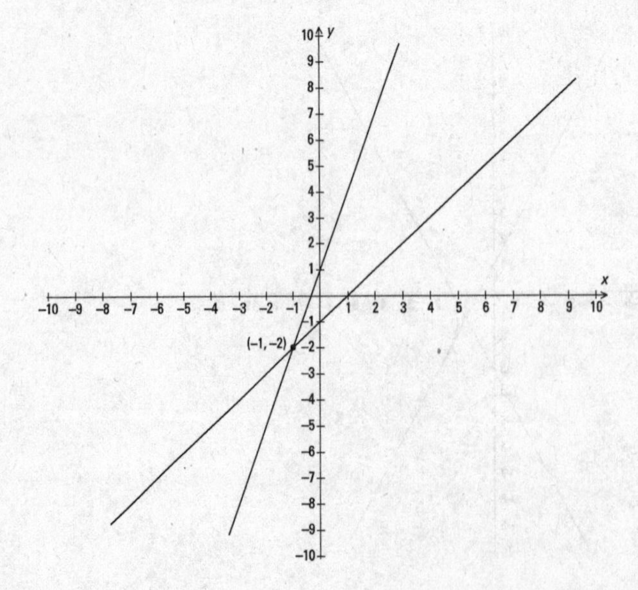

970. (3, 4)

The line $y = 4$ is a horizontal line, and the line $x = 3$ is a vertical line. The two lines intersect at the point (3, 4).

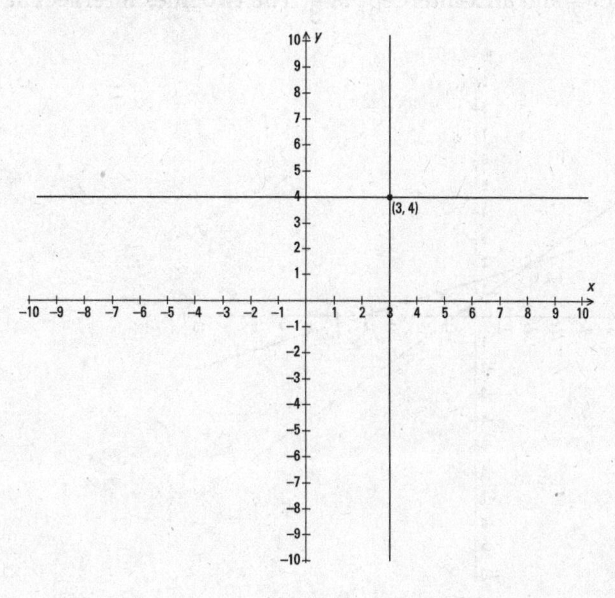

971. (0, –2)

The line $y = -2$ is a horizontal line, and the line $x = 0$ is a vertical line, the y-axis. The two lines intersect at the point (0, –2).

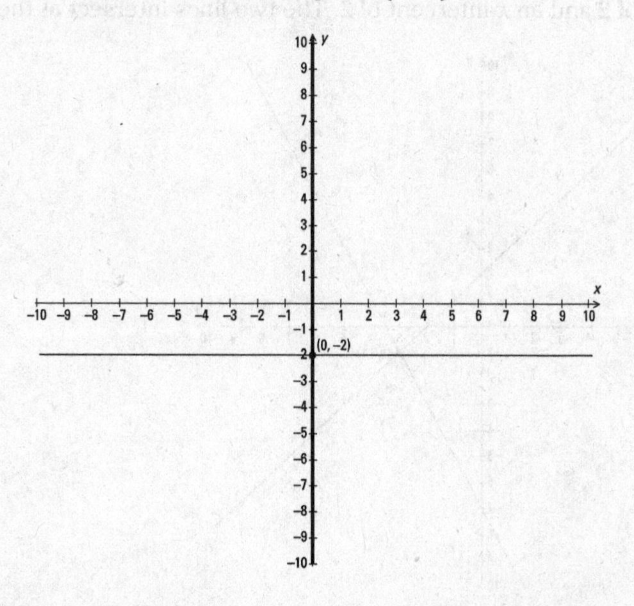

972. (–2, 2)

The line $x + 3y = 4$ has a y-intercept of $\frac{4}{3}$ and an x-intercept of 4. The line $3x + 4y = 2$ has a y-intercept of $\frac{1}{2}$ and an x-intercept of $\frac{2}{3}$. The two lines intersect at the point (–2, 2).

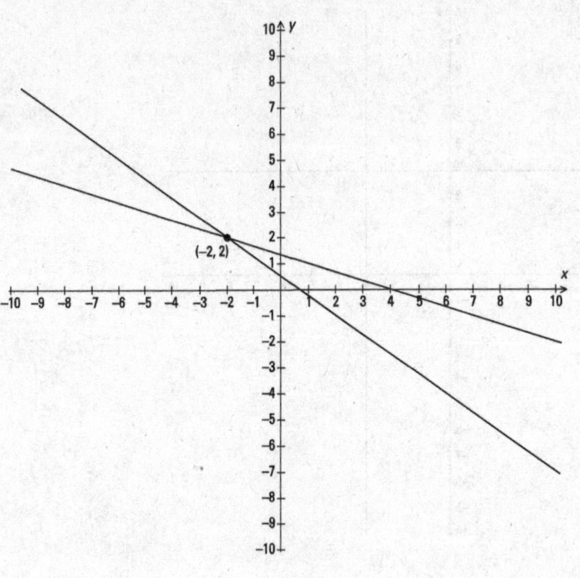

973. (3, –1)

The line $2x - y = 7$ has a y-intercept of –7 and an x-intercept of $\frac{7}{2}$. The line $x + y = 2$ has a y-intercept of 2 and an x-intercept of 2. The two lines intersect at the point (3, –1).

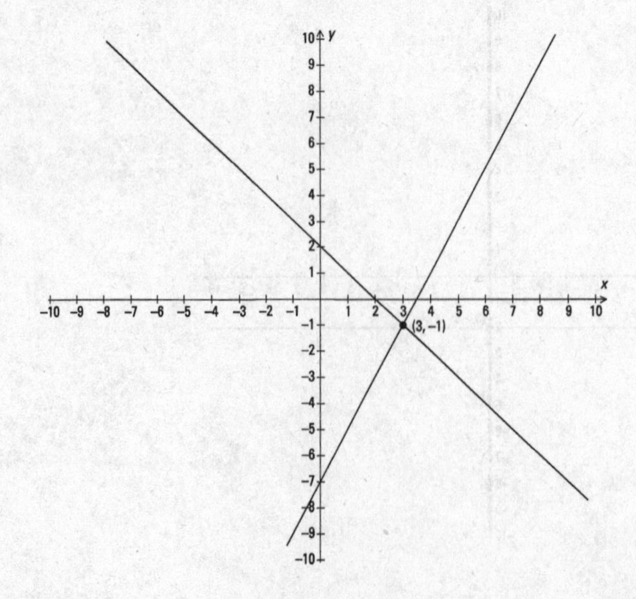

974. (0, 0); $r = 2$

Writing the equation in the standard form for a circle, $(x-h)^2 + (y-k)^2 = r^2$, where (h, k) is the center and r is the radius, you have:

$$(x-0)^2 + (y-0)^2 = 2^2.$$

The center is at (0, 0), and the radius is 2.

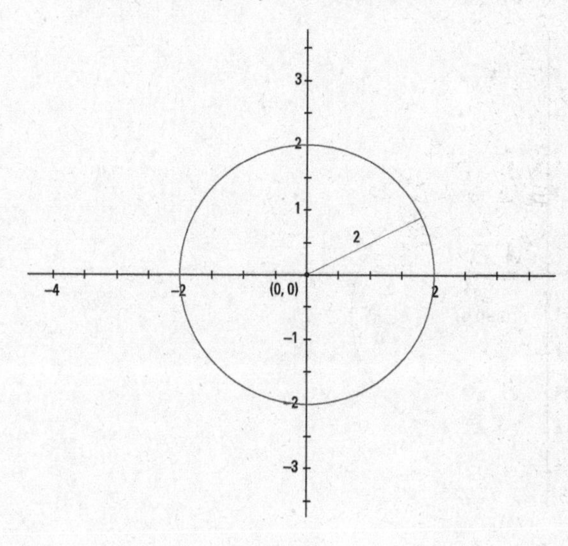

975. (0, 0); $r = \frac{1}{3}$

Writing the equation in the standard form for a circle, $(x-h)^2 + (y-k)^2 = r^2$, where (h, k) is the center and r is the radius, you have:

$$(x-0)^2 + (y-0)^2 = \left(\frac{1}{3}\right)^2$$

The center is at (0, 0), and the radius is $\frac{1}{3}$.

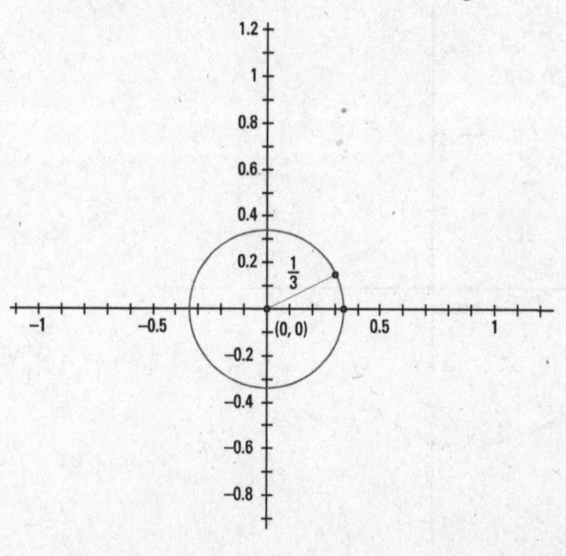

976. (2, –1); $r = 3$

Write the equation in the standard form for a circle, $(x-h)^2+(y-k)^2=r^2$, where (h, k) is the center and r is the radius.

$$(x-2)^2+(y-(-1))^2=3^2$$

The center is at (2, –1), and the radius is 3.

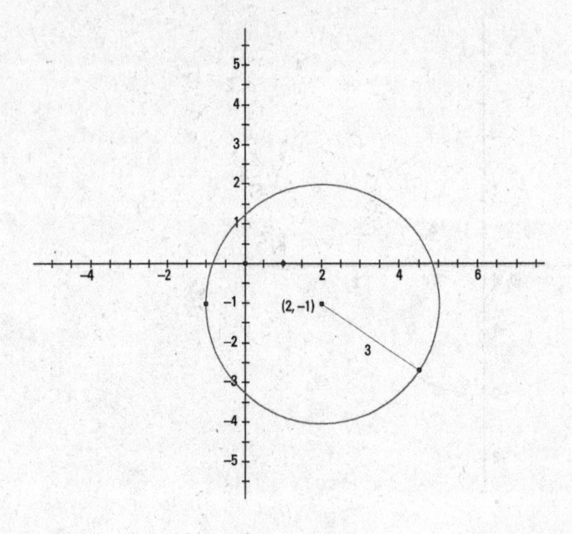

977. (–1, 0); $r = 4$

Write the equation in the standard form for a circle, $(x-h)^2+(y-k)^2=r^2$, where (h, k) is the center and r is the radius.

$$(x-(-1))^2+(y-0)^2=4^2$$

The center is at (–1, 0), and the radius is 4.

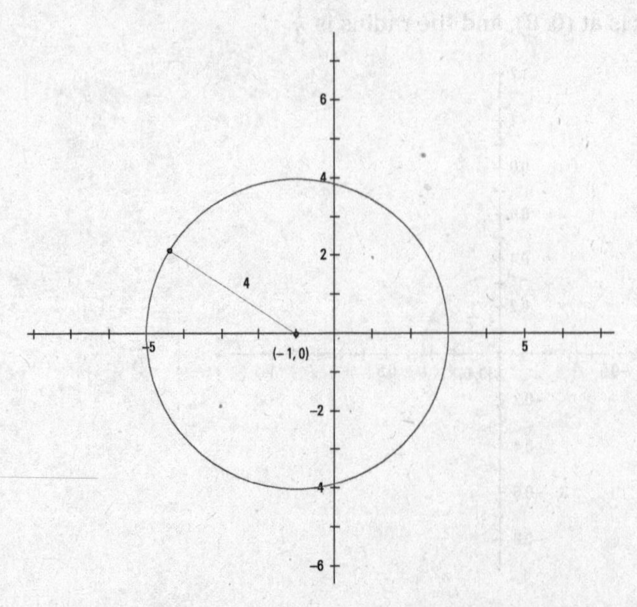

978. (4, 4); $r = 1$

Write the equation in the standard form for a circle, $(x-h)^2 + (y-k)^2 = r^2$, where (h, k) is the center and r is the radius.

$$(x-4)^2 + (y-4)^2 = 1^2$$

The center is at (4, 4), and the radius is 1.

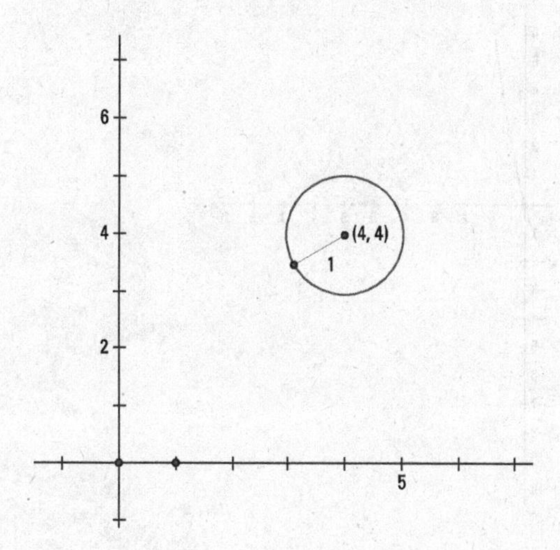

979. (0, 2); $r = 5$

Write the equation in the standard form for a circle, $(x-h)^2 + (y-k)^2 = r^2$, where (h, k) is the center and r is the radius, is written.

$$(x-0)^2 + (y-2)^2 = 5^2$$

The center is at (0, 2), and the radius is 5.

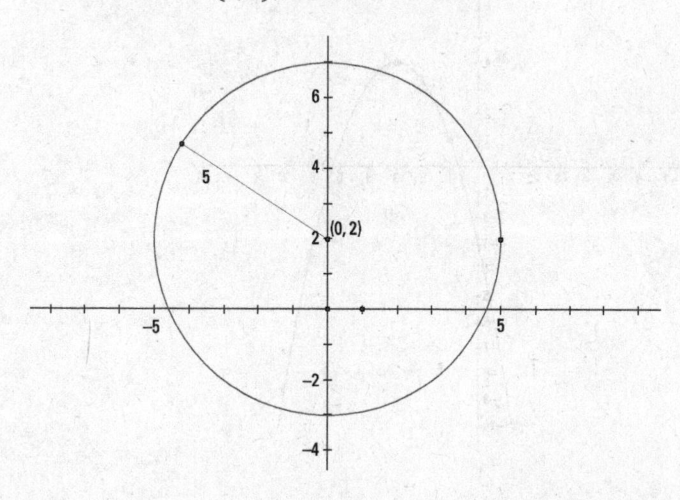

980. Also goes through (–2, 0)

The parabola opens upward and is symmetric about the line $x = -1$.

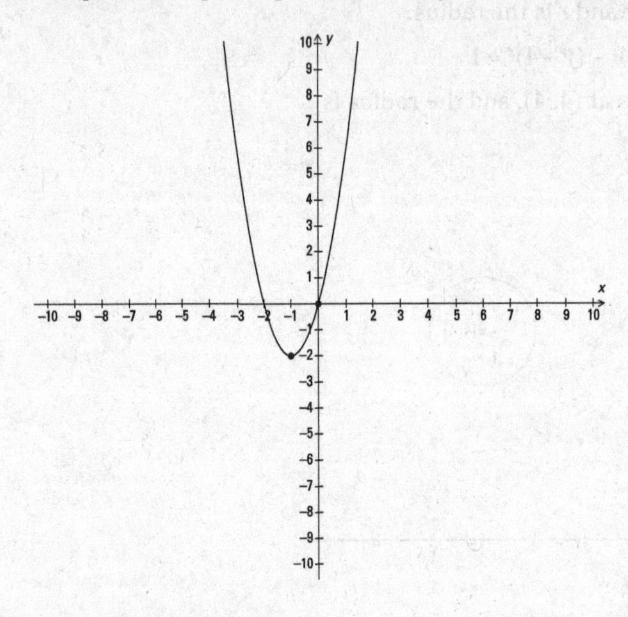

981. Also goes through (5, 1)

The parabola opens downward and is symmetric about the line $x = 3$.

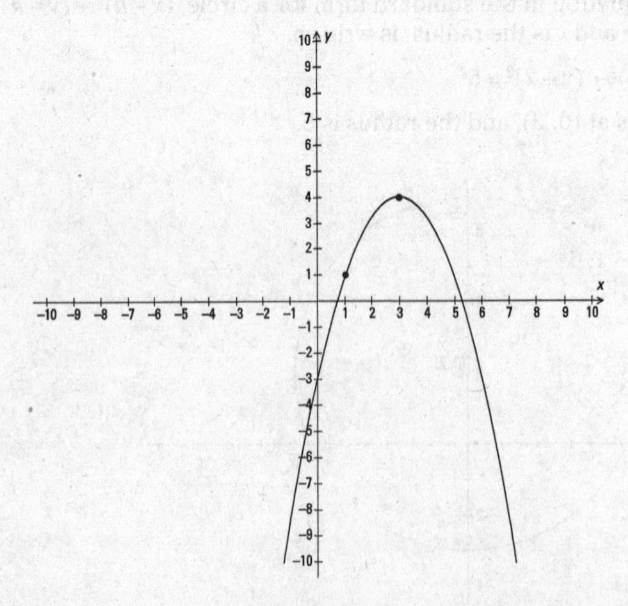

982. **Also goes through (–1, 0)**

The parabola opens downward and is symmetric about the line $x = 0$.

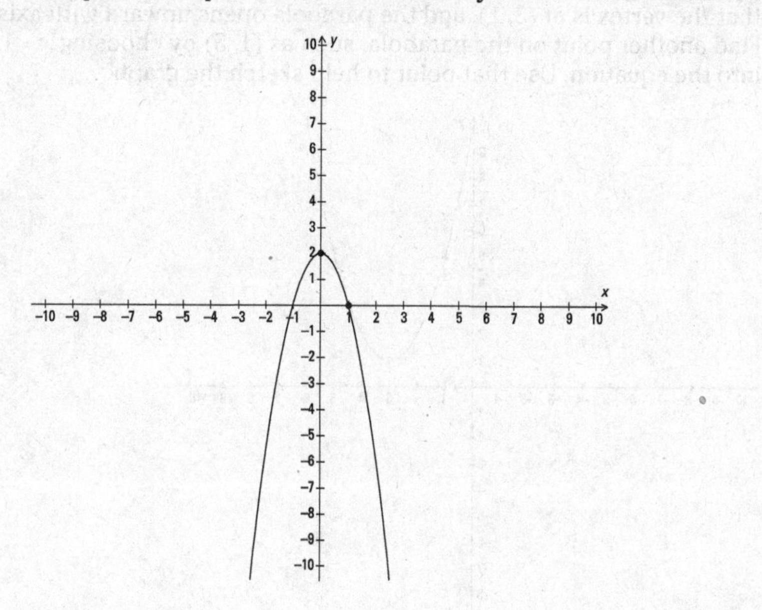

983. **Also goes through (7, 3)**

The parabola opens upward and is symmetric about the line $x = 3$.

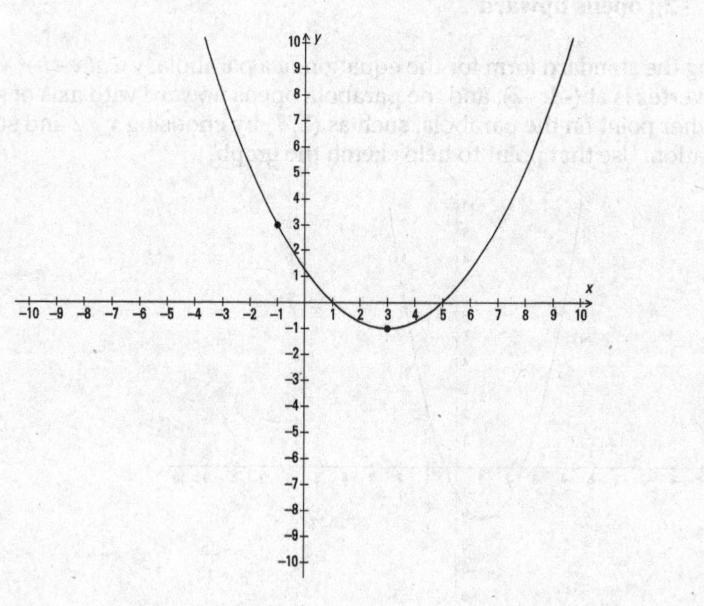

984. **Vertex (3, 1); opens upward**

Using the standard form for the equation of a parabola, $y = a(x - h)^2 + k$, you determine that the vertex is at (3, 1), and the parabola opens upward with axis of symmetry $x = 3$. Find another point on the parabola, such as (1, 5) by choosing $x = 1$ and substituting into the equation. Use that point to help sketch the graph.

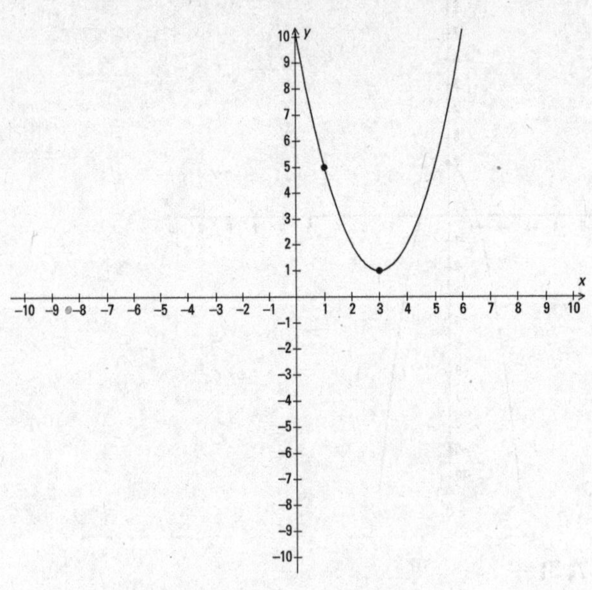

985. **Vertex (–1, –2); opens upward**

Using the standard form for the equation of a parabola, $y = a(x - h)^2 + k$, you determine that the vertex is at (–1, –2), and the parabola opens upward with axis of symmetry $x = -1$. Find another point on the parabola, such as (2, 7) by choosing $x = 2$ and substituting into the equation. Use that point to help sketch the graph.

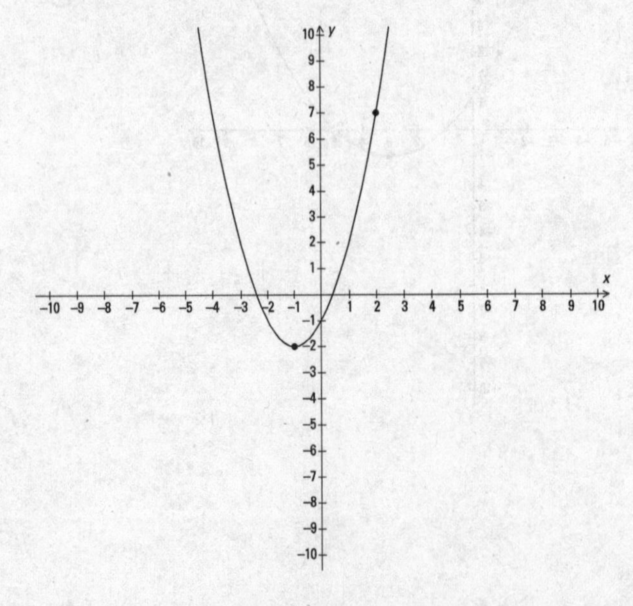

986. Vertex (2, 6); opens downward

Using the standard form for the equation of a parabola, $y = a(x - h)^2 + k$, rewrite the given equation as $y = -(x - 2)^2 + 6$. You determine that the vertex is at (2, 6), and the parabola opens downward with axis of symmetry $x = 2$. Find another point on the parabola, such as (1, 5) by choosing $x = 1$ and substituting into the equation. Use that point to help sketch the graph.

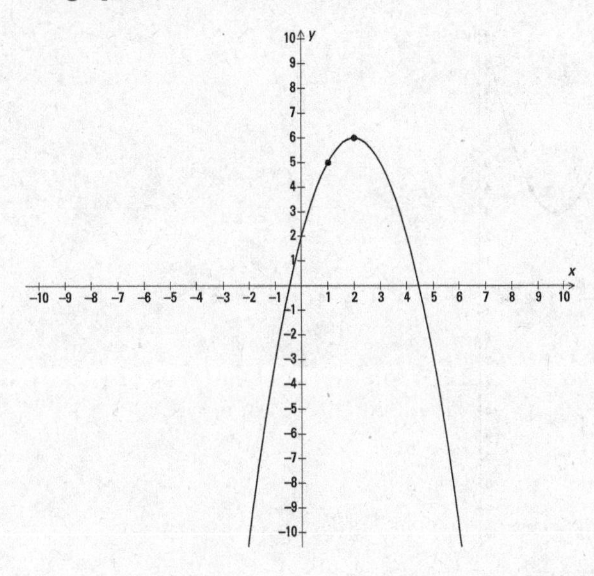

987. Vertex (–1, –1); opens downward

Using the standard form for the equation of a parabola, $y = a(x - h)^2 + k$, rewrite the given equation as $y = -(x + 1)^2 - 1$. You determine that the vertex is at (–1, –1), and the parabola opens downward with axis of symmetry $x = -1$. Find another point on the parabola, such as (0, –2) by choosing $x = 0$ and substituting into the equation. Use that point to help sketch the graph.

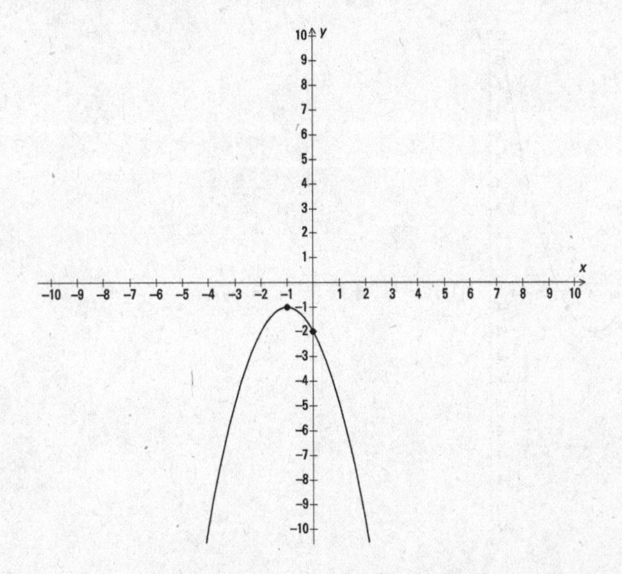

988. Vertex (–3, 3); opens upward

Using the standard form for the equation of a parabola, $y = a(x - h)^2 + k$, rewrite the given equation as $y = (x + 3)^2 + 3$. You determine that the vertex is at (–3, 3), and the parabola opens upward with axis of symmetry $x = -3$. Find another point on the parabola, such as (–2, 4) by choosing $x = -2$ and substituting into the equation. Use that point to help sketch the graph.

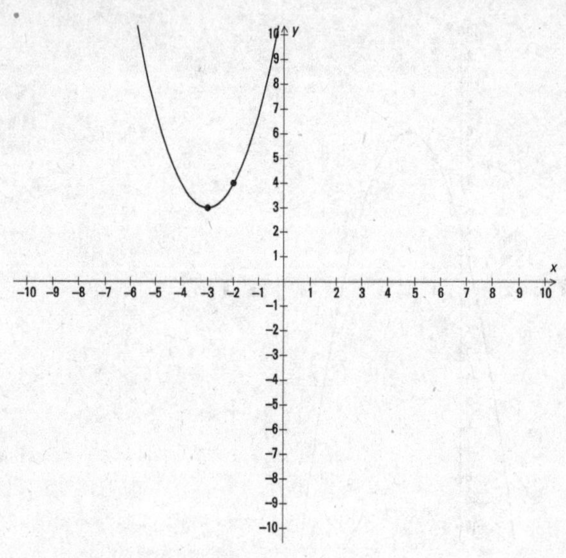

989. Vertex (–4, –2); opens upward

Using the standard form for the equation of a parabola, $y = a(x - h)^2 + k$, rewrite the given equation as $y = (x + 4)^2 - 2$. You determine that the vertex is at (–4, –2), and the parabola opens upward with axis of symmetry $x = -4$. Find another point on the parabola, such as (–3, –1) by choosing $x = -3$ and substituting into the equation. Use that point to help sketch the graph.

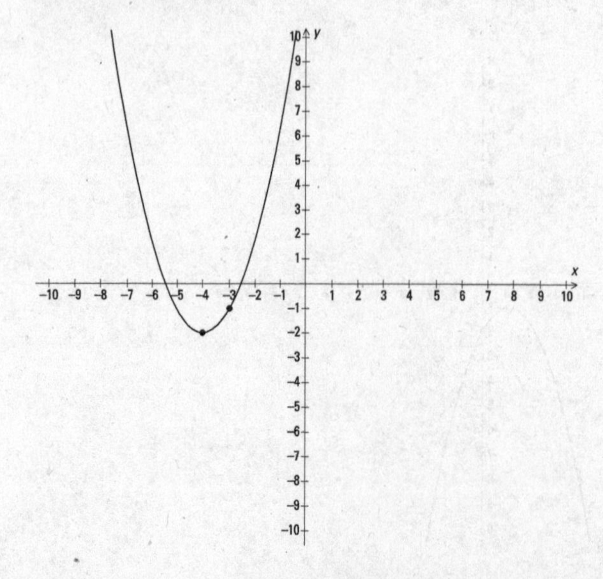

990. **Moves up 3**

The graph of $y = |x|$ is V-shaped with its lowest point at (0, 0). The transformation raises the graph by 3 units.

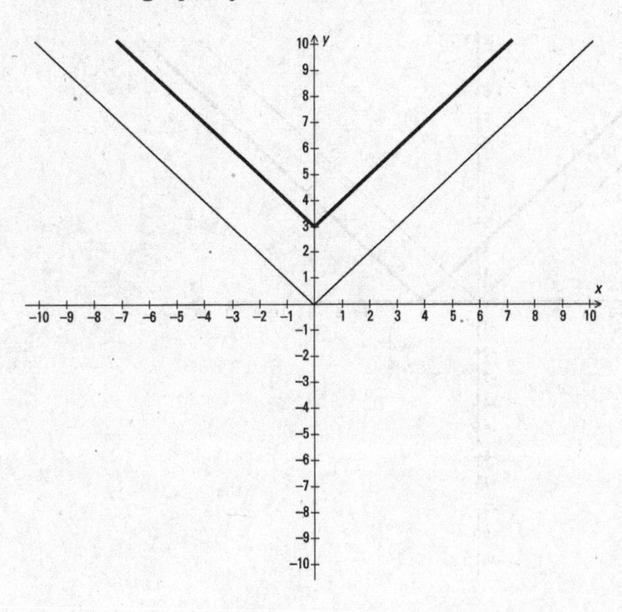

991. **Moves down 1**

The graph of $y = |x|$ is V-shaped with its lowest point at (0, 0). The transformation lowers the graph by 1 unit.

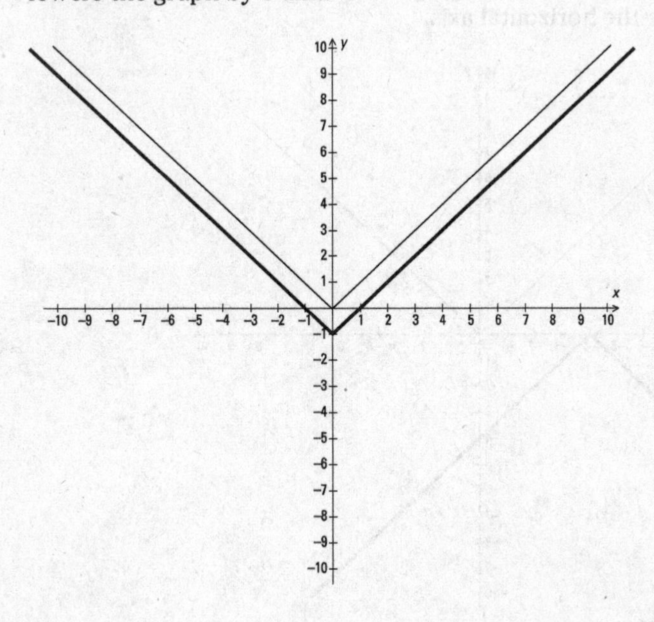

992. **Moves right 2**

The graph of $y = |x|$ is V-shaped with its lowest point at (0, 0). The transformation slides the graph two units to the right.

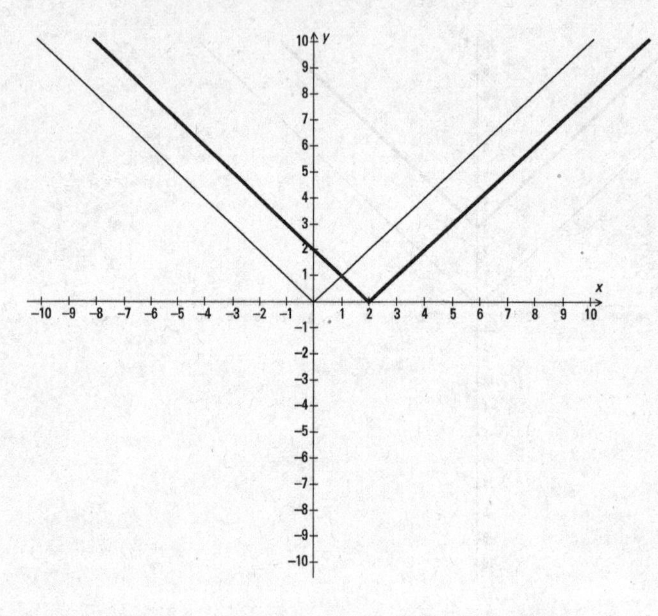

993. **Moves left 4 and flips over horizontal**

The graph of $y = |x|$ is V-shaped with its lowest point at (0, 0). There are two transformations. The + 4 slides the graph four units to the left, and the negative sign flips the graph over the horizontal axis.

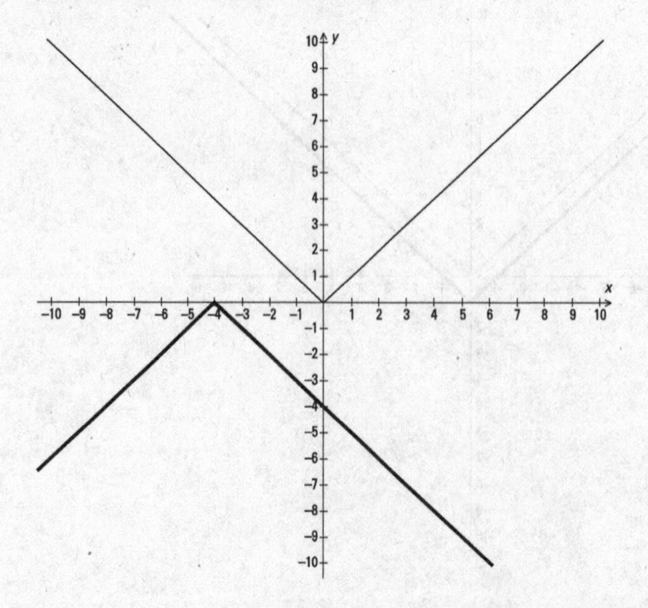

994. **Moves up 2**

The graph of $y = x^3$ is a flattened out S-shape moving up from left to right. The *flexing-point* (where it flattens out) is at (0, 0). The transformation raises the graph by two units.

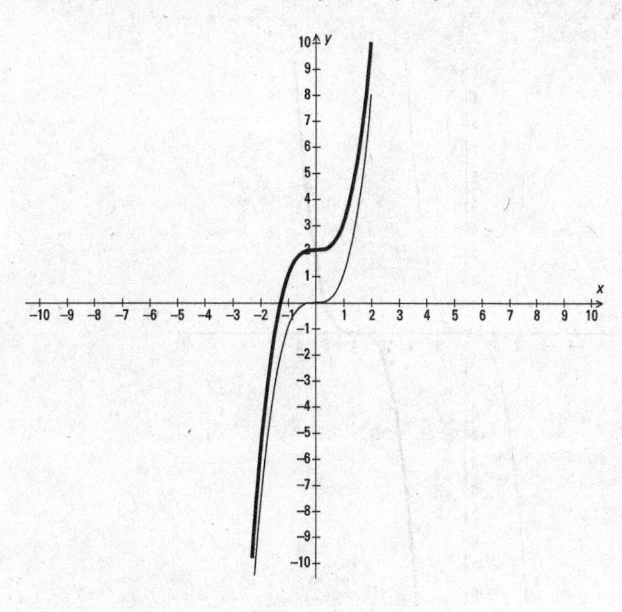

995. **Moves down 1 and flips over horizontal**

The graph of $y = x^3$ is a flattened out S-shape moving up from left to right. The *flexing-point* (where it flattens out) is at (0, 0). There are two transformations involved in this graph. The –1 lowers the graph by one unit, and the negative sign in front of the variable flips the graph horizontally over the flexing-point.

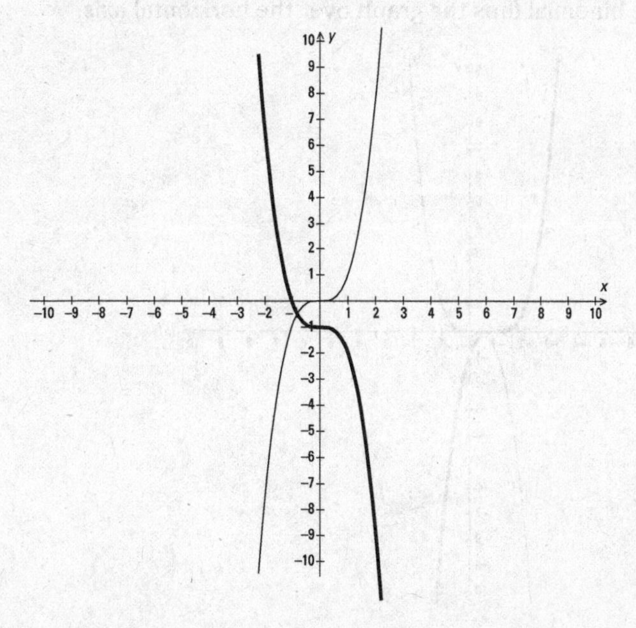

996. **Moves right 4**

The graph of $y = x^3$ is a flattened out S-shape moving up from left to right. The *flexing-point* (where it flattens out) is at (0, 0). The transformation sides the graph to the right by four units.

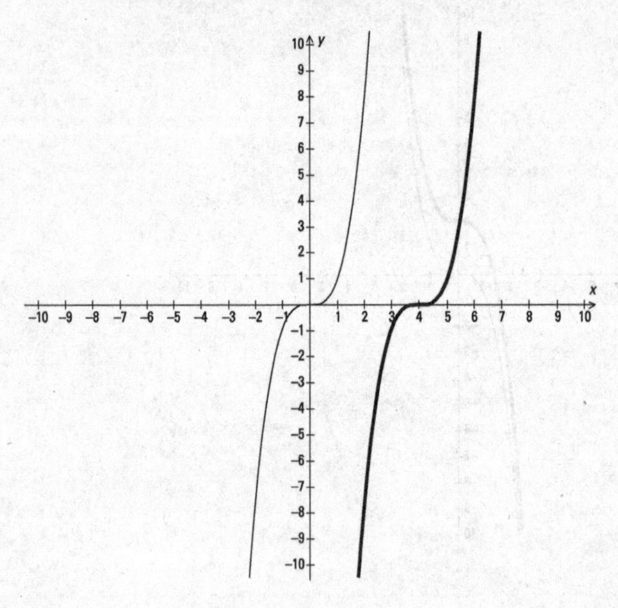

997. **Moves left 1 and flips over horizontal**

The graph of $y = x^3$ is a flattened out S-shape moving up from left to right. The *flexing-point* (where it flattens out) is at (0, 0). There are two transformations involved in the new graph. The + 1 slides the graph to the left by one unit, and the negative sign in front of the binomial flips the graph over the horizontal axis.

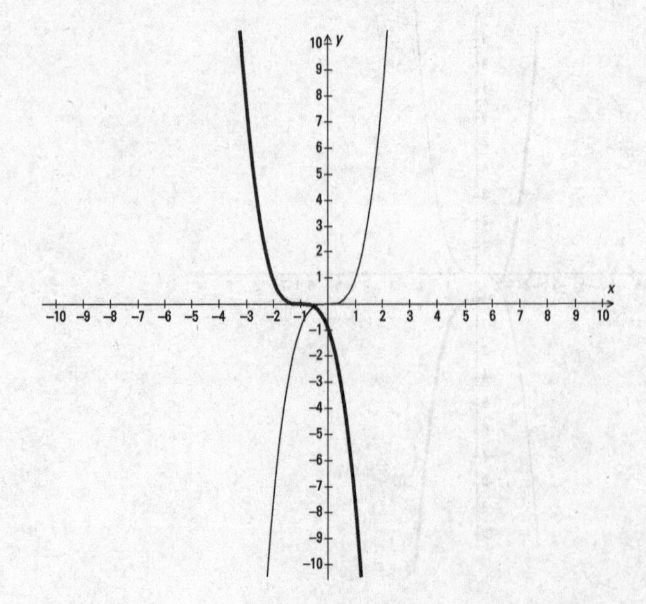

998. Moves up 1 and flips over horizontal

The graph of $y = (x - 1)^2$ is a U-shape with the lowest point (vertex) at (1, 0). There are two transformations involved in the new graph. The + 1 raises the graph one unit, and the negative sign in front of the binomial flips the graph over a horizontal line at the new vertex.

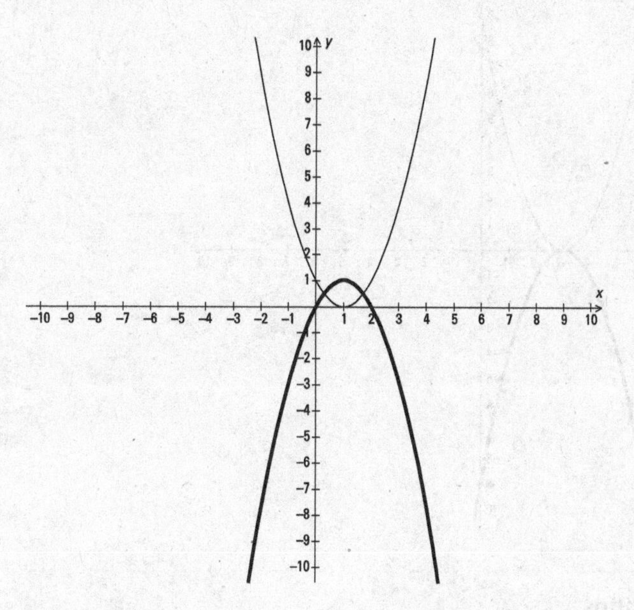

999. Moves down 4 and steepens

The graph of $y = (x - 1)^2$ is a U-shape with the lowest point (vertex) at (1, 0). There are two transformations involved in the new graph. The – 4 lowers the graph four units, and the multiplier of 2 in front of the binomial makes the graph steeper.

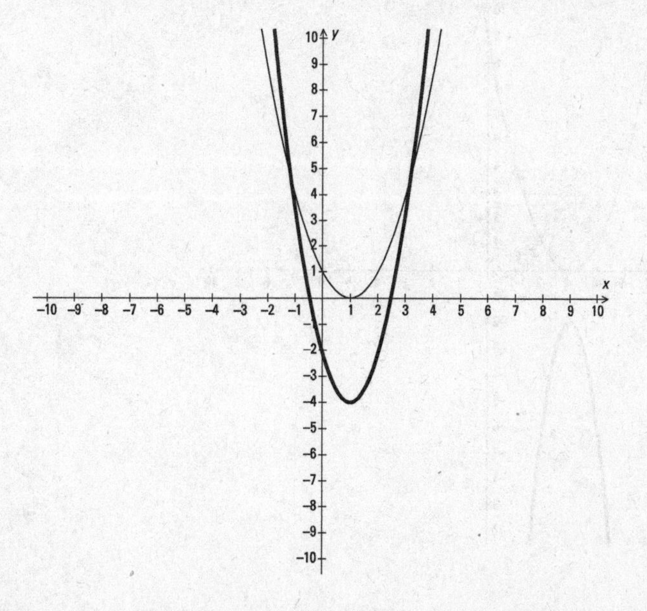

1000. Flips over horizontal

The graph of $y = (x + 3)^2$ is a U-shape with the lowest point (vertex) at (–3, 0). The transformation flips the graph over the horizontal axis.

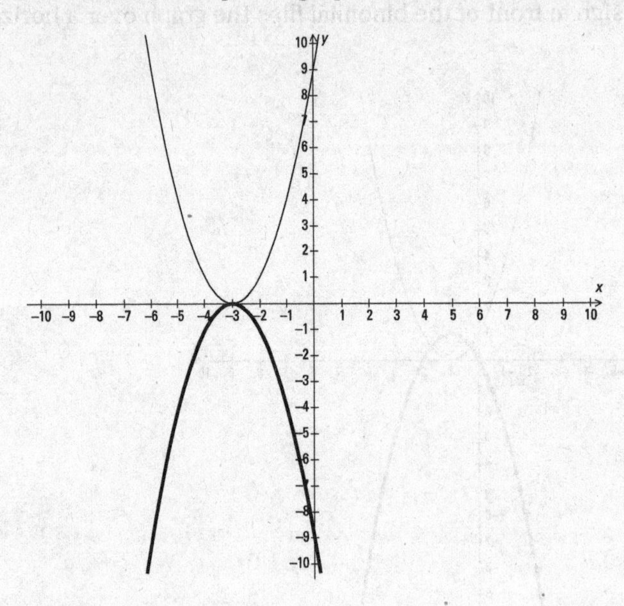

1001. Lowers, steepens, flips

The graph of $y = (x + 3)^2$ 'is a U-shape with the lowest point (vertex) at (–3, 0). There are three transformations involved in the new graph. The – 2 lowers the graph two units; the multiplier of –3 in front of the binomial makes the graph steeper, and the fact that it's a negative number flips the graph over a horizontal line at the new vertex.

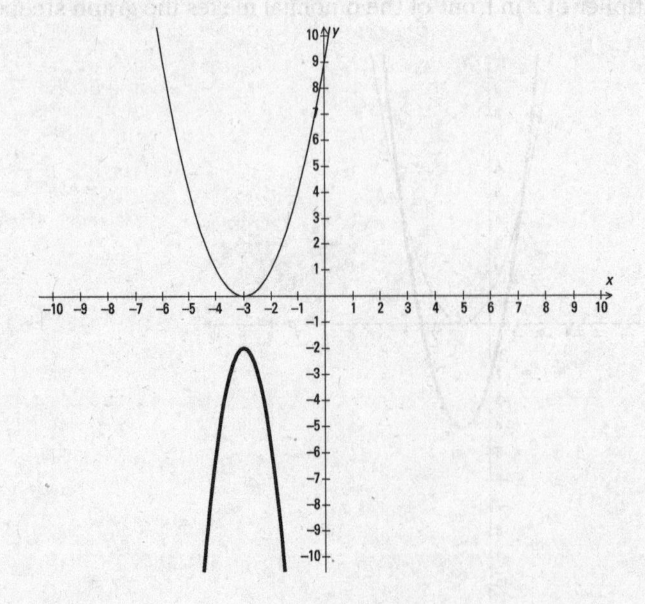

Index

• E •

• F •

• P •

• Q •

• R •

• S •

• T •

• V •

• W •

Workspace

Workspace

Workspace

Workspace

Workspace

Workspace

Workspace

Take Dummies with you everywhere you go!

Whether you're excited about e-books, want more from the web, must have your mobile apps, or swept up in social media, Dummies makes everything easier .

Visit Us

Like Us

Follow Us

Watch Us

Join Us

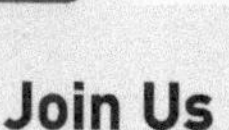

Pin Us

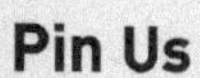

Circle Us

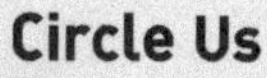

Shop Us